T0317721

CAMEL

CAMEL
INTELLIGENT NETWORKS FOR THE GSM, GPRS AND UMTS NETWORK

Rogier Noldus

Ericsson Telecommunications,
The Netherlands

John Wiley & Sons, Ltd

Other Wiley Editorial Offices

John Wiley & Sons Inc., 111 River Street, Hoboken, NJ 07030, USA

Jossey-Bass, 989 Market Street, San Francisco, CA 94103-1741, USA

Wiley-VCH Verlag GmbH, Boschstr. 12, D-69469 Weinheim, Germany

John Wiley & Sons Australia Ltd, 42 McDougall Street, Milton, Queensland 4064, Australia

John Wiley & Sons (Asia) Pte Ltd, 2 Clementi Loop #02-01, Jin Xing Distripark, Singapore 129809

John Wiley & Sons Canada Ltd, 22 Worcester Road, Etobicoke, Ontario, Canada M9W 1L1

Wiley also publishes its books in a variety of electronic formats. Some content that appears
in print may not be available in electronic books.

Library of Congress Cataloging-in-Publication Data:

Noldus, Rogier.
    CAMEL : intelligent networks for the GSM, GPSR and UMTS
network / Rogier Noldus.
      p. cm.
    Includes bibliographical references and index.
    ISBN-13: 978-0-470-01694-7 (cloth : alk. paper)
    ISBN-10: 0-470-01694-9 (cloth : alk. paper)
    1. Computer networks. 2. Artificial intelligence. 3. Global
system for mobile communications.   I. Title.
TK5105.84.N65 2006
621.382′1 – dc22

                           2005032765

British Library Cataloguing in Publication Data

A catalogue record for this book is available from the British Library

ISBN-13: 978-0-470-01694-7
ISBN-10: 0-470-01694-9

Typeset in 9/11pt Times by Laserwords Private Limited, Chennai, India

to Renée, Marc and Robyn

Contents

Foreword by Keijo Palviainen

In the 1990's the INAP (Intelligent Network Application Part) protocol was the dominant IN protocol. The INAP was mainly used in the fixed network environment and it worked well. However, the main issue was that the INAP deployments were vendor- and operator-specific since the INAP specification was lacking in some details. For example, many parameters are octet strings – leaving it up to the vendor to specify the precise encoding.

The other key functionality missing from INAP was mobility. The GSM system was becoming the dominant mobile network, and allowed for mobility between countries. The mobile operators were now seeing a real need to provide services to their subscribers when they were roaming.

To address these needs, ETSI started a project called CAMEL in late 1995. First, someone invented a distinctive name and then the words were filled in later. In fact, very few people actually remember what the 'abbreviation' actually stands for, including myself. As a result of this activity, CAMEL phase 1 was developed. CAMEL phase 1 is a very simple standard, but is tailored to the GSM-based core networks. One could claim that CAMEL is a child of INAP.

The CAMEL phase 2 extended CAMEL phase 1, the main focus being prepaid services. Then CAMEL and other GSM/UMTS works were moved to 3GPP responsibility, as the development of the 3G network was starting to become a global exercise. CAMEL phase 3 expands the service to include Short Message Service (SMS) as well as GPRS. Leading the pack, CAMEL phase 4 is the most advanced of the phases. It has about the same level of functionality as the Core INAP CS2 for fixed networks. The CAMEL phase 4 is the last CAMEL phase but it is extensible for any enhancements. In particular, the CAMEL phase 4 Call Party Handling has raised much interest among operators.

The original scope of CAMEL was the mobility but CAMEL has also been deployed for intra-network use in multi-vendor cases. Its deployment has begun in the large countries, such as India, China and the USA.

The main principle of CAMEL is that it is a toolkit that will enable many services. For example, when standardization was working on prepaid service, it was ensured that we have toolkits for online charging. However, nothing will now prevent us from using these tools for other services as well.

Much effort has been put into specification and testing specification work. However, the effort has proven to be money well spent, as CAMEL will continue to serve the circuit switched networks for many years to come.

Keijo Palviainen
Former ETSI SMG3 WPC and 3GPP CN2 chairman.
Nokia

Foreword by Gerry Christensen

When I started my career almost 18 years ago, I never envisioned the impact that mobile communications would have on telecom, IT, and for that matter, consumer lifestyles and business as a whole. The Yankee Group recently predicted that worldwide mobile operator revenue will reach $698 billion by 2009 with a unique user base of 2.4 billion individuals.

The exceptional growth of the customer base and usage of mobile communications raises some very important questions including "how will operators most cost effectively and efficiently deliver services?" and "how will service providers leverage common infrastructure to deploy new and innovative value-added services (VAS)?" In addition, IP Multimedia Subsystem (IMS) will have a profound effect on service creation and delivery for all service and content providers. While not the only answer, utilization of intelligent network technologies such as CAMEL will gain increasing importance as a tool in the mobile operator toolkit for voice and data applications.

While most consumers' top reasons for owning and using a cellular phone continue to be convenience and safety, most people will at least investigate new features if they add value to their daily lives. This is critical. Service providers must create and deliver VAS that generates incremental revenue as basic voice service becomes increasingly marginalized. In addition, momentum is gaining for wireless to be more than a medium for voice communications. The success in recent years of mobile personalization and entertainment applications and content such as ringtones, graphics, and games has proven the importance of non-voice applications to meet customer interests and derived new revenue for network operators.

In the book *Wireless Intelligent Networking*, I predicted five years ago that the future of CAMEL (and WIN) would be largely determined by its ability to evolve to support wireless data. The introduction of CAMEL phase III into mobile networks is beginning to make this a reality through its support of triggering and signalling within the core network infrastructure for SMS and GPRS control. However, there are also many emerging voice and voice/data hybrid services. A partial listing includes:

- **Calling Name Presentation:** The ability to provide the name of the calling party to the called party, allowing the called party to decide how to handle the call (e.g. the subscriber decides either to answer the call or let it go to voice mail). CAMEL is used to query a database that contains name information, which allows for a network-based service rather than programming the GSM phone to recognize caller names.
- **Prepay and Account Spending Limit (ASL):** Prepay and ASL utilize CAMEL to allow for metering usage on a prepaid basis and post-paid basis respectively. ASL has applications for those markets that are not debit based or credit-challenged but rather want to just manage usage. Markets include parental controls and corporate resource management.
- **Incoming Call Management (ICM):** CAMEL is leveraged to manage call termination attempts to customize subscriber's inbound calling experience. The subscriber can decide how inbound calls will be automatically managed. Features include automatic call handling (example: route all calls except boss to voice mail for the next hour) fixed-to-mobile convergence capabilities such as routing to mobile when a fixed network number is called.

- **Virtual Private Network (VPN):** CAMEL enables a mobile VPN that replicates PBX-like dialling in a mobile environment. For example, this (typically) group-based feature allows one to hit the digits "2706" and then SEND to actually place a call to Gerry Christensen at 650-798-2706.
- **Call Redirect Services (CRS):** CAMEL is utilized to provide a variety of CRS services including redirecting international roamers to their own customer care when they dial "611"
- **Location-based Services (LBS):** CAMEL has been used in the United States to support FCC mandates for wireless emergency calling (e.g. dialling 9-1-1 from a mobile phone. CAMEL thus allows for call control, information to be passed to databases, call assistance for routing to a Public Safety Answer Point, and for query of LBS infrastructure such as the Gateway Mobile Location Center (GMLC) for more precise positioning data based on A-GPS or TDOA. Commercial (non-regulatory) LBS applications are emerging that will rely on CAMEL include directory services and location-based search and information.

CAMEL also enables hybrid applications that allow for both voice and data interaction. For example, CAMEL is utilized in Teleractive mobile direct response marketing applications to allow the end-user to obtain information about products and services and to interact with brand and advertising agencies using data, voice, or both. CAMEL enables a simple and standard user interface for the end-user to engage in wireless data including SMS, MMS, and WAP.

An interesting thing to note is that the majority of the aforementioned services are subscriber-based and a few are group-based. This means that an end-user or group must subscribe in advance to be able to use the service. The mobile operator customer care department processes the request and instructs the engineering and operations department to provision the Home Location Register (HLR). The HLR is configured to utilize CAMEL functionality to recognize triggering events that occur typically on a per-subscriber/group, per-call basis.

CAMEL services may also be office-based, which means that any mobile phone user may use the service, whether in their home system or while roaming, without pre-subscription. CAMEL application triggering is based on events recognized by the Mobile Switching Center (MSC) rather than relying on communication and instruction from the HLR/VLR to arm a trigger detection point. For example, the Teleractive mobile direct response marketing applications are accessible to anyone with a mobile phone that dials a particular sequence of digits that follow "**" (example: **12345). The MSC recognizes "**" as a trigger to formulate a CAMEL message to be sent to a Service Control Point (SCP) for more information.

I have only scratched the surface with the few reference voice, data, and hybrid applications discussed in this foreward. The market for voice and data services for mobile is large and growing dramatically. Network operators, developers, service and content providers must focus on both market needs and the most effective and efficient creation and delivery mechanisms. The importance of CAMEL to fulfill this role cannot be ignored.

Until the availability of *CAMEL: Intelligent Networks for the GSM, GPRS, and UMTS Network*, there has been no book focused specifically on CAMEL. Rogier Noldus has really nailed the subject matter. I expect that, through use of this book, there will be more effective implementation of CAMEL-based applications and a lot more discussion about services heretofore unimagined.

CAMEL: Intelligent Networks for the GSM, GPRS, and UMTS Network is simply a must-have reference and instructional resource for anyone involved in planning and/or engineering applications and services within GSM voice and data networks. We use CAMEL in our mobile direct response marketing applications at Teleractive. I have declared Rogier's book to be must-reading for our engineering team.

Gerry Christensen
Chief Technology Officer
Teleractive, Inc.

Preface

This book provides an in-depth description of CAMEL. CAMEL is the embodiment of the Intelligent Networks (IN) concept, for the mobile network. The mobile networks for which CAMEL is specified, includes the GSM Network, the GPRS Network and the UMTS Network. This book is based mainly on the ETSI standards and the 3GPP specifications. Where appropriate, references to input document from other organizations, such as ITU-T, ISO, IETF are also included.

This book is not a GSM tutorial. However, since CAMEL is an integral part of GSM, the first chapter provides a rudimentary introduction into GSM. The remainder of the book will regularly fall back on the principles presented in that chapter. It will become clear, in the subsequent chapters of this book that CAMEL interacts mainly with the GSM **Core Network** (the Network Switching Subsystem). The entities that are part of the GSM Core Network, such as MSC, HLR, will be dealt with in detail. It should be emphasised that for general and in-depth background on GSM, a plethora of other text books are available.

This book is meant as reference material. For people who are new to IN, chapter two provides an introduction into IN. A brief history of IN is also included in that chapter. Chapters three to six describe the individual CAMEL Phases, i.e. CAMEL Phase 1 up to CAMEL Phase 4. Chapter seven describes some of the main charging principles related to CAMEL. And finally, chapter eight gives the reader a preview of the CAMEL features that are developed in 3GPP releases Rel-6 and Rel-7.

Few people will know the exact expansion of CAMEL: **C**ustomized **A**pplications for **M**obile networks **E**nhanced **L**ogic. The concept that CAMEL stands for, on the other hand, is now widely known within the telecommunications industry.

The present book has grown partly out of a personal desire to spread the knowledge about CAMEL, to those who work in the fields of Mobile Networks (GSM, GPRS, UMTS) and Intelligent Networks. The main drive, however, is a response to the question, "Where can I read up about CAMEL?" Hopefully, this book puts that question to rest! The present book aids those who are busy implementing CAMEL, developing CAMEL services, evaluating CAMEL etc.

CAMEL is the result of years of standardization work by ETSI and 3GPP. CAMEL development started in 1996, in the ETSI working groups SMG3-WPB and SMG1. I started participating in the SMG3-WPB meetings in September 1998. At that stage, development of the CAMEL Phase 2 standard was nearing completion. A "feet first" approach to the standardization work has resulted in years of active involvement in CAMEL development. A time which I thoroughly enjoyed.

With the finalizing of CAMEL Phase 4 in 3GPP Rel-7, the work on CAMEL may be considered complete. CAMEL is now deployed in most regions in the world, for pre-paid, VPN and many other services. It is expected that CAMEL will continue to serve mobile network operators for a vast number of years.

The IP Multimedia System (IMS) is currently gaining momentum. Whereas CAMEL is grafted on principles of the Circuit Switched (CS) technology (the "old world"), IMS is based on the Internet Protocol (IP) and is considered to represent the "new world". IP-based communication technology will eventually replace CS-based communication technology, both for wireline networks and for mobile networks. Full-scale IMS deployment within the UMTS network for speech services, will, however, take a couple of years to materialise. There are various estimates of the exact number of years that CS will remain the dominant technology for mobile speech services. IMS and CAMEL will co-exist for this transition period.

As goes for all major standards world wide, CAMEL is the product of a group of enthusiastic professionals. Without the commitment of the colleagues in ETSI and 3GPP, CAMEL would not have seen the light. It is therefore appropriate to thank those who have helped create CAMEL, both "the workers of the first hour" and those who continued to develop the later CAMEL phases. This group includes, in random order, Paul Martlew, Ian Park, Keijo Palviainen, Stanislav Dzuban, Jeremy Fuller, Noel Crespi, Michel Grech, Christian Homann, Sumio Miyagawa, Ruth Jones (nee Hewson), Veronique Belfort, Georg Wegmann, Nick Russell, Andrijana Jurisic, Angelica Remoquillo, Steffen Habermann, Isabelle Lantelme, Iris Moilanen, Kazuhiko Nakada and David Smith. Each person brought in his or her own expertise to the group. Especially those colleagues that were linked through the "humps" discussion group deserve special credit for their hard work on CAMEL. The above list does not pretend to be exhaustive. Hence, credit is due also to those whose names are not mentioned, but who have nevertheless contributed to the CAMEL standard. I also thank Gerry Christensen for supporting me during the initial stages of this book and during the process of writing the text. Richard Davies, from Wiley, has provided useful comments on style, grammar and layout for the book. I also thank my Ericsson colleagues of the "CAMEL team" for their support, expertise and commitment.

It further goes without saying that main credit is due to my wife Renee as well as to Marc and Robyn for being without husband and dad during the many hours, days and weeks spent on travelling and writing.

Rogier Noldus
February 2006

About the author

Rogier Noldus is senior specialist at Ericsson Telecommunicatie B.V. in Rijen, The Netherlands. He has been actively involved in Intelligent Networks (IN) standardization for six years and has driven the development of CAMEL within Ericsson. He advises customers worldwide about the implementation of CAMEL and about CAMEL service development.

Rogier is currently working in the area of Service Layer (for GSM, UMTS and IMS) system development. He has filed a large number of patent applications in the area of GSM and IN.

He holds a B.Sc. degree (electronics) from the Institute of Technology in Utrecht (The Netherlands) and a M.Sc. degree (telecommunications) from the University of The Witwatersrand (Johannesburg, South Africa). He joined Ericsson in 1996. Prior to that, he has worked for several companies in South Africa, in the area of telecommunications.

1

Introduction to GSM Networks

Figure 1.1 is a schematic overview of the main components in a GSM network. The various interface labels are the formal names given to these interfaces. More details about these interfaces are found in GSM TS 03.02 [26].

The GSM network consists mainly of the following functional parts:

- *MSC* – the mobile service switching centre (MSC) is the core switching entity in the network. The MSC is connected to the radio access network (RAN); the RAN is formed by the BSCs and BTSs within the Public Land Mobile Network (PLMN). Users of the GSM network are registered with an MSC; all calls to and from the user are controlled by the MSC. A GSM network has one or more MSCs, geographically distributed.
- *VLR* – the visitor location register (VLR) contains subscriber data for subscribers registered in an MSC. Every MSC contains a VLR. Although MSC and VLR are individually addressable, they are always contained in one integrated node.
- *GMSC* – the gateway MSC (GMSC) is the switching entity that controls mobile terminating calls. When a call is established towards a GSM subscriber, a GMSC contacts the HLR of that subscriber, to obtain the address of the MSC where that subscriber is currently registered. That MSC address is used to route the call to that subscriber.
- *HLR* – the home location register (HLR) is the database that contains a subscription record for each subscriber of the network. A GSM subscriber is normally associated with one particular HLR. The HLR is responsible for the sending of subscription data to the VLR (during registration) or GMSC (during mobile terminating call handling).
- *CN* – the core network (CN) consists of, amongst other things, MSC(s), GMSC(s) and HLR(s). These entities are the main components for call handling and subscriber management. Other main entities in the CN are the equipment identification register (EIR) and authentication centre (AUC). CAMEL has no interaction with the EIR and AUC; hence EIR and AUC are not further discussed.
- *BSS* – the base station system (BSS) is composed of one or more base station controllers (BSC) and one or more base transceiver stations (BTS). The BTS contains one or more transceivers (TRX). The TRX is responsible for radio signal transmission and reception. BTS and BSC are connected through the Abis interface. The BSS is connected to the MSC through the A interface.
- *MS* – the mobile station (MS) is the GSM handset. The structure of the MS will be described in more detail in a next section.

A GSM network is a *public land mobile network* (PLMN). Other types of PLMN are the time division multiple access (TDMA) network or code division multiple access (CDMA) network. GSM uses the following sub-division of the PLMN:

CAMEL: Intelligent Networks for the GSM, GPRS and UMTS Network   Rogier Noldus
© 2006 John Wiley & Sons, Ltd

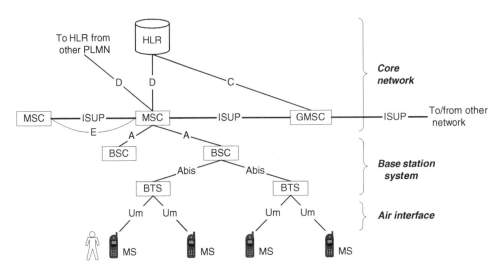

Figure 1.1   GSM network architecture

- *Home PLMN (HPLMN)* – the HPLMN is the GSM network that a GSM user is a subscriber of. That implies that GSM user's subscription data resides in the HLR in that PLMN. The HLR may transfer the subscription data to a VLR (during registration in a PLMN) or a GMSC (during mobile terminating call handling). The HPLMN may also contain various service nodes, such as a short message service centre (SMSC), service control point (SCP), etc.
- *Visited PLMN (VPLMN)* – the VPLMN is the GSM network where a subscriber is currently registered. The subscriber may be registered in her HPLMN or in another PLMN. In the latter case, the subscriber is *outbound roaming* (from HPLMN's perspective) and *inbound roaming* (from VPLMN's perspective). When the subscriber is currently registered in her HPLMN, then the HPLMN is at the same time VPLMN.[1]
- *Interrogating PLMN (IPLMN)* – the IPLMN is the PLMN containing the GMSC that handles mobile terminating (MT) calls. MT calls are always handled by a GMSC in the PLMN, regardless of the origin of the call. For most operators, MT call handling is done by a GMSC in the HPLMN; in that case, the HPLMN is at the same time IPLMN. This implies that calls destined for a GSM subscriber are always routed to the HPLMN of that GSM subscriber. Once the call has arrived in the HPLMN, the HPLMN acts as IPLMN. MT call handling will be described in more detail in subsequent sections. When basic optimal routing (BOR) is applied, the IPLMN is not the same PLMN as the HPLMN.

The user of a GSM network is referred to as the *served subscriber*; the MSC that is serving that subscriber is known as the *serving MSC*. Examples are:

- *mobile originated call* – the MSC that is handling the call is the *serving MSC* for this call; the calling subscriber is the *served subscriber*;
- *mobile terminated call* – the GMSC that is handling the call is the *serving GMSC* for this call; the called subscriber is the *served subscriber*.

---

[1] The CAMEL service requirement, GSM TS 02.78 [12] uses this strict definition. The term VPLMN is, however, commonly used to denote any network other than the HPLMN.

1.1 Signalling in GSM

The various entities in the GSM network are connected to one another through signalling networks. Signalling is used for example, for subscriber mobility, subscriber registration, call establishment, etc. The connections to the various entities are known as 'reference points'. Examples include:

- *A interface* – the connection between MSC and BSC;
- *Abis interface* – the connection between BSC and BTS;
- *D interface* – the connection between MSC and HLR;
- *Um interface* – the radio connection between MS and BTS.

Various signalling protocols are used over the reference points. Some of these protocols for GSM are the following:

- *mobile application part (MAP)* – MAP is used for call control, subscriber registration, short message service, etc.; MAP is used over many of the GSM network interfaces;
- *base station system application part (BSSAP)* – BSSAP is used over the A interface;
- *direct transfer application part (DTAP)* – DTAP is used between MS and MSC; DTAP is carried over the Abis and the A interface. DTAP is specified in GSM TS 04.08 [49];
- *ISDN user part (ISUP)* – ISUP is the protocol for establishing and releasing circuit switched calls. ISUP is also used in landline Integrated Services Digital Network (ISDN). A *circuit* is the data channel that is established between two users in the network. Within ISDN, the data channel is generally a 64 kbit/s channel. The circuit is used for the transfer of the encoded speech or other data. ISUP is specified in ITU-T Q.763 [137].

When it comes to call establishment, GSM makes a distinction between *signalling* and *payload*. Signalling refers to the exchange of information for call set up; payload refers to the data that is transferred within a call, i.e. voice, video, fax etc. For a mobile terminated GSM call, the signalling consists of exchange of MAP messages between GMSC, HLR and visited MSC (VMSC). The payload is transferred by the ISUP connection between GMSC and VMSC. It is a continual aim to optimize the payload transfer through the network, as payload transfer has a direct cost aspect associated with it. Some network services are designed to optimize the payload transfer. One example is optimal routing.

1.2 GSM Mobility

Roaming with GSM is made possible through the separation of *switching capability* and *subscription data*. A GSM subscriber has her subscription data, including CAMEL data, permanently registered in the HLR in her HPLMN. The GSM operator is responsible for provisioning this data in the HLR. The MSC and GMSC in a PLMN, on the other hand, are not specific for one subscriber group. The switching capability of the MSC in a PLMN may be used by that PLMN's own subscribers, but also by *inbound roaming* subscribers; see Figure 1.2.

In Figure 1.2, the GSM user who is a subscriber of PLMN-A roams to PLMN-B. The HLR in PLMN-A transfers the user's subscription data to the MSC in PLMN-B. The subscriber's subscription data remains in the MSC/VLR as long as she is served by a BSS that is connected to that MSC. Even when the user switches her MS off and then on again, the subscription data remains in the MSC. After an extended period of the MS being switched off, the subscription data will be purged from the MSC. When the subscriber switches her MS on again, the subscriber has to re-register with the MSC, which entails the MSC asking the HLR in the HPLMN to re-send the subscription data for that subscriber.

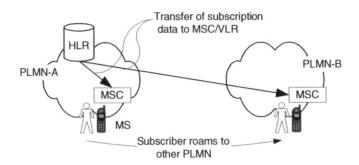

Figure 1.2   Transfer of GSM subscription data for a roaming subscriber

When the subscriber moves from one MSC service area (MSC-1) to another MSC service area (MSC-2), the HLR will instruct MSC-1 to purge the subscription data of this subscriber and will send the subscription data to MSC-2.

1.3 Mobile Station

The MS, i.e. the GSM handset, is logically built up from the following components:

- *mobile equipment (ME)* – this is the GSM terminal, excluding the SIM card;
- *subscriber identification module (SIM)* – this is the chip embedded in the SIM card that identifies a subscriber of a GSM network; the SIM is embedded in the SIM card. When the SIM card is inserted in the ME, the subscriber may register with a GSM network. The ME is now effectively personalized for this GSM subscriber; see Figure 1.3. The characteristics of the SIM are specified in GSM TS 11.11. The SIM card contains information such as IMSI, advice of charge parameters, operator-specific emergency number, etc. For the UMTS network an enhanced SIM is specified, the universal subscriber identity module (USIM); refer 3GPP TS 31.102.

1.4 Identifiers in the GSM Network

GSM uses several identifiers for the routing of calls, identifying subscribers (e.g. for charging), locating the HLR, identifying equipment, etc. Some of these identifiers play an important role for CAMEL.

1.4.1 International Mobile Subscriber Identity

The international mobile subscriber identity (IMSI) is embedded on the SIM card and is used to identify a subscriber. The IMSI is also contained in the subscription data in the HLR. The IMSI is used for identifying a subscriber for various processes in the GSM network. Some of these are:

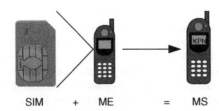

SIM   +   ME   =   MS

Figure 1.3   Components of the mobile station

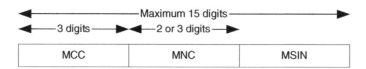

Figure 1.4 Structure of the IMSI

- *location update* – when attaching to a network, the MS reports the IMSI to the MSC, which uses the IMSI to derive the global title (GT) of the HLR associated with the subscriber;
- *terminating call* – when the GSM network handles a call to a GSM subscriber, the HLR uses the IMSI to identify the subscriber in the MSC/VLR, to start a process for delivering the call to that subscriber in that MSC/VLR.
- *roaming charging* – a VPLMN uses the IMSI to send billing records to the HPLMN of a subscriber.

Figure 1.4 shows the format of the IMSI.

- *mobile country code (MCC)* – the MCC identifies the country for mobile networks. The MCC is not used for call establishment. The usage of MCC is defined in ITU-T E.212 [129]. The MCC values are allocated and published by the ITU-T.
- *mobile network code (MNC)* – the MNC identifies the mobile network within a mobile country (as identified by MCC). MCC and MNC together identify a PLMN. Refer to ITU-T E.212 [129] for MNC usage. The MNC may be two or three digits in length. Common practice is that, within a country (as identified by MCC), all MNCs are either two or three digits.
- *mobile subscriber identification number (MSIN)* – the MSIN is the subscriber identifier within a PLMN.

The IMSI is reported to the SCP during CAMEL service invocation. The IMSI may be needed, for example, when identifying a country; countries in North America have equal country code (country code = 1), but different MCC (e.g. Canada = 303; Mexico = 334).

1.4.2 Mobile Station Integrated Services Digital Network Number (MSISDN Number)

The MSISDN is used to identify the subscriber when, among other things, establishing a call to that subscriber or sending an SMS to that subscriber. Hence, the MSISDN is used for routing purposes. Figure 1.5 shows the structure of the MSISDN.

- *country code (CC)* – the CC identifies the country or group of countries of the subscriber;
- *national destination code (NDC)* – each PLMN in a country has one or more NDCs allocated to it; the NDC may be used to route a call to the appropriate network;
- *subscriber number (SN)* – the SN identifies the subscriber within the number plan of a PLMN.

Figure 1.5 Structure of the MSISDN

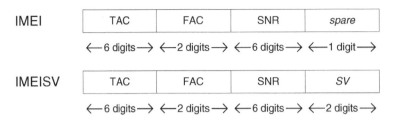

Figure 1.6 Structure of IMEI and IMEISV

The MSISDN is not stored on the subscriber's SIM card and is normally not available in the MS.[2] The MSISDN is provisioned in the HLR, as part of the subscriber's profile, and is sent to MSC during registration. The MSISDN is also reported to SCP when a CAMEL service is invoked. One subscriber may have multiple MSISDNs. These MSISDNs are provisioned in the HLR. At any one moment, only a single MSISDN is available in the MSC/VLR for the subscriber.

1.4.3 International Mobile Equipment Identifier

The international mobile equipment identifier (IMEI) is used to identify the ME [or user equipment (UE) in UMTS network]. Each ME has a unique IMEI. The IMEI is hard-coded in the ME and cannot be modified. Figure 1.6 shows the structure of the IMEI. The IMEI is not used for routing or subscriber identification.

Refer to GSM TS 03.03 [27] for the type approval code (TAC), final assembly code (FAC) and serial number (SNR). The software version (SV) may be included in the IMEI ('IMEISV') to indicate the version of software embedded in the ME. The IMEI is always encoded as an eight-octet string. As from CAMEL Phase 4, the IMEI(SV) may be reported to the SCP.

1.4.4 Mobile Station Roaming Number

The mobile station roaming number (MSRN) is used in the GSM network for routing a call to a MS. The need for the MSRN stems from the fact that the MSISDN identifies a subscriber, but not the current location of that subscriber in a telecommunications network. The MSRN is allocated to a subscriber during MT call handling and is released when the call to that subscriber is established. Each MSC in a PLMN has a (limited) range of MSRNs allocated to it. An MSRN may be allocated to any subscriber registered in that MSC. The MSRN has the form of an E.164 number and can be used by the GMSC for establishing a call to a GSM subscriber. An MSRN is part of a GSM operator's number plan. The MSRN indicates the GSM network a subscriber is registered in, but not the GSM network the subscriber belongs to. Figure 1.7 shows how the MSRN is used for call routing. The MSRN is not meant for call initiation. GSM operators may configure their MSC such that subscribers cannot dial numbers that fall within the MSRN range of that operator.

1.5 Basic Services

All activities that may be done in the GSM network, such as establishing a voice call, establishing a data call, sending a short message, etc., are classified as *basic services*. In order for a subscriber to use a GSM basic service, she must have a subscription to that service.[3] The handling of a basic

---

[2] GSM subscribers may program their MSISDN into the phone; this has, however, no significance for the network.

[3] Exceptions are Tele Service 12 (emergency call establishment) and Tele Service 23 (Cell Broadcast). Subscribers do not need a subscription to these Tele Services to use them.

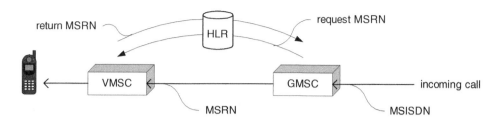

Figure 1.7 Usage of MSRN during call establishment to a GSM subscriber

service is fully standardized. Hence, a subscriber may use a basic service in any GSM network she roams to, provided that that basic service is supported in that network. The HLR will send a list of subscribed basic services to the MSC/VLR, during registration. When a GSM subscriber initiates a call, the MS supplies the serving MSC with a set of parameters describing the circuit-switched connection that is requested. These parameters are the bearer capability (BC), low-layer compatibility (LLC) and high-layer compatibility (HLC), as will be described below. The MSC uses the BC, LLC and HLC to derive the basic service for this call. The rules for deriving the basic service from LLC, HLC and BC are specified in GSM TS 09.07 [55]. The MSC then checks whether the subscriber has a subscription to the requested basic service, i.e. whether the subscription data in the VLR contains that basic service. If the service is not subscribed to, then the MSC disallows the call. The basic service is not transported over ISUP.

When a CAMEL service is invoked, the MSC reports the requested basic service to the SCP. The SCP may use the indication of the requested basic service for call service processing. Examples include:

- video calls may be charged at a higher rate than speech calls;
- for data calls and fax calls, the CAMEL service shall not play any announcements or tones.

Basic services are divided into two groups: tele services and bearer services.

1.5.1 Tele Services

Table 1.1 provides an overview of the available tele services (TS); see also GSM TS 02.03 [3].

1.5.2 Bearer Services

Table 1.2 provides an overview of the available bearer services (BS). The two bearer service groups are sub-divided into a variety of bearer services with different characteristics. Refer to GSM TS 02.02 [2].

1.5.3 Circuit Bearer Description

Bearer capability, low-layer compatibility and high-layer compatibility are descriptors of a circuit-switched (CS) connection. When a GSM subscriber initiates a call, the BC, LLC and HLC are transported from MS to MSC over DTAP. The MSC includes the parameters in the ISUP signal to the destination. These parameters are also reported to the SCP during CAMEL service invocation. That enables a CAMEL service to adapt the service logic processing to the type of call. Figure 1.8 shows the relation between LLC, HLC and BC on the DTAP and the corresponding parameters on ISUP.

Table 1.1 Tele services

Tele service	Description	Comment
11	Telephony	This TS represents the normal speech call
12	Emergency calls	The emergency call uses the characteristics of telephony (TS11), but may be established without subscription and bypasses various checks in the MS and in the MSC
21	Short message MT	This TS relates to receiving an SMS. This TS is not sent to the MSC/VLR. When an SMS is sent to the subscriber, the HLR checks whether the destination subscriber has a subscription to TS 21
22	Short message MO	This TS relates to the sending of an SMS
23	Cell broadcast	This TS relates to the capability of an SMS that is sent as a broadcast SMS
61	Alternate speech and fax group 3	This TS relates to the capability to establish a speech and fax (group 3) call
62	Automatic fax group 3	This TS relates to the capability to establish a fax (group 3) call
91	Voice group call	This TS relates to the capability to participate in a group call as specified in GSM TS 03.68 [35]
92	Voice broadcast	This TS relates to the capability to receive a voice broadcast as specified in GSM TS 03.68 [35]

Table 1.2 Bearer services

Tele service	Description	Comment
20	Asynchronous data bearer services	May be used for asynchronous services from 300 bit/s to 64 kbit/s.
30	Synchronous data bearer services	May be used for synchronous services from 1.2 to 64 kbit/s. This BS may be used, amongst other things, for multimedia services such as video telephony.[4]

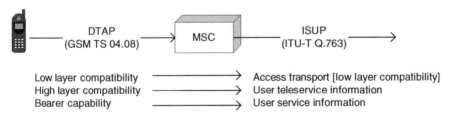

Figure 1.8   Transfer of LLC, HLC and BC through DTAP and ISUP

- *Low-layer compatibility* – the LLC is transported transparently between the calling entity and called entity; it may be used by the respective entities to adapt codecs for interworking purposes. LLC describes mainly characteristics related to the data transfer.

---

[4] 3GPP Rel-7 may include a dedicated bearer service for video telephony.

- *High-layer compatibility* – the HLC is also transported transparently between the calling entity and called entity; it is used to describe the requested service, such as telephony, Fax, video telephony, etc.
- *Bearer capability* – the BC describes the characteristics of the 64 kbit/s circuit requested for the call.

1.6 Supplementary Services

Supplementary services (SS) in GSM are a means of enriching the user experience. An SS may, for example, forward a call in the case of no reply from the called party, bar certain outgoing or incoming calls, show the number of the calling party to the called party, etc. In order to use an SS, a GSM user needs a subscription to that SS. The subscription to supplementary services is contained in the HLR and is sent to the MSC/VLR during registration. The supplementary services are fully standardized. A GSM subscriber can therefore use her supplementary services in any GSM network, provided that the network supports these supplementary services, and have the same user experience.

Table 1.3 GSM supplementary services

SS group	Supplementary services	GSM TS
Line identification	Calling line identification presentation (CLIP) Calling line identification restriction (CLIR) Connected line presentation (COLP) Connected line restriction (COLR)	02.81 [13]
Name identification	Calling name presentation (CNAP)	02.96 [24]
Call forwarding	Call forwarding – unconditional (CFU) Call forwarding – busy (CFB) Call forwarding – no reply (CFNRY) Call forwarding – not reachable (CFNRC)	02.82 [14],
	Call deflection (CD)	02.72 [11]
Call offering	Explicit call transfer (ECT)	02.91 [22]
Call completion	Call waiting (CW) Call hold (CH)	02.83 [15],
	Call completion to busy subscriber (CCBS) Multi-call (MC)	02.93 [23], 22.135 [69][a]
Multi-party	Multi-party call (MPTY)	02.84 [16]
Community of interest	Closed user group (CUG)	02.85 [17]
Charging	Advice of charge – information (AOCI) Advice of charge – charge (AOCC)	02.86 [18]
Additional information transfer	User-to-user signalling – service 1 (UUS1) User-to-user signalling – service 2 (UUS2) User-to-user signalling – service 3 (UUS3)	02.87 [19]

Call barring	Barring of all outgoing calls (BAOC)
	Barring of outgoing international calls (BOIC)
	Barring of outgoing international calls except to the
	  home country (BOIC-exHc)
	Barring of all incoming calls (BAIC)
	Barring of all incoming calls when roaming
	  (BICROAM)
		02.88 [20]

Call priority	enhanced multi-level precedence and pre-emption (eMLPP)	02.67 [10]

[a] For the multi-call service, there is no GSM TS available, but only a 3GPP TS (22.135).

Supplementary services may be provisioned for an individual basic service or for a group of basic services, e.g. a subscriber may have barring of all outgoing calls for all tele services and all bearer services, except SMS (tele service group 20). Such a subscriber is barred from establishing outgoing calls (except emergency calls), but may still send short messages. Some supplementary services may be activated or deactivated by the user. Examples include call forwarding and call barring. An operator may decide to bar certain subscribers or subscriber groups from modifying their supplementary services.

Table 1.3 shows the Supplementary Services. They are combined in service groups. Subscriptions are per individual Supplementary Service. The right-most column indicates the GSM technical specifications (TS) that specify the service requirement for the respective Supplementary Service(s).

The chapters on CAMEL Phases 1–4 describe the interaction between CAMEL and the various supplementary services. Not all GSM networks support all supplementary services. Many of the supplementary services in GSM have equivalent supplementary services in ISDN. The ISDN supplementary services are described in ITU-T recommendations.

GSM TS 02.03 [3] describes how the supplementary services may be activated, deactivated and invoked.

2

Introduction to Intelligent Networks

Intelligent networks (IN) is a technique that augments digital telecommunication networks with a method to lift the control over CS calls to a higher-layer control platform. These digital networks, which are based on signalling principles defined by ISUP, may include networks such as the integrated service digital network (ISDN), the public switched telephone network (PSTN) and the PLMN. Applying IN to any of these networks has in common that call establishment is intercepted at a designated node in the network. Control over the call is handed over to a *control platform*. The control platform determines how the establishment of this call shall continue. This is depicted in Figure 2.1.

The SCP forms the control platform for IN. The IN control protocol is the capability set that enables the operator to assert control over the call. Various IN standards have defined an IN protocol; CAMEL is one such standard. The exchange is located in the *core network* and may be a node such as a local exchange (LE), transit exchange (TE) or MSC. The SCP is located in the *service layer*. The service layer may contain a multitude of nodes, but for IN the SCP is the main entity through which control over the call may be asserted.

2.1 History of Intelligent Networks

Development of IN in the form that it is currently known started in the mid-1980s. One of the first IN standards was Bellcore's[1] advanced IN (AIN). AIN was developed as an IN standard for landline digital networks. A later IN standard was the wireless IN (WIN), which targets the mobile networks, amongst which is the personal communication system (PCS). WIN was later also applied to the TDMA and code division multiple access (CDMA) networks.

In the early-1990s, the International Telecommunication Unit – Telecommunications (ITU-T)[2] developed its first capability set (CS) standard, CS1. CS1 is the IN control protocol from which further IN standards were derived. IN application part (INAP) is often used as generic term to denote the IN control protocol between SCP and the core network. The ITU-T has subsequently published CS2, CS3 and CS4, all of which are successors and enhancements to CS1.

The European Telecommunication Standardization Institute (ETSI) has used the work from ITU-T to endorse IN standards for the European market. The ETSI CS standards are referred to as Core INAP CS1, Core INAP CS2 and Core INAP CS3.

The present book does not aim to provide in-depth description of the original IN standards as developed by Bellcore, ITU-T and ETSI. Rather, the chapters in this book describe how CAMEL,

---

[1] Bell Communications Research, North American laboratory, providing support to the Bell Companies.
[2] Other ITU sectors include radio communications (ITU-R) and telecom development (ITU-D).

CAMEL: Intelligent Networks for the GSM, GPRS and UMTS Network   Rogier Noldus
© 2006 John Wiley & Sons, Ltd

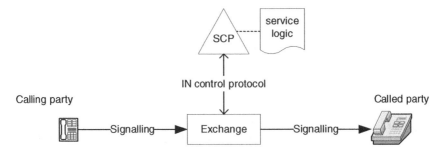

Figure 2.1 IN control to basic call

being an IN standard, is developed for the GSM network specifically. The reader who is acquainted with the original IN standards will recognize many of the fundamental principles of these standards, when reading the present book. However, CAMEL has been developed specifically for GSM (and subsequently for GPRS and UMTS); therefore, it contains many capabilities that are not found in any of the traditional IN standards.

For an in-depth description of traditional IN, the reader is referred to *The Intelligent Network Standards* [173] and *The Intelligent Network* [174].

2.2 Principles of Intelligent Networks

One pivotal principle of IN is the interaction between the core network signalling protocol (e.g. ISUP) and the IN control protocol (INAP). This is depicted in Figures 2.2–2.4. Figure 2.2 shows the network components for a typical mobile-to-mobile call in the GSM network; Figure 2.3 shows the ISUP signal sequence flow of this call; Figure 2.4 shows how IN interacts with this signalling, at designated points in the sequence flow.

In Figure 2.2, the call is established by the mobile station of the A-party (MS-A), through the VMSC of the A-party (VMSC-A), the GMSC for the B-party (GMSC-B), the VMSC of the B-party (VMSC-B) to the MS of the B-party (MS-B). The ISUP messages between VMSC-A, GMSC-B and VMSC-B are listed in the Appendix. Direct transfer application part (DTAP) is the call control protocol used between MS and the MSC.

At designated points in the message sequence flow, interaction between the MSC and the SCP takes place. These interactions enable the SCP to influence the processing of the call. In the example in Figure 2.4, interaction takes place at the following points:

- *Call establishment* – when MSC-A starts call establishment, as a result of receiving a setup message over the air interface from the A-party, it invokes an IN service in the SCP. The

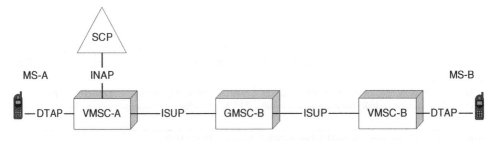

Figure 2.2 Network architecture for mobile-to-mobile call

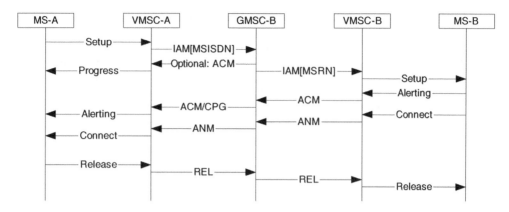

Figure 2.3 Example of ISUP message sequence flow

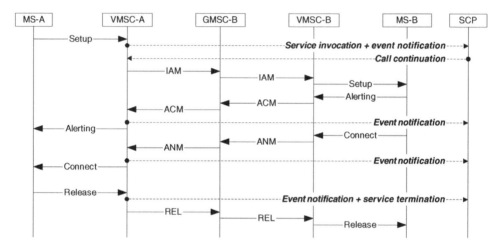

Figure 2.4 IN control for basic mobile-to-mobile call

invocation of an IN service entails the establishment of an *IN dialogue* between the MSC and the SCP. It is through this dialogue that the SCP can control the call.

- *Alerting* – when the MSC-A receives an indication that the B-party's terminal is alerting ('ringing'), it sends a notification to the SCP.
- *Answer* – when the MSC-A receives an indication that the B-party's terminal has answered the call, it sends a notification to the SCP.
- *Disconnect* – when the MSC-A receives an indication that one of the parties has released the call, it sends a notification to the SCP and terminates the IN dialogue. Closing the IN dialogue also has the effect of terminating the IN service.

At service invocation and event notification, the MSC copies information elements from the signalling message (i.e. the ISUP message) to the IN control message. The SCP decides how to control this call, based on the received information. The SCP may decide to allow the call to continue unmodified or to allow the call to continue with modified information. The latter may be done by providing the MSC with specific information elements that replace corresponding information elements in designated ISUP messages. The SCP may retain control over the call for the entire call duration or may relinquish control at an earlier moment. When the SCP relinquishes control over the call, i.e. terminates the IN service, the call may continue without IN control.

The IN service invocation, depicted in Figure 2.4 by the first arrow from VMSC-A to SCP, is based on criteria present in the exchange. In this example, VMSC-A has determined that, for this call, an IN service shall be invoked. Traditional IN does not define stringent triggering criteria. An operator may define these criteria in an MSC as deemed suitable. Examples include:

- *number-based triggering* – the MSC triggers an IN service for certain numbers or number ranges, e.g. calls to numbers starting with 0800 trigger a free-phone service;
- *trunk-based triggered* – calls that arrive over a particular trunk ('trunk' is generic term for the transmission channel between two switching nodes) trigger an IN service, e.g. all calls arriving from another network trigger an incoming call-screening service;
- *subscription-based triggering* – calls from a particular subscriber trigger an IN service, e.g. all calls from subscribers belonging to a certain company trigger a virtual private network (VPN) service.

The exchange from where an IN service is invoked needs configuration for various other characteristics of the IN service. These characteristics include:

- the address of the SCP where the IN service resides; the service invocation will be sent to that address;
- the protocol that will be used for this IN service;
- the information elements that will be provided to the IN service.

All of these aspects of the IN service are configured in the exchange from where the IN service may be invoked. The operator owning the exchanges may decide on this configuration, to suit that operator's IN services.

2.3 Service Switching Function

The IN control protocol at the exchange is handled by the service switching function (SSF). The SSF passes call control from the exchange to the SCP and relays instructions from the SCP back to the exchange. All IN protocol aspects are handled by the SSF. Figure 2.5 depicts the SSF in an MSC.

At IN service invocation, the SSF copies information from the access protocol (e.g. ISUP or DTAP) onto the INAP message that is used to invoke the IN service. When the SSF receives instruction from SCP, it copies information received from the SCP on to the call control protocol.

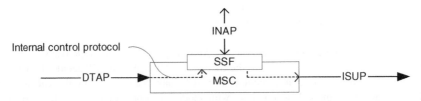

Figure 2.5   SSF inside an MSC

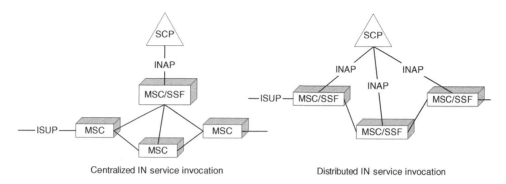

Figure 2.6 Centralized vs distributed IN control

In a GSM network, each MSC may be equipped with an SSF or only designated MSCs may be equipped with an SSF (Figure 2.6).

A network may have a mix of centralized IN control and distributed IN control, depending on the type of IN service. Centralized IN control requires less investment in SSF, but may lead to more ISUP signalling since calls need to be routed through a designated MSC for IN service invocation. Centralized IN control may also be applied when the designated MSC has specific IN control capability that cannot be offered by the other MSCs in the network. This method may, for example, be applied in a network with core network equipment from different vendors.

2.4 Service Control Function

The service control function (SCF) is the functional entity residing in the SCP. It forms an application in the SCP that facilitates the execution of IN services. The SCP is an *addressable* node in the SS7 network. Other nodes in the network may communicate with the SCP through the SS7 signalling protocol.

For both CAMEL and INAP, the behaviour of the SCF is specified in less detail than the behaviour of the SSF. The rationale is that, once the SCF has gained control over a call, it may decide how the call shall continue. The SCF supports the IN protocol (e.g. INAP), but the behaviour of the service logic is operator-specific.

2.5 Basic Call State Model

A fundamental concept for IN control is the basic call state model (BCSM). When a call is processed by an exchange, the call goes through a number of pre-defined phases. These phases of the call are described in the BCSM. The BCSM generally follows the ISUP signalling of a call. ISUP messages received by the exchange result in the transition from one BCSM state to another. The definition of the BCSM enables the MSC to interact with the SSF at defined points in the call. The SSF may in its turn contact the SCP at these points in the call.

The BCSM contains detection points (DP) and points in call (PIC). This is reflected in Figure 2.7.

The PIC indicates the state of the call, i.e. analysis, routing, alerting and active. A DP is associated with a *state transition*. When the call reaches a certain PIC, the BCSM first processes the DP that is associated with the transition to that PIC, e.g. when the call is in the alerting phase and an answer event is received over ISUP, the BCSM processes the DP that is associated with the answer event. After the processing of the DP is complete, the BCSM transits to the active PIC.

The BCSM describes a *model* according to which an exchange may handle the establishment of a call. For each call that is handled by an exchange a process is started that behaves as defined by

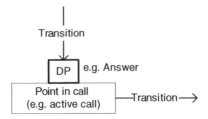

Figure 2.7 Elements of the BCSM: PIC and DP. Reproduced from GSM TS 03.78 v5.8.0 Figure 7.1/1, by permission of ETSI

the BCSM. This is commonly described as 'an instance of the BCSM is created' or 'the BCSM is instantiated'. ISUP messages passing through this exchange may have the result that the BCSM instance for this call transits from one state to another state, e.g. when a call is in the setup phase and the exchange receives the ISUP answer message, the BCSM instance transits to DP answer. The SSF in that exchange may notify the SCP and, depending on the IN service logic behaviour, continue processing the ISUP answer message. Practically, this means the forwarding of the answer message in the backwards direction to the originator of the call.

Core INAP CS1 has defined two types of BCSM: the originating call BCSM and the terminating call BCSM. These BCSMs are based on the ISUP messages used for call establishment and on the digital subscriber signalling 1 (DSS1) protocol. DSS1 is the access protocol used between ISDN terminal and ISDN network. The BCSMs that are defined in CAMEL are derived from the Core INAP CS1 BCSMs. These CAMEL BCSMs are described in Chapters 3–6.

IN defines four DP types:

- *Trigger detection point – request (TDP-R)*: when the BCSM instance for a call transits to a DP that is defined as TDP-R, an IN service may be started at that point. This entails the internal SSF notifying the SCF and waiting for further instructions. The call processing in the MSC is halted until the SSF has received instructions from the SCF. TDPs are statically defined in an exchange. By defining different DPs in the BCSM as TDP, the exchange may invoke an IN service at different points in the call.
- *Trigger detection point – notify (TDP-N)*: the TDP-N is a variant of the TDP-R. An IN service may be triggered from a DP that is defined as TDP-N as opposed to TDP-R. The SSF will in that case not wait for instructions from the SCP, but will return the call control immediately to the MSC. As a result the call processing is not halted. The SCP has not gained control over the call; the SCP was merely notified about the occurrence of the call event. The use of TDP-N is not very common for IN. CAMEL defines TDP-R, but not TDP-N.
- *Event detection point – request (EDP-R)*: When an IN service is invoked, it may *arm* DPs within the BCSM as an event detection point (EDP). Arming a DP entails that the IN service instructs the SSF to monitor for the occurrence of the event associated with the DP. When the event occurs, the SSF notifies the SCP. If the DP is armed as EDP-R, the SSF halts call processing after the notification and waits for instructions from the SCP. The reporting of an event that was armed as an EDP-R is referred to as *interrupt mode*.
- *Event detection point – notify (EDP-N)*: The IN service may arm an EDP in *interrupt* mode (EDP-R) or in *notify* mode (EDP-N). When a DP is armed as an EDP-N, the SSF reports the occurrence of the event associated with the DP, but the SSF does not halt call processing. Instead, the SSF instructs the MSC to continue call processing.

An IN service normally keeps a mirror image of the BCSM instance in the SSF for the call that the IN service is controlling. In this way, the IN service knows the phase of the call and which events

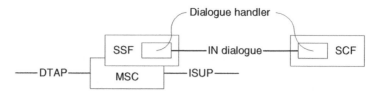

Figure 2.8   IN dialogue handler

may occur. In order to keep this mirror image of the BCSM, the IN service will arm the DPs in the BCSM, so as to receive a notification when a state transition occurs in the BCSM. When a DP is not armed, the DP is said to be *transparent*.

2.6 Dialogue Handling

The invocation of an IN service involves the establishment of an IN dialogue between SSF and SCF. SSF and SCF start a process that governs this dialogue (Figure 2.8).

The IN dialogue between SSF and SCF facilitates the exchange of instructions and notifications between SSF and SCF. When the IN service terminates, the IN dialogue is closed. Two methods exist for closing the IN dialogue:

- *Pre-arranged end* – when communication has taken place between SCF and SSF and both entities can deduce that, for this call, there will not be any further communication through this IN dialogue, then both entities may terminate the dialogue without informing the other entity.
- *Basic end* – an entity may explicitly terminate the IN dialogue by sending a dialogue closing notification to the other entity.

Figure 2.9 contains examples that reflect both methods for dialogue termination. The transaction capability (TC) messages (TC_Begin, TC_Continue, TC_End) are explained in a later section. The IN service is started by the SSF by sending the *initial DP* operation to the SCF. The IN service responds by sending the *continue* operation, which instructs the SSF to continue call establishment. In the pre-arranged end example, the SCF does not explicitly close the IN dialogue. However, since the SCF did not arm any of the DPs in the BCSM, there will not be any further communication between SSF and SCF through this IN dialogue. The SSF and SCF therefore decide to close the IN dialogue. In the basic end example, the SCF instructs the SSF to continue call establishment and at the same time instructs the SSF to close the IN dialogue. Section 2.9 presents further details related to the signalling between SSF and SCF.

2.6.1 DP Arming/Disarming Rules

As described above, the DPs in the BCSM are the defined *contact points* between SSF and SCF. Arming and disarming DPs in the BCSM is a tool used by the IN service to keep informed about

Figure 2.9   Pre-arranged end vs basic end

the phase of the call and to maintain or close the IN dialogue. A set of DP arming and disarming rules are defined below.

- *TPD arming* – TDPs are statically armed in the exchange. The operator may decide for which calls an IN service is invoked and at which DP in the BCSM for that call.
- *EDP arming* – when an IN service is invoked from a particular TDP in the BCSM, the IN service may dynamically arm DPs in the BCSM as EDP-N or EDP-R. The arming of a DP as EDP is valid only for the duration of the IN service. The IN protocol that is used for the IN service determines which DPs are available in the BCSM and whether these DPs may be armed as EDP-N or EDP-R.
- *EDP disarming* – when a DP is armed as EDP, it may be disarmed in various ways: (1) the IN service may explicitly instruct the SSF to disarm the DP; (2) when the DP occurs, the SSF disarms the DP; the IN service may re-arm the DP, if needed;[3] (3) the occurrence of a particular DP in the SSF may result in the implicit disarming of other DPs in the BCSM; CAMEL specifies strict rules for this form of *implicit disarming*; and (4) when a call or call leg is released, all DPs associated with that call or call leg are disarmed.

2.6.2 Control vs Monitor Relationship

The SSF and SCF maintain a *relationship* through the IN dialogue. The relationship is a means of describing the level of control the SCF has over the call. A relationship exists between SSF and SCF under the following conditions:

- the SSF has reported a TDP-R or EDP-R to the SCP and is waiting for instructions from the SCP; or
- at least one DP in the BCSM is armed as EDP-N or EDP-R; or
- the SCP has requested the SSF to send a charging report; or
- the SCP has requested the SSF to send a call information report.

The charging report and call information report are described in Chapter 4. Two forms of relationship are defined.

2.6.2.1 Control Relationship

When a *control relationship* exists between SSF and SCF, the IN service may take actions like releasing the call. A control relationship exists under the following conditions:

- the SSF has reported a TDP-R or EDP-R to the SCP and is waiting for instructions from the SCP; or
- at least one DP is armed as EDP-R.

When the BCSM transits to a DP that is armed as EDP-R, the SSF automatically disarms that DP. The control relationship between the SSF and SCF remains at least until the end of the processing of this DP. For example, IN service may arm the disconnect event (indicating that the calling or called party has terminated the call) as EDP-R. When the disconnect event occurs, the SSF reports the event to the SCP and waits for instructions. The SSF has, in accordance with DP disarming rules, disarmed the disconnect event. Hence, there is currently no DP armed for this call. However, as long as the SCP is busy processing the disconnect event, which was reported in interrupt mode, the control relationship exists.

---

[3] In CAMEL Phase 4, 'automatic re-arming' is introduced for selected DPs.

2.6.2.2 Monitor Relationship

When a monitor relationship exists between SSF and SCF, the IN service may keep informed about the call progress, but cannot assert any control over the call. It cannot, for example, order a follow-on call when call establishment fails. When a relationship exists between SSF and SCF, but does not qualify for a control relationship, it is a monitor relationship. A control relationship may downgrade to a monitor relationship, but not vice versa.

2.7 Evolution of the CAMEL Standard

CAMEL is a natural evolution of the IN standards that were defined by Bellcore, the ITU-T and ETSI. Many of the concepts that are defined for Core INAP CS1 also apply to CAMEL. Hence, CAMEL is by definition an IN standard.[4] The need for CAMEL grew during the development of the GSM network standard. When GSM development started in the late 1980s, the concept of IN was already in place. When operators started to deploy GSM in the early 1990s, IN was still mainly used for fixed line networks, such as PSTN and ISDN. When the need arose for more advanced services than were available in the GSM network, operators started using the existing IN standards. Over and above, vendors introduced their specific enhancements to the IN standards.

This practice had the following aspects:

- IN capability related to charging is largely undefined in the existing standards; these capabilities may be defined by the equipment vendor that implements the IN standard.
- Triggering methods are not defined; hence, no unified set of rules exists that indicates when an exchange such as an MSC will invoke an IN service for a subscriber.
- The existing IN standards were developed for wireline networks. Many GSM-specific network aspects are not supported in the existing IN.
- The existing IN standard does not support the mobility aspect of GSM, i.e. subscribers roaming to other GSM networks and using their basic services and supplementary services in those other networks.

To address the above issues, ETSI introduced an IN standard specifically for the GSM network. This GSM-specific IN standard, i.e. CAMEL, forms an integral part of the ETSI GSM standards. The first version of CAMEL was included in the GSM phase 2+ release 96 (GSM R96). Table 2.1 shows the relation between GSM releases and CAMEL phases. The table also shows the evolution of the GSM standard into the third generation network standard. Table 2.1 does not consider 3GPP releases beyond Rel-7.

Major distinctive aspects of the CAMEL IN standard, compared with traditional IN standards, include:

- The IN control protocol from CAMEL, the CAMEL application part (CAP), is fully specified, including charging capability.
- CAMEL may be used in GSM networks consisting of equipment from multiple vendors.
- CAMEL services may be used for subscribers in their home network and for roaming subscribers.
- The CAMEL IN standard is tailored to the GSM network.
- CAMEL is evolving with further development in the GSM network.

2.7.1 Third-generation Partnership Project

Although ETSI is a *regional* (European) standardization institute, the GSM network standard developed by ETSI, including the CAMEL standard, is used by operators worldwide. The GSM network

---

[4] IN is often used as a term to refer to non-CAMEL IN standards such as ETSI CS1.

Table 2.1 Overview of GSM releases and CAMEL phases

GSM release	Organization	Year	CAMEL phase	Comment
GSM phase 1	ETSI	1992	–	
GSM phase 2	ETSI	1994	–	
GSM phase 2+ R96	ETSI	1996	Phase 1	
GSM phase 2+ R97	ETSI	1997	Phase 2	
GSM phase 2+ R98	ETSI	1998	Phase 2	CAMEL phase 2 in R98 is identical to CAMEL Phase 2 in R97
3G network R99	3GPP	1999	Phase 3	
3G network Rel-4	3GPP	2000	Phase 3	CAMEL phase 3 in Rel-4 is identical to CAMEL phase 3 in R99
3G network Rel-5	3GPP	2002	Phase 4	
3G network Rel-6	3GPP	2004	Phase 4	CAMEL phase 4 in Rel-6 is enhanced, compared to Rel-5
3G network Rel-7	3GPP	2006	Phase 4	CAMEL phase 4 in Rel-7 is enhanced, compared with Rel-6

development was taken over by the Third Generation Partnership Project (3GPP) from release R99 onwards. 3GPP also took over the maintenance of the existing GSM specifications. 3GPP is a consortium of various organizations, jointly developing the specifications for the Third Generation (3G) mobile network. The 3G mobile network is a global standard. Table 2.2 shows the organizations that form 3GPP. 3GPP has a well-defined structure of working groups, each of which carries responsibility for developing specific aspects of the mobile network. Table 2.3 lists the working groups responsible for the CAMEL standardization.

Table 2.2 3GPP organizations

Organization	Acronym	Region
Association of Radio Industries and Businesses	ARIB	Japan
Alliance for Telecommunications Industry Solutions	ATIS	North America
China Communications Standards Association	CCSA	China
European Telecommunications Standardisation Institute	ETSI	Europe
Telecommunications Technology Association	TTA	Korea
Telecommunications Technology Committee	TTC	Japan

Table 2.3 3GPP working groups for CAMEL standardization

Working group	Task
Service architecture – working group 1 (SA1)	Defining service requirements
Service architecture – working group 2 (SA2)	Overall network architecture
Service architecture – working group 5 (SA5)	Network management and charging aspects
Core network – working group 2 (CN2)	Overall CAMEL development responsibility
Core network – working group 4 (CN4)	Core network protocols

CAMEL was mainly developed by working group CN2. In 2005, the tasks of the CN2 working group and the CN4 working group were bundled into a new group, 'Core network and terminals – working group 4' (CT4).

2.7.2 CAMEL Standards and Specifications

CAMEL is defined in the following set of technical specifications (TS).

2.7.2.1 For GSM R96, R97 and R98

- GSM TS 02.78 [12] – this TS specifies the service requirements; it is also known as 'stage 1'.
- GSM TS 03.78 [38] – this TS specifies the technical implementation, information flows, subscription data etc; this TS is also referred to as 'stage 2'.
- GSM TS 09.78 [56] – this TS specifies the CAMEL application part (CAP) which is the IN protocol used between SCF and GSM/3G core network entities such as MSC/SSF. This TS is known as 'stage 3'.

The distinction between stage 1 specification, stage 2 specification and stage 3 specification is a common method in 3GPP.

2.7.2.2 For 3G R99, Rel-4, Rel-5, Rel-6 and Rel-7

- 3GPP TS 22.078 [66] – this TS specifies the service requirements; it is the 3G successor of GSM TS 02.78 [12].
- 3GPP TS 23.078 [83] – this TS specifies the technical implementation; it is the 3G successor of GSM TS 03.78 [38].
- 3GPP TS 29.078 [106] – this TS specifies CAP; it is the 3G successor of GSM TS 09.78 [56].
- 3GPP TS 23.278 [93] – this TS specifies the technical implementation of CAMEL control of IMS; this TS is applicable from 3G Rel-5 onwards. There is no separate stage 1 specification for CAMEL control of IMS.
- 3GPP TS 29.278 [111] – this TS specifies the CAP that is used for CAMEL control of IMS; this TS is applicable from 3GPP Rel-5 onwards.

The different GSM and 3G releases use distinctive version numbers for the specifications. This helps designers and implementers identify the specific document that is needed for the CAMEL specification in a specific GSM or 3G release. Table 2.4 lists the document version number per GSM or 3GPP release.

Table 2.4 GSM and 3G specification versions

GSM release	TS version	3G release	TS version
GSM R96	5.y.z	3GPP R99	3.y.z
GSM R97	6.y.z	3GPP Rel-4	4.y.z
GSM R98	7.y.z	3GPP Rel-5	5.y.z
		3GPP Rel 6	6.y.z
		3GPP Rel-7	7.y.z

When amendments are approved for a particular TS, the suffix of the version number (.y.z) is increased. For technical corrections, the .y part of the version number is normally stepped; for editorial corrections, the .z part of the version number is normally stepped.

2.7.2.3 Relation Between Stage 2 TS and Stage 3 TS

The CAMEL stage 2 specification (03.78/23.078) specifies information flows and the CAMEL stage 3 specification (09.78/29.078) specifies the syntax of CAP. There is often confusion about the priority of these specifications where it comes to optionality of information elements that may be carried in CAMEL operations. The rule is as follows:

- stage 2 defines semantics for the information flows (IF) and information elements (IE) that may be sent between gsmSSF and gsmSCF;
- stage 3 defines the syntax of conveying these IEs, encoded as operation parameters, through the CAP protocol.

When stage 2 and stage 3 *seem* to be contradicting for a particular IE/parameter, then stage 2 is leading. For example:

(1) GSM TS 03.78 [38] specifies for the Initial DP (IDP) IF that IMSI is a mandatory IE. GSM TS 09.78 [56] specifies, however, that the corresponding parameter in the initial DP operation is syntactically *optional*. Stage 2 is leading in this regard, meaning that the initial DP operation shall always contain IMSI.

(2) GSM TS 03.78 [38] specifies that an IN service may provide a no answer timer value in the range 10–40 s. The corresponding parameter in GSM TS 09.78 [56], application timer, has a range of 0–2047 s, but again, stage 2 is leading with a range of 10–40 s. A parameter that is specified in stage 3 may be used by different stage 2 information flows. These stage 2 information flows may specify different restrictions on the use of this parameter.

2.8 Principles of CAMEL

Possibly the most important characteristic of CAMEL is its mobility aspect. A GSM operator may offer an IN service to its subscribers; this IN service may be used in identical manner in the home network and when roaming in other networks. This is accomplished by strict specification of the gsmSSF, the SSF entity for GSM networks, in combination with CAMEL subscription data. In order for a subscriber to use a CAMEL service, the subscriber shall have CAMEL subscription data in her GSM profile in the HLR. When the subscriber registers in a PLMN, the HLR may transfer the CAMEL subscription data to the MSC/VLR. When that subscriber originates a call from that PLMN, the serving MSC may invoke a CAMEL service for that subscriber. The transfer of CAMEL subscription data from HLR to MSC/VLR is in line with the mobility aspect of GSM. A GSM network supports various basic services and supplementary services. Subscribers of the GSM network may subscribe to these services. That implies that the subscriber has subscription data in the HLR for those services; the HLR transfers the subscription data to the MSC/VLR. In that way, the supplementary service is *personalized* for that subscriber. Similarly, the transfer of CAMEL subscription data from HLR to MSC/VLR *personalizes* the CAMEL service invocation for a subscriber.

2.8.1 Location Update Procedure

When a GSM CAMEL subscriber registers in an MSC/VLR, *CAMEL capability negotiation* takes place between HLR and VLR. This negotiation entails that the HLR determines whether the sub-scriber is allowed to register in that VLR and which CAMEL data shall be sent to that VLR. This negotiation relates to the fact that different GSM networks have different levels of CAMEL support, i.e. the HPLMN of a GSM subscriber may support different CAMEL phases than the VPLMN. Examples include:

- HPLMN of a subscriber supports CAMEL phase 1 + CAMEL phase 2; VPLMN supports CAMEL phase 1 only;
- HPLMN of a subscriber supports CAMEL phase 1 + CAMEL phase 2; VPLMN does not support CAMEL.

Hence, a GSM subscriber who subscribes to a CAMEL phase 2 service for mobile-originating (MO) calls may roam to a PLMN that does not support CAMEL phase 2. If the subscriber registers in that PLMN, the HLR is not allowed to send that subscriber's CAMEL phase 2 subscription data to the VLR and, as a result, she cannot use her CAMEL phase 2 service. In this situation, the HLR shall take a *fallback* action during registration. This fallback action may be one of the following:

(1) The HLR allows normal registration, without sending CAMEL data to the VLR. This option may be used for GSM subscribers who subscribe to, for example, a CAMEL phase 2 VPN service and the operator does not have a CAMEL phase 1 VPN service. The subscriber will not have the VPN features available in this network, such as short number dialling.
(2) The HLR allows normal registration and sends CAMEL data of a lower phase, provided that that lower CAMEL phase is supported in the VLR. This option may be used for CAMEL pre-paid GSM subscribers. If the VPLMN does not support CAMEL phase 2, but supports CAMEL phase 1, then the subscriber may register with CAMEL phase 1. The service level of the CAMEL phase 1 service will be lower than that of CAMEL phase 2, but at least the pre-paid subscriber can register in the network and make outgoing calls.
(3) The HLR allows restricted registration. This option entails the HLR sending barring of all outgoing calls (BAOC) to the VLR. BAOC prevents the subscriber from establishing outgoing calls or forwarding calls. The ability to receive calls is not affected. The subscriber may use USSD Callback[5] to establish voice calls.
(4) The HLR disallows registration. This option may, for example, be used for CAMEL phase 2 pre-paid GSM subscribers, when the operator does not have CAMEL phase 1 pre-paid or USSD Callback service. The MS of the subscriber will attempt to register with another PLMN.

The above list of options is not mandatory for CAMEL, but is a common implementation in HLRs that support CAMEL. The fallback option may normally be set per subscriber; see Figure 2.10 for an example.

As more and more CAMEL subscription elements are introduced in later CAMEL phases, the algorithm in HLR to decide what CAMEL data to send to VLR gets more complex. The exact HLR behaviour, to determine which CAMEL data to send to VLR, remains operator-specific. However, the following rules are used by VLR and HLR:

- When an MSC/VLR does not indicate its supported CAMEL phases, the HLR may assume that the MSC/VLR supports CAMEL phase 1.
- When the HLR sends CAMEL data to the MSC/VLR, the MSC/VLR will, in response, indicate its supported CAMEL phases.

---

[5] USSD Callback is a service whereby a subscriber uses a USSD service code to request a service node in the HPLMN to establish a call-back call. See Chapter 4 for a description of USSD.

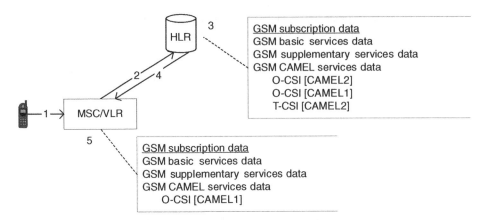

Figure 2.10 Registration procedure with fallback to CAMEL phase 1. (1) Subscriber with CAMEL phase 2 service registers with an MSC (in VPLMN). The MSC supports CAMEL phase 1, but not CAMEL phase 2; (2) MSC requests HLR (in HPLMN) for subscription data. MSC indicates to HLR that it (the MSC) supports CAMEL phase 1; (3) HLR cannot send CAMEL phase 2 data (O-CSI), but decides to send CAMEL phase 1 data instead; (4) HLR sends CAMEL phase 1 data (O-CSI) to MSC; (5) the subscriber is now registered in MSC; MSC will invoke a CAMEL phase 1 service for MO and mobile-forwarded (MF) calls.

- An MSC/VLR that supports a particular CAMEL phase will support the entire capability set of that CAMEL phase, in as far as the capability relates to MSC/VLR.[6]
- An MSC/VLR that supports a particular CAMEL phase will support all previous CAMEL phases. This rules guarantees that CAMEL services that are operational may continue to be used in a VPLMN when that VPLMN operator upgrades its core network to become compliant with the next CAMEL phase.

2.8.1.1 Selective CAMEL Support

A GSM operator may decide to offer the CAMEL capability in its core network to selected roaming partners, e.g. Vodafone UK offers CAMEL phase 2 capability to inbound roaming subscribers from Vodafone Germany, but not to inbound roaming subscribers from E-Plus Germany. This distinction may be made through IMSI analysis during registration in MSC. The operator may configure per IMSI range what capabilities are offered by the MSC. Further refinement is possible. An MSC that supports CAMEL phase 3 may offer CAMEL phase 3 capability to certain IMSI ranges, CAMEL phase 2 capability to certain other IMSI ranges, CAMEL phase 1 capability to some other IMSI ranges and no CAMEL capability to remaining IMSI ranges. It should be borne in mind that, when the MSC offers a particular CAMEL phase for a certain IMSI range, previous CAMEL phases are also offered to that IMSI range.

2.8.2 CAMEL Application Part

Although CAMEL includes a wide range of functionalities related to deploying IN in the GSM network, one major part of CAMEL is the IN control protocol, used between the gsmSSF and the gsmSCF. The CAMEL application part (CAP) is derived from Core INAP CS1. The capability of CAP is defined by means of 'operations'. An operation may be regarded as a mechanism for one entity to start a procedure in the peer entity. For example, the gsmSSF in an MSC invokes a

---

[6] This convention applies up to CAMEL phase 3; refer to Chapter 6 for CAMEL phase 4 subsets.

CAMEL service by sending the initial DP (IDP) operation to the SCP. The sending of IDP to SCP means that the gsmSSF starts a procedure in the SCP. The SCP may, in turn, send an operation to the gsmSSF; by doing so, the SCP starts a procedure in the gsmSSF. The entity receiving an operation may send a response to the sender of the operation. The sending of a response depends on the specific operation and on the outcome of the processing of the operation. Three types of information may be specified for each operation:

- *Argument* – the sender of an operation may include an argument in the operation. The argument contains parameters that shall be used as input for the procedure call. For example, the argument of the IDP operation contains a set of parameters that are used for service logic processing.
- *Result* – for some operations, a result is defined. The receiver of an operation may report the outcome of the processing of the operation in the result.
- *Errors* – for most operations, the receiver of the operation may return an error. An error is sent when the receiver encounters a problem in processing the operation. If the sender of an operation does not receive an operation error within a defined time period, then the sender assumes that the operation was executed successfully. This time period (known as 'operation time') is specified per CAP operation.

The concept of the operations is defined by ITU-T, in recommendations X.880 [155], X.881 [156] and X.882 [157]. Figure 2.11 shows an example of a CAP v1 operation (connect); this is extracted from GSM TS 09.78 [56].

For the connect operation, argument and errors are defined. The argument consists of a sequence of *parameters*. Each parameter in the connect argument, except for destination routing address, is

```
Connect                        ::= OPERATION
      ARGUMENT
          ConnectArg
      ERRORS {
          MissingParameter,
          SystemFailure,
          TaskRefused,
          UnexpectedComponentSequence,
          UnexpectedDataValue,
          UnexpectedParameter
          }

ConnectArg                          ::= SEQUENCE {
      destinationRoutingAddress     [0]  DestinationRoutingAddress,
      originalCalledPartyID         [6]  OriginalCalledPartyID        OPTIONAL,
      extensions                    [10] SEQUENCE SIZE(1..numOfExtensions) OF
                                              ExtensionField OPTIONAL,
      genericNumbers                [14] GenericNumbers               OPTIONAL,
      callingPartysCategory         [28] CallingPartysCategory        OPTIONAL,
      redirectingPartyID            [29] RedirectingPartyID           OPTIONAL,
      redirectionInformation        [30] RedirectionInformation       OPTIONAL,
      suppressionOfAnnouncement     [55] SuppressionOfAnnouncement    OPTIONAL,
      oCSIApplicable                [56] OCSIApplicable               OPTIONAL,
      ...
      }
```

Figure 2.11 Connect operation. Reproduced from GSM TS 09.78 v5.7.0 Section 6.1, definition of Connect OPERATION, by permission of ETSI

optional. That implies that the argument may or may not contain that parameter. The formats of the various parameters are specified in CAP. The errors definition for CAP connect indicates which error values may be returned to the SCP. Each error value (missing parameter, system failure etc.) is specified in CAP.

2.8.3 Abstract Syntax Notation

GSM uses a formal language to describe CAP. This formal language is the Abstract Syntax Notation 1 (ASN.1[7]). ASN.1 is defined in ITU-T X.680 [150], X.681 [151], X.682 [152] and X.683 [153]. ASN.1 is also used for most of the protocols specified for GSM, including, for example, MAP. ASN.1 facilitates the rigid definition of a protocol, in a compact manner. Extensive tutorials on ASN.1 include the works of J. Larmouth [170] and O. Dubuisson [171].

ASN.1 has mechanisms that allow for extending a protocol definition. CAP uses two of these mechanisms.

2.8.3.1 Ellipsis

Many data type definitions in CAP consist of a SEQUENCE of elements. Figure 2.12 contains an example (extracted from 3GPP TS 29.078 [106] Rel-5). The three dots at the end of the SEQUENCE definition are known as an 'ellipsis' or 'extension marker'. The ellipsis allows for future extension of the data type definition. Later CAMEL phases may, for example, add a new parameter to the Burst definition by placing a parameter after the ellipsis. Placing the new parameter after the ellipsis may be done without impacting the protocol version. If the receiver does not recognize any parameter after the ellipsis, then the receiver ignores that parameter.

A practical use case could be the addition of a frequency indicator in Burst. Currently, the CAMEL flexible warning tone uses the MSC built-in 900 Hz tone generator. A future CAMEL release could add a frequency indicator after the ellipsis. An MSC/gsmSSF that supports that new functionality would use that parameter for its flexible tone generation; an MSC/gsmSSF that does not support that new functionality ignores the parameter and uses the standard 900 Hz tone generator.

The ellipsis is used, for example, in 3GPP Rel-6 for adding new functionality to CAMEL phase 4 without having to introduce CAP v5.

2.8.3.2 Extension Container

The extension container is a data type definition that facilitates the transfer of operator-specific or vendor-specific information in an operation. Extension containers are included in most CAP

```
Burst ::= SEQUENCE {
    numberOfBursts          [0] INTEGER (1..3)       DEFAULT 1,
    burstInterval           [1] INTEGER (1..1200)    DEFAULT 2,
    numberOfTonesInBurst    [2] INTEGER (1..3)       DEFAULT 3,
    toneDuration            [3] INTEGER (1..20)      DEFAULT 2,
    toneInterval            [4] INTEGER (1..20)      DEFAULT 2,

    ...

    }
```

Figure 2.12 ASN.1 definition of CAMEL phase 4 flexible tone. Reproduced from 3GPP TS 29.078 v5.8.0, Section 5.1, definition of Burst SEQUENCE, by permission of ETSI

---

[7] At the time of defining ASN.1, it was envisaged that a successor, ASN.2, would be developed. However, there exists no ASN.2 at present.

operation arguments. The operator may decide to place designated information elements in the extension containers. Each extension container that is included in a CAP operation has an *identifier*. The identifier identifies the type of data that is contained in the extension container. The identifier will be unique for an operator; extension containers are identified by means of a global object identifier, which, if used properly, guarantees global uniqueness of a data type definition.

Examples of the use of extension containers include:

(1) An operator has configured the MSC/gsmSSF to place network-specific charging parameters in an extension container in CAP IDP. The CAMEL service uses this information to adapt its service processing, e.g. adapt the charging rate for the call.
(2) That same operator may include network-specific information in an extension container in CAP connect. This network-specific information may, for example, be an information element to be copied to ISUP initial address message (IAM).

Extension containers shall be used only between entities that are configured to use these specific extension container definitions. The extension container is therefore used only within an operator's own network or between networks of different operators when special agreements are in place.

2.8.3.3 Basic Encoding Rules

CAP protocol elements are encoded in accordance with the basic encoding rules (BER). BER defines a set of encoding rules specifically for formal language defined in ASN.1. A basic principle of BER is that data elements are encoded in the format, as presented in Figure 2.13.

The tag indicates the parameter that is encoded. If the data element to be encoded is, for example, the *numberOfBursts* from Figure 2.12, then the Tag takes the value 0; 0 is the value that is used as tag for the *numberOfBursts* parameter in the Burst data type. The length part of the BER-encoded data element indicates the number of octets contained in the data part. The data part contains the actual data that is conveyed. The type of data element that is contained in the data part, e.g. INTEGER, OCTET STRING or BOOLEAN, follows from the tag value.

The data part of the BER-encoded data element may itself be a constructed data type, such as a SEQUENCE. The encoded data element then takes the form indicated in Figure 2.14. BER is defined in ITU-T X.690 [154]. A good tutorial on BER is contained in *ASN1* [172]. Normally, when analysing the data transfer over a CAP dialogue, an analyser is used that performs

Figure 2.13 Structure of BER-encoded data element

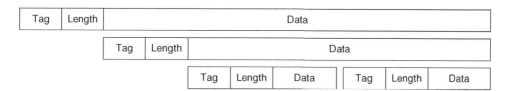

Figure 2.14 BER encoding of a constructed data type

the data decoding (BER decoding) and presents the CAP operations, results, errors, etc., in textual form.

2.8.4 Application Context

The application context (AC) is the mechanism for identifying the protocol and the protocol version of an application part (AP). Since various versions of CAP exist (CAP v1–CAP v4), the AC is especially needed. When an SCP receives a CAMEL service invocation request, it uses the AC to determine which protocol manager to use for this service, in other words, whether a CAMEL phase 1, CAMEL phase 2, etc., service is invoked. The definition of the AC for CAMEL is included in the specification of CAP. Figure 2.15 shows an example of an AC definition (extracted from 3GPP TS 29.078 [106] Rel-5). In this example, the AC is identified by 'id-ac-CAP-gsmSSF-scfGenericAC'. This AC name represents an object identifier, with value:

{itu-t(0) identified-organization(4) etsi(0) mobileDomain(0) umts-network(1) cap4OE(23) ac(3) 4}.

The object identifier is used to assign a globally unique identifier to a protocol object, such as an AC for CAP. When the gsmSSF starts the CAMEL service, by sending the initial DP operation to the SCP, it includes the AC in the service invocation request. Only the numerical values of the elements within the object identifier are transported, not the element tags. For the above example, the AC is represented by the following sequence of numbers:

0 4 0 0 1 23 3 4.

2.9 Signalling for CAMEL

The transfer of the CAP operations between the MSC/gsmSSF (or other applicable core network entities) and the SCP is done through the signalling system no. 7 (SS7) network, which is also used for the other application parts used in GSM, such as MAP or BSSAP. (Figure 2.16).

The signalling transfer points (STP) in the SS7 network provide signalling connection between the nodes in the SS7 network. SS7 defines a layered communication protocol, in accordance with the seven-layer open system interconnection (OSI) reference model, developed by the International Standards Organization (ISO). The OSI reference model is described in ITU-T X.200 [149]. Figure 2.17 depicts how the SS7 protocol stack, when used for CAP, relates to the OSI reference model.

SS7 allows for the transport of signalling data (e.g. ISUP, MAP, CAP) and user data (e.g. speech, data) through a common network. As such, SS7 is a common channel signalling (CCS) network. The SS7 layers that are relevant for CAP are described in the following sections.

```
capssf-scfGenericAC APPLICATION-CONTEXT ::= {
        CONTRACT                      capSsfToScfGeneric
        DIALOGUE MODE                 structured
        ABSTRACT SYNTAXES             { dialogue-abstract-syntax |
                                      gsmSSF-scfGenericAbstractSyntax }
        APPLICATION CONTEXT NAME      id-ac-CAP-gsmSSF-scfGenericAC }
```

Figure 2.15 Example of application context definition. Reproduced from 3GPP TS 29.078 v5.8.0, Section 6.1.2.1, definition of capssf-scfGenericAC APPLICATION-CONTEXT, by permission of ETSI

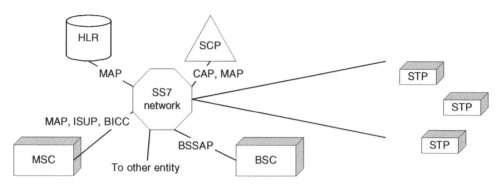

Figure 2.16   SS7 network

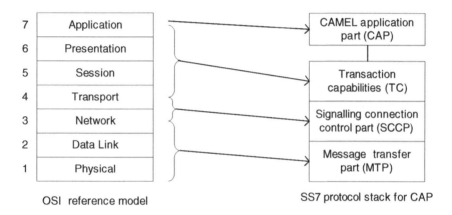

OSI reference model   SS7 protocol stack for CAP

Figure 2.17   Relation between OSI reference model and SS7 protocol stack for CAP

2.9.1 Message Transfer Part

The message transfer part (MTP) layer is responsible for transporting messages between signalling points in the SS7 network. A signalling point may, for example, be an STP, MSC/SSF, HLR or SCP. MTP is defined in ITU-T Q.701 [133]. In Figure 2.17, the signalling connection control part (SCCP) layer is indicated as the MTP user. Other protocols such as ISUP also run over MTP.

2.9.2 Signalling Connection Control Part

The SCCP part of the SS7 stack provides the signalling connection between two signalling end-points in the SS7 network. An MSC may, for example, address a message to the HLR of a subscriber. The MSC and HLR are in this case signalling end-points. The SCCP layer takes care of transporting the message to the correct HLR. The SCCP connection may run through one or more STPs. An STP determines the next signalling point for an SCCP message; an STP may also apply address translation (Figure 2.18). The SCCP link may span several networks, e.g. when MSC and SCP are located in different networks. When a signalling connection spans different regions, e.g. Europe and America, then one STP in the signalling connection interconnects between the European SCCP message format and the American SCCP message format.

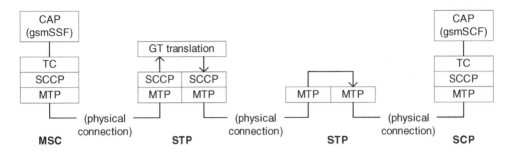

Figure 2.18   Message transfer through SCCP

SCCP is defined in ITU-T Q.711 [134]. Two releases of SCCP are commonly used:

(1) CCITT SCCP definition; this release of SCCP is known as 'Blue SCCP';
(2) ITU-T Q.711 – Q.716, which is the successor of the CCITT SCCP; this SCCP is known as 'White SCCP'.

The colour reference of the SCCP recommendations relates to the colour of the printed books in which these recommendations were published. White SCCP supports larger message length than Blue SCCP. This is achieved, amongst others, through message segmentation and re-assembly. In order to use this functionality of White SCCP, the entire link connection needs to support White SCCP. The increased message size of White SCCP may be needed, for example, when an operator uses the extension container mechanism in CAP, resulting in an SCCP message size that exceeds the maximum length for Blue SCCP.

For the transport of MAP messages that exceed the maximum Blue SCCP message length, an additional mechanism may be used: *MAP segmentation*. When an entity determines that the amount of data to be transported in an SCCP message exceeds the maximum length, it may transport the data in a series of individual MAP messages, each one within Blue SCCP message length constraint. The receiving entity combines the data received in the successive MAP messages. Examples where this mechanism is applied include:

- insert subscriber data (ISD) – from HLR to VLR;
- send routing information result (SRI-Res) – from HLR to GMSC;
- resume call handling (RCH) – from MSC to GMSC.[8]

Exceeding the Blue SCCP maximum message length may be caused by the inclusion of a full set of O-CSI conditional triggering or a full set of D-CSI.

2.9.2.1 Global Title Translation

Each entity in the SS7 network is addressable with a signalling point code (SPC), which is a locally unique number. Hence, to invoke a CAMEL service in the SCP, the SPC of the SCP is needed to send the first SCCP message, containing the CAP IDP operation, to this SCP. However, the gsmSCF address in the O-CSI is not the SPC of the SCP but a global title (GT) of the CAMEL service. One rationale of this principle is that the operator of the network containing the SCP may alter the SS7 network configuration in that network. Altering the network configuration may involve

---

[8] MAP segmentation for MAP ISD and MAP SRI-Res is introduced in GSM R97; for MAP RCH it is introduced in GSM R98.

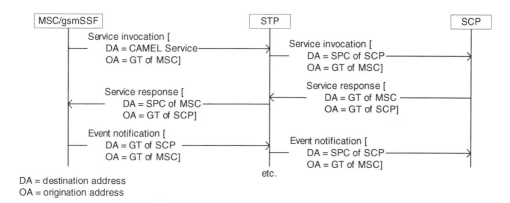

DA = destination address
OA = origination address

Figure 2.19 Global title translation

changes in SPC allocation. However, the GTs that are used to address the entities in the network in which the SPC was changed are not affected. This is accomplished by GT translation in the STP (Figure 2.19).

The STP translates the GT that is used to address the CAMEL service into the SPC of the corresponding SCP. When the MSC/gsmSSF receives the first response from the SCP, it stores the address of the responding SCP. This address will be the GT of that SCP. From that moment onwards, the MSC/gsmSSF uses this internally stored address of the responding SCP for further communication with that SCP, for the remainder of the CAMEL dialogue. The TC layer in the SS7 protocol stack in the MSC/gsmSSF does not report this GT of the responding SCP to the application in the MSC/gsmSSF, i.e. the CAP protocol. The application only knows the GT that it used to invoke the CAMEL service. Rationale of this restriction is that the entity that initiates a global service should not receive information about the network configuration of the other operator.

2.9.2.2 Subsystem Number

Entities in the GSM network have a subsystem number (SSN). The SSN is used to address a particular subsystem within an SS7 node. One node in a GSM network may contain, for example, MSC and HLR, or MSC, HLR and gsmSCF. Such a node would have one SPC in the SS7 network, but its internal subsystems would have different SSN. SSN for the GSM network are defined in GSM TS 03.03 [27]. Table 2.5 contains some SSNs that are relevant for CAMEL.

Table 2.5 Subsystem numbers for CAMEL

Entity	Protocol	SSN
HLR	MAP	6
VLR	MAP	7
MSC	MAP	8
	CAP	146
SCP	MAP	147
IM-SSF	MAP	147
SGSN	MAP	149
GGSN	MAP	150

The SSN may also be used in the STP, when performing GT translation, to derive the SPC of the destination entity. The CAP SSN is not allocated to a *node* but to a *protocol*. This SSN is used by any entity that talks CAP, such as gsmSCF, gsmSSF (in MSC, GMSC), gprsSSF (in SGSN), assisting gsmSSF, etc. When MSC/gsmSSF and gsmSCF reside in the same node in the SS7 network, then the SSN of an incoming SCCP message cannot be used to select the entity for which the message is destined. Instead, the node may have to analyse, for example, the application context of the protocol or the operation code. The SSN is used during dialogue establishment only. When MSC/gsmSSF and gsmSCF have established a CAMEL relationship, the two entities have exchanged a TC *transaction identifier*. The transaction identifier is unique within an SS7 node and relates to a specific (CAMEL) dialogue. The SS7 node allocates a new TC transaction identifier for each TC dialogue it initiates or accepts.

When CAMEL was first released (1996), SSN 5 was allocated to CAP. It turned out later that this SSN was already allocated to another entity in the SS7 network. For that reason, the SSN was changed to 146. At the same time, a separate SSN for MAP messaging with the SCP (SSN 147) was introduced.

2.9.3 Transaction Capabilities

The TC layer in the SS7 communication layer is responsible for establishing, maintaining and closing a (CAMEL) dialogue between two entities. This is done by the transfer of TC messages between the entities. The TC messages are also used to carry the CAP components, such as operation invoke, operation return or operation error. A TC dialogue runs from signalling end point to signalling end point. STPs in the signalling link are not involved in the TC dialogue. CAP operations are embedded in TC messages. The TC messages are encapsulated in SCCP messages and then transported to the destination entity by the SCCP and MTP.

TC is defined in ITU-T Q.771 [140]. The following TC messages are used for the dialogue management.

- *TC_Begin* – this TC message is used to initiate a TC dialogue; TC_Begin may contain operations like initial DP, initiate call attempt or assist request instruction.
- *TC_Continue* – this TC message is used to continue or to terminate a TC dialogue (with pre-arranged end) and to transfer one or more CAP operations, CAP error or CAP result.
- *TC_End* – TC_End is used to explicitly close a TC dialogue; this TC message may also contain one or more CAP operations.
- *TC_Abort* – TC_Abort is used to abort a TC dialogue when an error has occurred requiring the closing of the TC dialogue, but the entity does not have the possibility to close the dialogue in a regular way. The TC_Abort may be issued by the TC itself ('P_Abort') or by the TC_User, i.e. the application ('U_Abort').

Figure 2.20 shows an example of TC signalling between MSC/gsmSSF and SCP. The names of the CAP Operations used in Figure 2.20 are listed in the Appendix.

The TC dialogue is *established* as soon as the gsmSSF has received the first TC_Continue. From then onwards, the gsmSSF addresses the CAMEL service logic instance in the SCP, with the transaction identifier. The transaction identifier is included in the first TC_Continue from the SCP.

Similar to SCCP, network entities can choose between Blue and White (and Pure White) communication services. CAMEL mandates the use of White (or Pure White) TC. That means that CAP should attach itself to TC as a White (or Pure White) TC_User. The reason is that White (and Pure White) TC supports the transport of the AC in the TC message. The TC_Begin needs to contain the AC, so the SCP knows which version of CAP is requested. The transfer of AC is done only during dialogue establishment. *Note*: the term 'Pure White TC', also known as 'White+ TC', is terminology used in the industry. It refers to a form of TC usage whereby the sending

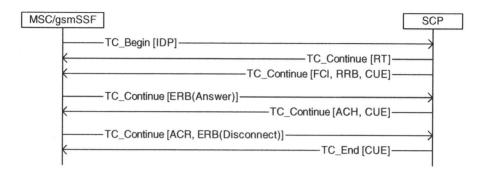

Figure 2.20 Example TC signal sequence

entity uses White TC and expects the receiving entity to support White TC as well. When, on the other hand, normal 'White TC' is used by the sending entity, it may establish a TC dialogue with an entity that supports Blue TC. However, the receiving entity will not receive the AC in that case.

MAP messages, on the other hand, may be transported through Blue TC or White TC. When the AC is not present in MAP dialogue establishment, the addressed subsystem uses a default AC version.

2.9.3.1 TC Components

CAP operation, errors and results are transported as TC components. A TC message may contain one or more TC components. Table 2.6 lists the types of TC components that may be transported in a TC message.

2.9.3.2 CAP Component Execution

The gsmSSF and gsmSCF, and other entities that talk CAP, may include multiple CAP components in a single TC message, instead of transporting these CAP components in separate TC messages. This method has two-fold purpose: (1) signalling efficiency and (2) error handling. Figure 2.21 shows an example of a TC buffer in the gsmSSF with multiple CAP operation components.

When the gsmSSF receives the TC_Continue message containing the operations apply charging (ACH), furnish charging information (FCI) and continue (CUE), it starts processing these operations

Table 2.6 TC Component types

Component type	Description
TC-Invoke	Used to invoke the execution of a CAP operation
TC-Return-L	Used to convey the result of a CAP operation. The 'L' suffix indicates that no further return components for this operation will follow
TC-Error	This component indicates that the processing of a CAP operation failed. It contains one of the error values that are defined for this operation
TC-Reject	This component indicates that TC rejected a previous TC-Invoke, TC-Return (*Last* or *Not last*) or TC-Error component
TC-Return-NL	Used to convey the result of an operation. The 'NL' suffix indicates that this return component forms part of a segmented operation result and that more return component(s) will follow. CAMEL does not use this TC component type

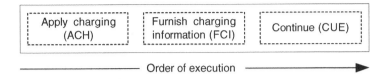

Figure 2.21 TC buffer with multiple CAP operations

in this order. Should the processing of any CAP operation fail, then the gsmSSF reports the execution error to the SCP (provided that the failed operation has error definition) and discards those operations from the TC buffer for which the execution has not yet started. If in the example in Figure 2.21 the processing of FCI fails, then the SCP concludes from the FCI error that the CUE was not executed. The SCP may now take corrective action, which may be sending CAP CUE or CAP release call (RC). Grouping CAP operations in this manner should be done only for CAP operations that have error definition, unless the CAP operation is the last one in the TC message. Otherwise, failure in execution of such operation would have the effect of other CAP operations from the same TC message being discarded without notice to the SCP.

2.10 Dynamic Load Sharing

Dynamic load sharing is a technique whereby an operator may distribute the load for a particular CAMEL service over two SCPs. Dynamic load sharing may also be used for other entities in the network, such as HLR, but that is not considered in this section. As described in an earlier section, the STP in the served subscriber's HPLMN translates the GT that is used to address the SCP where the CAMEL service resides into a corresponding SPC. When dynamic load sharing is applied, the STP translates the GT of the CAMEL service into <u>one of two SPCs</u>. These SPCs are both associated with an SCP where the required CAMEL service is located. The related architecture is shown in Figure 2.22.

The STP applies a dynamic translation from GT to SPC. This dynamic translation is applied only to GTs that address a *service*, such as an IN service (or an HLR). Every time a gsmSSF initiates a CAMEL dialogue to such GT, the STP selects between two SPCs associated with this GT. The two SPCs belong to two SCPs that contain the requested CAMEL service. The SCP that receives a particular service invocation request returns its own GT in the first TC_Continue message. This GT is used in further signalling from gsmSSF to gsmSCF. This GT does not indicate a *service*, but a node. Hence, the STP translates that GT when handling an SCCP message destined for that GT into

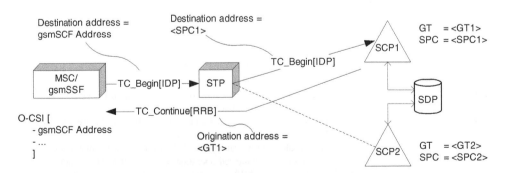

Figure 2.22 Dynamic load sharing

the SPC for the corresponding SCP. According to this configuration, the HPLMN operator needs to allocate three GTs for addressing the CAMEL service: one GT to indicate the CAMEL service and one GT for each SCP. A convention, when using dynamic load sharing, is that the CAMEL service that is shared between two SCPs has no subscriber-specific data. Any subscriber-specific data resides in an external node, such as an SDP, which may be addressed by either SCP. The addressing of this external node by the SCP may be based on subscriber's IMSI, MSISDN or other vendor-specific means. As a result, when a subscriber initiates a call and a CAMEL service is started for that call, the CAMEL service may be handled by one of two SCPs, but the subscriber-specific data is retrieved from the designated node for that subscriber.

If the SCPs in load sharing configuration do contain subscriber-specific data, then the SCPs will apply a method whereby the subscriber data is updated over the two SCPs as and when the subscriber data changes.

The load sharing method as described in the present section applies in principle to any CAMEL dialogue establishment towards a gsmSCF. Other scenarios where load sharing may be applied include CAP v3 or CAP v4 between smsSSF and gsmSCF, CAP v3 between gprsSSF and gsm-SCF, etc. Chapter 5 explains that the CAP v3 signalling between gprsSSF and gsmSCF has some CAMEL-specific aspects for load sharing.

The load sharing based on the dynamic GT translation allows for evenly distributed load from one STP to two SCPs. The STP may also apply another load sharing ratio. The dynamic load sharing over multiple STPs is a matter of network configuration. When more than two SCPs are used, an operator may use multiple gsmSCF addresses in the CAMEL subscription data in HLR.

2.11 Using Signalling Point Code for Addressing in HPLMN

This section specifies a possible signalling optimization, although not formally specified by CAMEL. When the gsmSSF sends TC_Begin to start a CAMEL service, it includes its own address in the form of a GT. When the gsmSCF receives this TC_Begin, it responds to the initiator of the dialogue and includes its own address, again in the form of a GT, in the response message. In this way, gsmSSF and gsmSCF exchange one another's GTs. These GTs are used for the remainder of the CAMEL dialogue for the sending of SCCP messages. The use of GT for the respective entities guarantees that the CAP dialogue may be established internationally. Every TC message sent between gsmSSF and gsmSCF transits through an STP, for the translation of the GT of the destination entity into that entity's SPC.

However, when the CAP signalling is taking place between a gsmSSF and a gsmSCF in the HPLMN (or in the VPLMN), CAMEL dialogue continuation could be based on SPC instead of GT. This could be achieved in the following manner:

Figure 2.23 Using SPC for CAP signalling in HPLMN

- when gsmSSF sends TC_Begin to initiate a CAMEL dialogue and the GT of the CAMEL service is associated with the same PLMN as the gsmSSF, then gsmSSF uses its SPC as an origination address in the SCCP message carrying the TC_Begin;
- gsmSCF notices that gsmSSF uses SPC as origination address and hence uses its SPC as an origination address in the SCCP message carrying the TC_Continue.

This is reflected in Figure 2.23 (gsmSSF and gsmSCF in HPLMN; STP is not reflected in this figure). In this manner, dynamic load sharing may still be used, since gsmSSF uses GT as a destination address when initiating the CAMEL dialogue. The effect is, however, that gsmSSF and gsmSCF continue the CAMEL dialogue, which is taking place in the HPLMN (or VPLMN), using one another's SPC. Using SPC reduces the load on the STP and increases signalling throughput.

3

CAMEL Phase 1

In the previous chapters, the concept of HPLMN, VPLMN and IPLMN was described briefly. These network domains play a crucial role in the CAMEL architecture.

3.1 Architecture for CAMEL Phase 1

Figure 3.1 depicts the network architecture for CAMEL phase 1.

3.1.1 Functional Entities

Chapter 1 has already provided an introduction into the various functional entities for the GSM network. The present section describes these entities from a CAMEL perspective.

Although Figure 3.1 includes a single HLR, a single gsmSCF etc., there may be multiple instances of each functional entity in a GSM network. In addition, in a particular network configuration, two or more entities may be located in the same physical node. In most networks, the MSCs may also function as GMSCs. In small networks, the gsmSCF may be co-located in an MSC/gsmSSF.

3.1.1.1 HLR

The HLR is the entity in the network that contains the CAMEL subscription data. CAMEL subscription data is grouped in *CAMEL service information (CSI)* elements. Chapter 2 has already described the principle of *subscribed IN services*. CAMEL services are in principle all 'subscribed' IN services. This means that a subscriber has a subscription to a CAMEL service. This means that the subscriber has the corresponding CSI in her GSM subscription profile in the HLR. The contents of the CSI relate to the subscribed CAMEL service, i.e. which CAMEL service will be invoked.

Subscription data, contained in the HLR, can be distributed to the various nodes in the GSM network, including MSC, GMSC, SGSN[1] and gsmSCF.[2] These nodes may be located in the HPLMN or in VPLMN. Refer also to Section 2.8.1 for a description of the distribution of the CAMEL data through the network.

The HLR always resides in the HPLMN of the served CAMEL subscriber. The HLR may send the CAMEL subscription data to the MSC/VLR and to the GMSC. These two cases differ as follows:

---

[1] The sending of CAMEL subscription data from HLR to SGSN is applicable from CAMEL phase 3 onwards.
[2] Refer to Section 5.8.

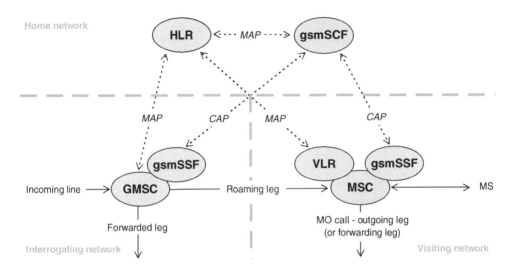

Figure 3.1 CAMEL phase 1 network architecture. Reproduced from GSM TS 03.78 v5.8.0, Figure 4/1, by permission of ETSI

- CSIs that are sent from HLR to MSC/VLR remain present in that MSC/VLR as long as the subscriber is registered in that MSC/VLR. When a subscriber switches her phone off, she remains registered in the MSC. When the phone is switched on again, the MSC/VLR does not need to fetch the CAMEL data from the HLR again.
- When a subscriber's phone is switched off for a long time, that subscriber's data, including the CAMEL subscription data, may be removed from the VLR. As a result, the subscriber is no longer registered in that MSC. This duration is operator-specific, e.g. 24 h.
- When the subscriber is attached, but the VLR does not have radio contact with a particular subscriber for a certain, operator-specific period, then the VLR may also decide to purge that subscriber's data, including the CAMEL subscription data.
- When the subscriber re-establishes contact with the MSC (e.g. switches on the phone), the MSC performs a location update, after which the subscriber is registered again in the MSC, including CAMEL data.
- CSIs that are sent from HLR to GMSC remain present in that GMSC only for the duration of terminating call handling. When the call is cleared, the CAMEL data is discarded in the GMSC. This means that, for every terminating call handling in the GMSC, the GMSC receives the CAMEL data (and other GSM data) from the HLR. Terminating calls for a subscriber may be handled by different GSMCs in the network.

This behaviour of the MSC/VLR, GMSC and HLR is in line with generic GSM principles.

Another function of the HLR is to apply CAMEL-specific handling for MT calls. As an example, the HLR may obtain subscriber state and subscriber location from the VLR, and send this information to the GMSC, so the GMSC can send the information to the gsmSCF, as input for MT CAMEL service processing. The HLR handling for MT calls is described below. A third CAMEL-related function of the HLR is any time interrogation, which is also described below.

3.1.1.2 gsmSCF

The gsmSCF is the entity where the CAMEL services reside; the gsmSCF resides in the HPLMN. All CAMEL service requests are directed to a gsmSCF. (Chapter 5 explains that, as from CAMEL phase 3, the possibility exists for the gsmSCF to reside in the VPLMN.) The gsmSCF is a *functional entity*. The node in which the gsmSCF resides is called the 'service control point' (SCP). The term 'gsmSCF' was introduced in the GSM CAMEL specifications to make a clear distinction between CAMEL standardized SCF and proprietary SCF. Hence, the following convention applies:

- *gsmSCF* – service control function, in accordance with GSM standardized protocol (i.e. CAMEL); the gsmSCF supports CAP and relevant parts of MAP. A gsmSCF may also support other protocols than CAP and MAP; however, support of other protocols by the gsmSCF is not specified by CAMEL.
- *SCF* – service control function, in accordance with ETSI/ITU-T INAP (e.g. CS1, CS2; see Chapter 2) or vendor-specific IN protocols. The SCF may also support relevant parts of MAP (or proprietary extensions to MAP), and other protocols.

An SCP may contain both a gsmSCF and an SCF. That means that such an SCP may run CAMEL services and CS1, or other non-CAMEL services at the same time, e.g. an SCP may offer a CS1 service to a subscriber in the HPLMN. When that same subscriber is registered in another PLMN, then the SCP may offer a CAMEL phase 2 service to that subscriber. To accomplish such service differentiation, the HLR needs to send different IN subscription data to MSC/VLRs in different networks. The sending of any non-CAMEL subscription data from HLR to MSC/VLR or GMSC is not specified in the CAMEL standard. However, extension mechanisms in the various MAP operations allow for the transfer of operator-specific information to MSC/VLR and GMSC.

An operator may use CS1 (or other non-CAMEL protocol) in the HPLMN for a pre-paid service. That operator may wish to allow its subscribers to roam abroad, using CAMEL, but may also wish to keep on using CS1 for the pre-paid service, for subscribers in the HPLMN. In that case, the operator would operate both a CS1 service and a CAMEL service, for the same subscriber group. These services may reside in the same SCP or in different SCPs.

3.1.1.3 MSC/gsmSSF

The MSC/gsmSSF is an MSC with integrated CAMEL capability. The CAMEL capable MSC has the ability to receive and store CAMEL data in the VLR. The MSC also has an integrated gsm service switching function (gsmSSF). The gsmSSF forms the relay between the MSC and the gsmSCF. When a call is established in the MSC, the MSC ascertains whether that call will be subject to CAMEL handling. The presence of O-CSI in the VLR for that subscriber indicates that the MSC will request the invocation of a CAMEL service for that call.

If the MSC has determined that the call shall be subject to CAMEL control, then the MSC instantiates a gsmSSF process.[3] The gsmSSF then establishes a *relationship* with the gsmSCF in the HPLMN of that subscriber, provided that any available trigger criteria are fulfilled.[4] *Conditional triggering* is described in Chapter 4. The details of the CAMEL relationship that may now be established are described in Chapter 2.

There is a similar distinction in naming for CAMEL SSF and non-CAMEL SSF:

- *gsmSSF* – service switching function, in accordance with GSM standardized protocol (i.e. CAMEL); a gsmSSF is always located in a visited MSC (VMSC, i.e. MSC connected to the radio access network) or gateway MSC (i.e. MSC that may interrogate an HLR as part of MT call handling). In CAMEL phase 4 in 3GPP Rel-7, the gsmSSF may also be located in a transit MSC.

---

[3] CAMEL specifies the gsmSSF as a *process*. When a MSC or GMSC establishes a CAMEL control relationship with the gsmSCF, an *instance* of a gsmSSF process is started.

[4] Conditional triggering is specified for CAMEL phase 2 onwards.

- *SSF* – service switching function, in accordance with non-CAMEL protocol; an SSF may be located in VMSC or GMSC but also in transit MSC or fixed network exchange.

An MSC or fixed network exchange with integrated SSF is referred to as a service switching point (SSP). The term SSP is generally not used for the combination of MSC or GMSC and gsmSSF.

An MSC may have an integrated gsmSSF and an integrated SSF at the same time. That means that that MSC can establish CAMEL dialogues and non-CAMEL dialogues at the same time. However, for a particular registered subscriber, the MSC would establish either a CAMEL dialogue or a non-CAMEL dialogue, e.g. an MSC may offer CAMEL or non-CAMEL services (e.g. CS1) to registered subscribers that belong to the same PLMN as the PLMN in which the MSC resides. For registered subscribers from other PLMNs, the MSC may offer CAMEL services only.

Mobile number portability (MNP) specifies a method whereby an MSC may invoke a CS1 service for a call, after a CAMEL service was invoked. See Chapter 6 for MNP. In addition, vendors may use a proprietary triggering mechanism to have an MSC invoke a CAMEL service and a CS1 service in the same call.

3.1.1.4 GMSC/gsmSSF

The GMSC/gsmSSF is a GMSC with integrated CAMEL capability. The CAMEL-capable GMSC has the capability to receive CAMEL data during terminating call handling. The GMSC also has an integrated gsmSSF, which forms a relay between the GMSC and the gsmSCF. When a call arrives at the GMSC (i.e. GMSC receives ISUP initial address message (IAM) for a subscriber from that network[5]) and the GMSC receives CAMEL data from the HLR (in the MAP SRI-Res message), then the GMSC instantiates a gsmSSF process. The gsmSSF then establishes a relationship with the gsmSCF in the HPLMN of the terminating subscriber. As described in a previous chapter, although the GMSC is conceptually located in the IPLMN, the IPLMN and HPLMN are normally one and the same network. Hence GMSC and SCP are normally in the same network.

Although CAMEL specifies a single gsmSSF process for the MSC and for the GMSC, the gsmSSF process instances behave differently, depending on the call case. Similar to the MSC, the GMSC may have an integrated gsmSSF and SSF at the same time, i.e. the GMSC may establish CAMEL relationships and non-CAMEL relationships at the same time, for different subscribers. Proprietary triggering mechanisms allow for having the GMSC invoke a CS1 service (or other non-CAMEL service) and a CAMEL service in the same call.

There is a slight difference between the co-existence of CAMEL and non-CAMEL relationships in the VMSC and the co-existence of CAMEL and non-CAMEL relationships in the GMSC. A subscriber from PLMN-A may receive a CAMEL service when registered in an MSC from PLMN-B (e.g. roaming abroad) and a proprietary, non-CAMEL service when registered in an MSC from PLMN-A (i.e. registered in her home network).

The GMSC, on the other hand, is normally handling terminating calls for subscribers from the same network only; this is the result of the fact that the GMSC normally resides in the HPLMN. As a result, terminating call handling is always done in the HPLMN of the called subscriber, independent of the origin of the call (i.e. calling network/subscriber) and independent of the location of the called subscriber. The operator can therefore offer a consistent IN service for MT calls to a particular subscriber group, i.e. either a proprietary IN service or a CAMEL service. This differentiation in IN services is reflected in Figure 3.2.

Should an operator have different subscriber groups (in HLR), whereby these subscribers have different IN services, then one group of subscribers may consistently receive a CAMEL service in

---

[5] MNP may have the effect that a GMSC initiates an HLR interrogation for a subscriber who belongs to another network.

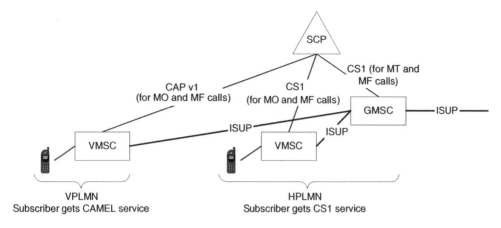

Figure 3.2 CAMEL service (in VPLMN) vs CS1 service (in HPLMN)

the GMSC, whereas the other group of subscribers may consistently receive a proprietary service in the GMSC.

Besides MT calls, the GMSC may also handle MF calls. When an MT call in the GMSC results in call forwarding, then the forwarded call may in turn also be subject to a CAMEL service. The HLR may send both O-CSI and T-CSI to the GMSC:

(1) *T-CSI* – the HLR sends T-CSI to the GMSC for the invocation of an MT call CAMEL service.
(2) *O-CSI* – the HLR sends O-CSI to the GMSC for the invocation of an MF call CAMEL service.[6]
   An HLR may be configured as follows:
   (a) the HLR sends O-CSI to the GMSC only in the case that GSM call forwarding is pending, i.e. when the HLR sends a forwarded-to number (FTN) to the GMSC; should a terminating call CAMEL service induce call forwarding when there is no GSM call forwarding pending, then the GMSC does not have O-CSI available to invoke a CAMEL service for the forwarded call;
   (b) The HLR sends O-CSI to the GMSC when GSM call forwarding is pending and also when it sends T-CSI to the GMSC; should the terminating call CAMEL service induce call forwarding, then the GMSC has O-CSI available to invoke a CAMEL service for the forwarded call.

The choice between method (1) and (2) is HLR-vendor specific.

As a result of call forwarding in the GMSC, it may occur that two CAMEL services are active simultaneously in the GMSC, for a subscriber: (1) a CAMEL service for the MT call; and (2) a CAMEL service for the MF call. In principle, the gsmSSF process instance for the MT call and the gsmSSF process instance for the MF call work independently of one another. That entails that these gsmSSF process instances have their own basic call state model (BCSM) and their own CAMEL dialogue. In addition, the actual CAMEL services, i.e. the MT CAMEL service and the MF CAMEL service, work independently of one another. These services may be running on the same or on different SCPs (Figure 3.3). The combination of MT call and MF call is further described in a later section.

---

[6] When call forwarding in the GMSC is the result of ORLCF, then the GMSC does not receive the O-CSI from HLR, but from VMSC.

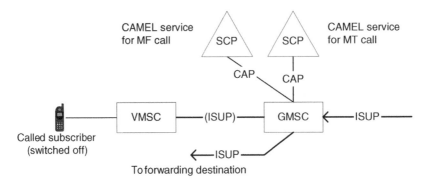

Figure 3.3   Combination of MT call CAMEL service and MF call CAMEL service from the GMSC

Table 3.1   CAMEL phase 1 interfaces

Interface	Protocol	Purpose
VLR – HLR	MAP v3	Transport of CAMEL trigger data from HLR to VLR; removal of trigger data from VLR Retrieval of subscriber information during mobile terminating call handling Retrieval of subscriber information using any time interrogation
GMSC – HLR	MAP v3	Terminating call handling
gsmSCF – HLR	MAP v3	Any time interrogation
MSC – GMSC	MAP v3, MAP v4[7]	Optimal routing of late call forwarding
MSC/gsmSSF – gsmSCF	CAP v1	CAMEL control of mobile originated calls CAMEL control of mobile forwarded calls (late forwarded calls)
GMSC/gsmSSF – gsmSCF	CAP v1	CAMEL control of mobile terminated calls CAMEL control of mobile forwarded calls – early forwarded calls CAMEL control of mobile forwarded calls – late forwarded calls in combination with optimal routing

3.1.2 Information Flows

Table 3.1 provides an overview of the network interfaces involved in CAMEL phase 1. For further details of the MAP messages, refer to GSM TS 03.18 (Basic Call Handling) [31], GSM TS 03.78 (CAMEL) [38], GSM TS 03.79 (Optimal Routing) [39] and GSM TS 09.02 (MAP) [54].

3.1.2.1 VLR–HLR

This interface is used for administrative purposes, terminating call handling and for the retrieval of subscriber data. This interface uses the MAP procedures listed in Table 3.2.

---

[7] MAP v4 for this interface is normally not needed for CAMEL phase 1.

Table 3.2 MAP messages for the VLR–HLR interface

Location update (LU)	When a subscriber registers with an MSC, the MSC uses MAP LU to request the HLR of that subscriber for GSM subscription data
	If the VLR supports CAMEL, then the VLR will indicate this in MAP LU; the VLR will indicate each individual CAMEL phase that it supports.[8] The indication of CAMEL support in MAP LU tells the HLR that it is allowed to send CAMEL subscription data to that VLR
Restore data (RD)	The VLR uses MAP RD when it needs to reload its subscription data. The RD procedure is not triggered by subscriber action, such as registration in MSC. The handling of the RD procedure is similar to the handling of the location update procedure. That is, the VLR indicates to HLR which CAMEL capability is supported in the VLR, so the HLR can decide whether it can send CAMEL data to the VLR for that subscriber
Insert subscriber data (ISD)	The HLR uses MAP ISD for the transfer of GSM subscription data to VLR. MAP ISD may be used in response to MAP LU or MAP RD. When the VLR has indicated that it supports CAMEL and the HLR contains CAMEL subscription data for the subscriber, then the HLR may include the CAMEL subscription data in MAP ISD
	The HLR keeps track of the CAMEL data that is transferred to that VLR; the HLR also keeps track of the CAMEL capability that is supported in that VLR. Refer to GSM TS 03.08 [28] for an overview of subscriber and VLR-related data that is stored in the HLR
	MAP ISD may also be used outside the context of a location update procedure. When subscriber data changes in the HLR, the HLR may use MAP ISD to update the VLR with the new subscription data. The HLR uses the internally stored subscriber and VLR-related data to determine whether the modified subscription data may be sent to the VLR or the HLR shall take fallback action
	When CAMEL data needs to be sent to the VLR outside the context of a location update procedure, then the HLR may have to use a combination of MAP DSD (to remove the CAMEL data from the VLR) and MAP ISD (to place the modified CAMEL data in the VLR)
Delete subscriber data (DSD)	The HLR uses MAP DSD to remove subscription data from the VLR. When the HLR uses MAP DSD to remove CAMEL phase 1 or 2 subscription data, then all CAMEL phase 1 and CAMEL phase 2 subscription data is removed from VLR. This is due to protocol limitation in MAP. From CAMEL phase 3 onwards, the HLR may remove individual CAMEL subscription elements
Provide subscriber information (PSI)	The HLR may use MAP PSI to obtain subscriber data from VLR; the subscriber data that may be obtained with MAP PSI includes location information and subscriber state. In CAMEL phase 3 and CAMEL phase 4, additional information may be obtained from the VLR
	Within the context of CAMEL, the HLR may use MAP PSI for the following procedures: (1) MT call handling (2) ATI
	The HLR may also use MAP PSI for optimal routing (OR); refer to GSM TS 03.79 [39] for a description of OR
Provide roaming number (PRN)	The HLR uses MAP PRN to obtain a mobile station roaming number (MSRN) from the VLR. Obtaining an MSRN from the VLR is part of normal MT call handling, even when CAMEL is not invoked
	For MT calls that are subject to CAMEL handling, the HLR may include the call reference number (CRN) and GMSC address (GMSCA) in MAP PRN. The HLR receives CRN and GMSCA from the GMSC, in MAP SRI. Placing the CRN and GMSCA in MAP PRN allows for the linking of call detail records (CDR) by post-processing systems. For CDR processing, including the linking of CDRs, refer to Chapter 7
	The inclusion of CRN and GMSC Address in MAP PRN is also done for the purpose of optimal routing of late call forwarding

---

[8] If a VLR supports CAMEL phase 4, then the VLR will also report which subset(s) of CAMEL phase 4 are supported by that VLR. See Section 6.1.2.

3.1.2.2 GMSC–HLR

This interface is used for terminating call handling. This involves the MAP procedure listed in Table 3.3.

3.1.2.3 gsmSCF–HLR

This interface is used for retrieval of subscriber information. This involves the MAP procedure listed in Table 3.4.

3.1.2.4 MSC–GMSC

This interface is used for optimal routing of late call forwarding (ORLCF). This involves the MAP procedure listed in Table 3.5.

Table 3.3 MAP messages for the GMSC–HLR interface

Send routing information (SRI)[9]	The GMSC uses MAP SRI for the purpose of MT call handling. When the GMSC sends MAP SRI to the HLR, it includes an indication of the CAMEL phases that are supported by that GMSC. If the GMSC supports CAMEL phase 4, then it also indicates which subset(s) of CAMEL phase 4 are supported
	The indication of CAMEL support in MAP SRI tells the HLR that it is allowed to send CAMEL subscription data to that GMSC. If the HLR sends CAMEL subscription data to the GMSC, in response to MAP SRI, then the GMSC establishes a CAMEL relationship with the gsmSCF. The CAMEL service in the gsmSCF can now control the MT call or the MF call (in the case of call forwarding) for this subscriber
	The GMSC–HLR interface is also used for forwarding interrogation; this feature is briefly described in Chapter 5.

Table 3.4 MAP messages for the gsmSCF–HLR interface

Any time interrogation (ATI)	The gsmSCF may use MAP ATI to request the HLR for subscriber information. ATI may be used at any moment, also outside the context of a call. The gsmSCF may use ATI to request the following subscriber data:
	• Subscriber location and • subscriber state
	CAMEL phases 3 and 4 specify enhanced capability for ATI, when used between gsmSCF and HLR
	CAMEL phase 3 and CAMEL phase 4 also specify the usage of ATI for the purpose of location services and mobile number portability

---

[9] GSM specifications use both the American spelling, 'Send Routing Info', and the UK spelling, 'Send Routing Info'.

Table 3.5 MAP messages for the MSC–GMSC interface

Resume call handling (RCH)	When the VMSC sends MAP RCH to the GMSC, in order to initiate ORLCF, it may include CAMEL subscription information, specifically O-CSI and D-CSI (CAMEL phase 3 onwards). The GMSC may use O-CSI and D-CSI in MAP RCH, to invoke a CAMEL service for the forwarded call that will be established in the GMSC. MAP RCH may use MAP v3 or MAP v4 The interworking between CAMEL and optimal routing is extensively described in Chapter 4

3.1.2.5 MSC/gsmSSF–gsmSCF

This interface is the CAP v1 interface for call control. A CAP v1 *dialogue* is established by the MSC/gsmSSF as a result of mobile originated call establishment or mobile forwarded call (late call forwarding) establishment.

The CAP v1 dialogue establishment from the MSC/gsmSSF is the result of the presence of O-CSI in the VLR. The address of the gsmSCF, and other protocol-related details, are retrieved from O-CSI. The same O-CSI is used for both MO and MF calls. As a result, the same CAMEL service is invoked for MO and MF calls. However, the MSC/gsmSSF signals to the gsmSCF the type of call (MO or MF call), so the CAMEL service can adapt its behaviour to the particular type of call.

3.1.2.6 GMSC/gsmSSF–gsmSCF

This interface is the CAP v1 interface for call control. A CAP v1 *dialogue* is established by the GMSC/gsmSSF as a result of a mobile terminated call or mobile forwarded call (early call forwarding or late call forwarding with optimal routing). The CAP v1 dialogue establishment from the GMSC/gsmSSF is the result of the reception of T-CSI/O-CSI from the HLR. The address of the gsmSCF, and other protocol related details, are retrieved from T-CSI and O-CSI.

The O-CSI that may be sent to the GMSC is normally the same as the O-CSI that is sent to the VLR. As a result, the three forms of call forwarding (early call forwarding, late call forwarding and ORLCF) will invoke the same CAMEL service.

All MAP interfaces mentioned above, except ATI, are specified for GSM call handling and are not CAMEL-specific. However, CAMEL specifies additional information elements that are transported via these MAP procedures. The CAP v1 interfaces are specific for CAMEL. The CAP v1 interfaces are specified in GSM TS 03.78 ('stage 2') [38] and GSM TS 09.78 ('stage 3') [56].

3.2 Feature Description

The main features offered by CAMEL Phase 1 include:

(1) control of mobile originated calls;
(2) control of mobile terminated calls;
(3) control of mobile forwarded calls;
(4) any time interrogation.

In this context, a 'call' refers to a CS call. The different call handling processes are described in more detail below.

3.2.1 Mobile-originated Calls

Mobile originated calls are calls that are established by a GSM subscriber, who is registered in an MSC. All MO calls that are established by a CAMEL subscriber may be subject to CAMEL control, except emergency calls (i.e. calls with tele service code TS12). TS12 calls bypass all CAMEL handling in the MSC.

Emergency calls are calls to '112' (Europe) or '911' (America), etc. The emergency number is normally programmed in the ME. When the user enters 112 on the keypad, a TS12 service, as specified in GSM TS 02.03 [3], is started. Multiple band MEs (e.g. GSM900/1800/1900) may have multiple emergency numbers programmed in, to cater for the various regions where these MEs may be used. Additional emergency numbers may be programmed in the (U)SIM by the operator. In either case, the resulting call to such number will be a TS12 call. The use of emergency numbers is further specified in 3GPP TS 22.101 [68].

When a subscriber establishes an MO call, the MSC will start a basic call handling process. Basic call handling entails checking whether the subscriber has a subscription to the requested basic service. If the requested basic service is not subscribed, i.e. the VLR has not received the required subscription data for that service from the HLR, then the MSC rejects the call attempt. The call establishment is in that case rejected even before a CAMEL service is invoked.

The subscribed basic services are provisioned in the HLR and are sent to the VLR at the time of location update.

In addition to a check on the basic service, the MSC may apply call barring to the call. Two forms of call barring may be applied to an MO call:

- Call barring (CB) supplementary service;
- Operator determined barring (ODB).

CB is a supplementary service that a GSM subscriber may subscribe to; supplementary services are provisioned in the HLR. A subscriber may modify her GSM SS settings, e.g. changing her call barring profile. A password may be needed to change call barring settings in HLR. A subscriber may subscribe to several CB categories, such as barring of all outgoing calls (BAOC) or barring of outgoing international calls (BOIC). CB categories may be defined for specific basic services or for entire basic service groups. Basic services are specified in GSM TS 02.01 [1]. The CB check may result in a rejection of the call attempt even before a CAMEL service is invoked. CB category BAOC is often used as a fallback option in the HLR, for the case that a CAMEL subscriber registers in a VPLMN that does not support the required CAMEL phase. When a subscriber has BAOC, she may receive calls and use USSD to establish a call-back call, provided that her operator offers USSD callback.

ODB is a network service; an operator may decide to apply ODB for a subscriber. ODB categories may be defined in the HLR for a subscriber and are transferred to the VLR at location update, together with other subscription data, such as basic services, supplementary services, CAMEL trigger elements etc. As the name ('operator determined barring') implies, a user does not have the ability to modify her ODB settings.

Here as well, the ODB check may result in a rejection of the call attempt, even before a CAMEL service is invoked. Figure 3.4 reflects the CB check for an MO call. When the HLR sends BAOC to the VLR, the HLR should not send O-CSI to the VLR. The presence of BAOC in VLR prevents the establishment of MO calls, hence O-CSI is not needed. The HLR may still send tele service 11 (speech service) to the VLR, since it is needed for MT calls.

Another check that forms part of basic call handling is whether the served subscriber has O-CSI in her subscription data in VLR. The presence of O-CSI in VLR for that subscriber serves as an indication to the basic call handling process that the call will be subject to CAMEL control, i.e. the MSC will invoke a CAMEL service for this call.

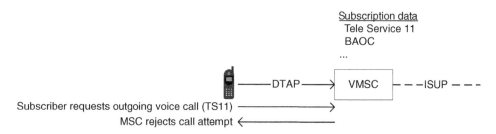

Figure 3.4 Effect of CB category BAOC on MO call attempt

When the subscriber has O-CSI and none of the subscription checks in the MSC has prevented call establishment, the MSC will hand over control of the call to the gsmSSF. The gsmSSF is the functional entity in the MSC that is responsible for the communication between MSC and gsmSCF.

When the basic call handling process has handed over control of the call to gsmSSF, the call handling process is suspended. The gsmSSF uses the information in the O-CSI to invoke the subscribed CAMEL service. The gsmSSF process is now also suspended, whilst waiting for instructions from the gsmSCF. Call establishment continues once the gsmSCF has sent a *call continuation operation* to the gsmSSF. The following operations are call continuation operations:

- *Continue (CUE)* – the gsmSCF instructs the gsmSSF to establish the call to the dialled destination; the MSC will compile an ISUP IAM towards this destination, using information that was received from the radio network and using subscriber-related information from the VLR.
- *Connect (CON)* – the gsmSCF instructs the gsmSSF to establish the call to the destination that is contained in CAP CON; CAP CON may also contain other parameters; the parameters in the CAP CON operation overwrite the corresponding parameters in ISUP IAM;
- *Release call (RC)* – the gsmSCF instructs the gsmSSF to release the call; the gsmSCF supplies the cause code to be returned to the calling subscriber.

Figure 3.5 depicts the call continuation operations.

If the gsmSCF uses CAP CUE or CAP CON, then the call processing in the MSC continues. The MSC analyses the called number, possibly modified by the SCP, and sends out ISUP IAM to establish the call. The CAMEL service may remain monitoring the call or may terminate at this point; this depends on the arming of DPs in the O-BCSM. If the gsmSCF had armed one or more DP prior to sending CAP CUE or CAP CON, then the CAMEL service remains active; if the gsmSCF had not armed at least one event, then the CAMEL service terminates at this point; see

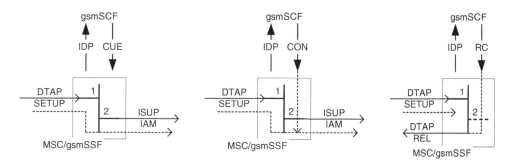

Figure 3.5 Call continuation operations

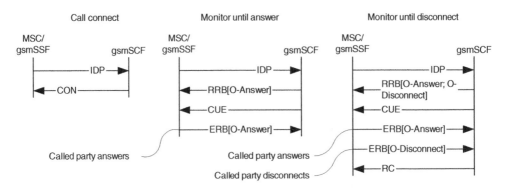

Figure 3.6   CAMEL phase 1 – MO call monitoring

Figure 3.6 for example sequence flows. The abbreviations of the CAP operations are listed in the Appendix.

In the 'call connect' example, the CAMEL service does not arm any DP. The CAMEL service terminates after CAP CON has been sent. In the 'monitor until answer' example, the CAMEL service arms the answer event. As a result, the CAMEL service remains active after sending CAP CUE. When the answer event has been reported, there are no more DPs armed in the O-BCSM. As a result, the CAMEL service terminates. Finally, in the 'monitor until disconnect' example, the CAMEL service arms both the answer event and the disconnect event. The CAMEL service remains active until it has received the disconnect event. The disconnect event may be armed in interrupt mode; in that case, the CAMEL service will send CAP CUE or CAP RC to the gsmSSF after the disconnect is reported.

The CAMEL phase 1 control capability is limited, as may be clear from Figure 3.6. A CAMEL phase 1 service may arm the answer event and the disconnect event, but not the call establishment failure events. This may lead to premature CAP dialogue termination. When call establishment failure occurs (called subscriber busy, no reply, etc.), then the gsmSSF will abort the CAMEL dialogue.

Figure 3.7 presents three examples of a CAMEL phase 1 service that is used for international VPN.[10] The VPN service translates the dialled number into the public number associated with the destination subscriber, who belongs to the same VPN group as the calling subscriber. Hence, subscribers of one VPN group, e.g. an enterprise, may call one another by GSM, by dialling the PABX extension number.

In case (1), the CAMEL service connects the calling subscriber to a mobile VPN subscriber. The mobile VPN subscriber belongs to the same VPN group as the calling subscriber. The call is routed to the GMSC of the HPLMN operator, from where the call is routed to the VMSC, where the called subscriber currently resides. In case (2), the calling subscriber is connected to an extension of the PABX at the company. In example (3), the calling subscriber is connected to a VPN colleague in an office in the visited country. In all three cases, the ISUP signalling may take the shortest possible path between the VMSC of the calling subscriber and the PSTN/PLMN of the called subscriber. In these examples, the VPN service may ensure that the called party receives the calling party's VPN number on her display, instead of the calling party's public number (Table 3.6). On-net calls are calls between users of the same VPN group; off-net calls are calls from a user of a VPN group to a user outside the VPN group or vice versa.

When the calling party (+31 6 516 34 567) dials 3341, the VPN service determines that this number belongs to a subscriber of the same VPN group. The VPN service translates the dialled

---

[10] If the international VPN service is also used for cost control, then CAMEL phase 2 is required.

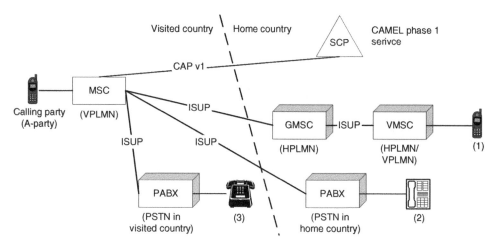

Figure 3.7   CAMEL phase 1 call routing examples

Table 3.6   Call routing for a VPN subscriber

Calling party number	Called number	Public number	Displayed number
+31 6 516 34 567	3341 (= on-net)	+31 6 516 34 568	3342
+31 6 516 34 568	3342 (= on-net)	+31 6 516 34 567	3341
+31 6 516 34 567	+31 70 456 6782 (= off-net)	+31 70 456 6782	+31 6 516 34 567

number into the public GSM number of the destination subscriber (+31 6 516 34 568). The VPN service also provides an additional calling party number (3342) in CAP CON. The called subscriber receives 3342 on her display, instead of +31 6 516 34 567.

Should the called VPN subscriber return the call, i.e. dial 3342, then the VPN service will connect the call to +31 6 516 34 567. If the calling subscriber dials an off-net number (e.g. +31 70 456 6782), then the VPN service allows the call to continue to the dialled destination, without affecting the routing of the call. VPN does not provide an additional calling party number, since the VPN number should not be presented to a called party that does not belong to the (same) VPN group.

When a call crosses an international boundary, it may occur that the calling party number or the additional calling party number is not transported in the ISUP signalling link.

A VPN subscriber may receive an on-net call when she is roaming in a non-CAMEL network. In that case, the VPN service for that called subscriber may remove the additional calling party number from the ISUP signalling flow. The rationale is that the called VPN subscriber might otherwise return a call to the displayed VPN number of the calling party. However, since the network where the called party is currently roaming does not support CAMEL, the VPN service does not have the capability to connect a call from that subscriber to a VPN destination.

3.2.2 Mobile-terminated Calls

Refer to Figures 3.8 and 3.9 for the MT call case with CAMEL control. When a call is established for a GSM subscriber, the ISUP IAM, containing the MSISDN of the destination GSM subscriber as the called party number, is routed to the HPLMN of the destination subscriber. This routing of

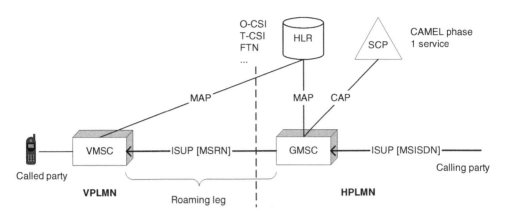

Figure 3.8   Node overview for CAMEL control of an MT call

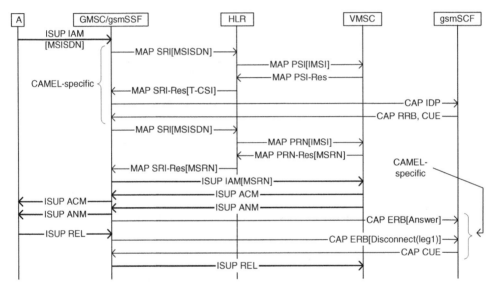

Figure 3.9   Sequence diagram for CAMEL control of MT call

the call to the HPLMN does not consider the current location of the called subscriber. If the called subscriber is a subscriber of Orange France, for example, but is currently registered in an MSC from Telefonica Spain, then the call will still be routed to her HPLMN, i.e. Orange France.[11] When the call arrives in the HPLMN of the called subscriber, a border GMSC in that network, handling the incoming call, analyses the destination of the call. The GMSC knows, by configuration, which number ranges represent MSISDN series of this operator. Hence, the GMSC can determine that this incoming call is destined for a subscriber of this network. Consequently, the border GMSC will behave as a GMSC for MT call handling.

---

[11] Unless BOR is used for this call; refer to Section 4.8.1.

If the call was established in an MSC of the same network, then the originating MSC, handling the MO call, may at the same time act as a GMSC for the MT call. Operators may configure their MSCs to function as VMSC and as GMSC at the same time.

It is in fact at this point, where the GMSC has determined that the call is destined for a subscriber belonging to this network, that MT call handling commences. The first step the GMSC takes is contacting the HLR, by sending MAP SRI to the HLR. The address of the HLR, in the form of a global title, is derived from the MSISDN of the called party. A STP in the HPLMN's SS7 network translates the HLR GT into the SPC of the HLR for this subscriber. Hence, a network may contain one or several HLRs; the GMSCs do not need configuration to select the HLR of a subscriber.[12]

The GMSC indicates in the MAP SRI that it supports CAMEL. The GMSC includes this indication in MAP SRI without knowing whether the called subscriber is a CAMEL subscriber or not. The GMSC copies various information elements from ISUP IAM to MAP SRI. When the HLR receives MAP SRI, it determines that the MAP SRI relates to a call for a CAMEL subscriber; the MSISDN that is included in MAP SRI is used to index the subscriber in HLR. The HLR will take various actions as part of the terminating call handling for the CAMEL subscriber. These actions include:

- Determine the requested BS for the call; the BS is derived from the BC, LLC and HLC. BC, LLC and HLC are parameters that are transported in ISUP IAM and are copied to MAP SRI. GSM TS 09.07 [55] specifies the rules for deriving the BS form BC, LLC and HLC. If the BS cannot be derived, e.g. because of missing information in MAP SRI, then the HLR may apply a default BS.
- Verify that the subscriber has a subscription to the requested basic service. For example, a subscriber may subscribe to speech calls (basic service TS11), but not to video calls (basic service BS30).
- Check whether any supplementary services apply, such as call barring and call forwarding.
- Check whether conditional triggering applies for T-CSI. Conditional triggering for T-CSI is part of CAMEL phase 2; see Chapter 4.

Vendor-specific actions in the HLR may be performed, such as:

- Suppress T-CSI when the called subscriber currently resides in HPLMN. This option may be used when the T-CSI relates to pre-paid service; presuming that MT calls in HPLMN are not charged, the MT CAMEL service (i.e. pre-paid service) does not need to be triggered when the called subscriber is in HPLMN.
- Suppress T-CSI when early call forwarding is pending. In the case of early call forwarding, a terminating charging service may not be required, hence T-CSI may be suppressed.

When the HLR has performed the various actions as listed above and has determined that CAMEL handling applies for the call, it sends the MAP PSI to the VLR. The MAP PSI is used to obtain the following subscriber information:

- subscriber location, including cell ID, location number, geographical information, etc.;
- subscriber state (busy, idle, not reachable).

The use of MAP PSI during MT call handling is optional; an HLR may be configured to use or not use it, depending on the subscriber. It may be required for pre-paid subscribers and for home zone VPN subscribers, but not for other CAMEL subscribers.

---

[12] MNP may affect the signalling sequence for MT call handling; refer to Section 6.8.

As a further implementation option, an HLR could have the capability to use MAP PSI when the called subscriber is currently in the HPLMN and to return the HLR-internally stored VLR address when the subscriber is roaming outside the HPLMN. The rationale would be that, when the subscriber is in the HPLMN, the detailed location information that may be obtained with MAP PSI may be used for real distance charging or Home Zone charging, for example. When that same subscriber is roaming outside the HPLMN, the VLR address may be accurate enough for these features.

The HLR does not normally keep track of the location and the state of the subscriber, so this information needs to be retrieved from VLR per terminating call. The information that is received from VLR is included in the MAP SRI-Res that is returned to the GMSC. The MAP SRI-Res also contains T-CSI, enabling the GMSC to invoke a CAMEL service for the MT call. If the above-listed HLR checks indicate that CAMEL does not apply for this MT call, e.g. because the called subscriber does not have T-CSI, or T-CSI is suppressed for this call, then the HLR would send the MAP PRN to the VLR, to obtain an MSRN. The HLR would then include the MSRN in MAP SRI-Res to GMSC. Hence, since the GMSC does not know whether the called subscriber is a CAMEL subscriber, the GMSC is prepared to receive either MSRN (non-CAMEL call) or T-CSI (CAMEL call).

The subscriber information that is retrieved from VLR and sent to GMSC is used in the CAMEL service invocation. The GMSC forwards the T-CSI and other data received from the HLR such as location information, subscriber state and IMSI, to the gsmSSF, which uses this data to invoke the CAMEL service. The MT call handling process is now suspended until the gsmSCF instructs the gsmSSF to continue call processing. The gsmSCF may use the location information, for example, to determine whether VPN restrictions or incoming call screenings apply.

The location information that is reported to the gsmSCF is, however, not the *current* location information. The called subscriber may have changed location to another cell within a defined location area. When changing location within the location area, the VLR is not updated. When the VLR responds to the MAP PSI from the HLR, it reads the internal register and returns the stored location information. This stored location information may therefore not indicate the current cell of the subscriber. The *age of location* parameter in the location information serves as an indication of the reliability of the location information.[13]

Presuming that the CAMEL service determines that the call may continue, the gsmSCF sends CAP CUE or CAP CON to the gsmSSF. Unlike MO call handling, the usage of CAP CUE and CAP CON is distinctively different in MT call handling:

(1) *CAP CUE* – the GMSC continues call establishment to the called MSISDN; this is described below;
(2) *CAP CON* – when the CAMEL service uses CAP CON, the following applies:
    (a) if the destination routing address in CAP CON argument is identical to the called party number in CAP IDP, then CAP CON is treated as CAP CUE with additional information. The GMSC uses the information carried in CAP CON to overwrite the corresponding parameters in ISUP. Examples are calling party category or additional calling party number. CAP CON may also be used to provide the alerting pattern; the alerting pattern is sent to VLR via MAP signalling through HLR. See Section 4.3.6 for a description of this feature.
    (b) If the destination routing address in CAP CON argument differs from the called party number in CAP IDP, then CAP CON is treated as a call forwarding instruction. This is described below.

---

[13] 3GPP has considered introducing the possibility of allowing the HLR to instruct the VLR to page the subscriber when sending MAP PSI during MT call handling. This has, however, not been introduced in CAMEL.

When the GMSC, upon receiving CAP CON, compares the destination routing address with the called party number it had sent in CAP IDP, it considers both the address digits and the number header; the format of the number header is defined in ITU-T Q.763 [137]. Service designers should take note of the following. An SCP platform may apply number normalization prior to passing the called party number on to the CAMEL service for MT call handling. The goal of such number normalization may be to provide the MT CAMEL service consistently with the called party number in international format. Normalization may include:

- removing the ST digit at the end of the address signals; ST is the 'end of pulsing signal' that may be present in the called party number in ISUP IAM; this digit may be present when the call originates from PSTN;[14]
- converting the MSISDN from national format to international format.

When the CAMEL service is not aware of the number normalization that is applied by the platform, then the CAMEL service may send CAP CON with 'unmodified number', with the intention that call handling to the called destination subscriber continues. However, the GMSC will receive a number that differs from the called party number in CAP IDP, resulting in call forwarding.

Before the CAMEL service sends CAP CUE or CAP CON, it may use CAP Request Report BCSM (RRB) to arm the answer event and the disconnect event. When the GMSC has received CAP CUE or CAP CON with unmodified destination address, it continues call handling. The GMSC has not yet received an MSRN from HLR, so it needs to contact the HLR again to obtain the required MSRN. This is where the *double HLR interrogation* mechanism comes in:

(1) *First interrogation* – the first MAP SRI sent from GMSC to HLR is the MAP SRI at the beginning of MT call handling. At this point, the GMSC does not know whether CAMEL will apply for the call. The GMSC may receive MSRN or T-CSI. The GMSC may also receive FTN from HLR; refer to Section 3.2.3.
(2) *Second interrogation* – the second MAP SRI sent from GMSC to HLR is the MAP SRI after the GMSC has received CAP CUE or CAP CON, with unmodified destination address, from gsmSCF.

The purpose of the second interrogation is to obtain MSRN from HLR, not T-CSI. The HLR is *stateless* with respect to processing MAP SRI. For that reason, the GMSC includes an indication in MAP SRI that this MAP SRI is a 'second SRI'. This indication takes the form of the 'suppress T-CSI' parameter in MAP SRI. As a consequence, the HLR bypasses CAMEL MT call handling. The HLR uses MAP PRN to obtain the MSRN from the VLR. The MSRN is returned to the GMSC in MAP SRI-Res. The GMSC then uses the MSRN to complete the call to the called subscriber; the MSRN contains the SS7 signalling address of the VMSC. The call leg between GMSC and VMSC is called the 'roaming leg', regardless of whether the VMSC is in HPLMN or in VPLMN.

Since the second HLR interrogation is processed independently of the first HLR interrogation, the HLR will again derive the basic service and apply subscription checks. These checks were already performed in response of the first interrogation.

The VLR is also *stateless* with respect to the handling of MAP PSI and MAP PRN. When the VLR has responded to MAP PSI during MT call handling, this does not affect the subsequent handling of MAP PRN for this call. The VLR does not correlate MAP PSI and MAP PRN.

Further call handling includes the sending of the ISUP address complete message (ACM) and the ISUP answer message (ANM) by the VMSC in the backwards direction. ISUP ANM may result

---

[14] Refer to ITU-T Q.763 for the End of Pulsing signal.

in the sending of CAP Event Report BCSM (ERB) to gsmSCF, to inform the CAMEL service that the called party has answered. The eventual ISUP Release (REL) may also result in the sending of CAP ERB to gsmSCF, to inform the CAMEL service that the calling party or called party has disconnected from the call.

3.2.2.1 Call set up Failure for MT Call

Various events or conditions may result in failure of the MT call establishment; some of these events occur before CAMEL invocation, the others occur after CAMEL invocation.

- *Incoming call barring* – when the HLR determines that the call may not be established to the subscriber due to call barring or ODB, then the HLR will not return CAMEL data to the GMSC. The HLR will instruct the GMSC to release the call.
- *Non-subscribed basic service* – when the subscriber does not have a subscription to the requested basic service for the call, e.g. data call, then the HLR will not return CAMEL data to the GMSC; the call will be released without CAMEL service being invoked.
- *Subscriber not registered in VLR* – presuming no call forwarding is active for this condition, the HLR will not return CAMEL data to the GMSC; the GMSC releases the call.
- *Subscriber detached* – the detached condition is normally not known in HLR; as a result, the HLR returns the T-CSI to the GMSC. The HLR may also have retrieved subscriber information from VLR, i.e. subscriber state and subscriber location. The subscriber state may indicate 'detached'. The HLR still returns T-CSI to the GMSC. When the CAMEL service is invoked and returns CAP CUE, the GMSC does the second interrogation. The HLR, however, fails to obtain an MSRN from VLR, since the subscriber is currently not attached to the MSC. Presuming no call forwarding is active for this condition, the HLR returns an error to the GMSC. Since the T-BCSM for CAMEL phase 1 does not contain DPs associated with call establishment failure, the gsmSSF in the GMSC terminates the CAMEL service by aborting the CAMEL dialogue. Figure 3.10 shows the signal sequence for this condition.
- *Busy, no Reply, no paging response* – when the VLR has allocated an MSRN for the call to the subscriber and the GMSC has routed the call to the VMSC, the VMSC attempts to deliver the call to the called subscriber. The conditions no paging response, subscriber busy[15] and no reply all lead to call establishment failure. The busy condition includes both network determined user busy (NDUB) and user determined user busy (UDUB); see GSM TS 02.01 [1] for descriptions of NDUB and UDUB. Presuming no call forwarding is active for the failure condition, the VMSC

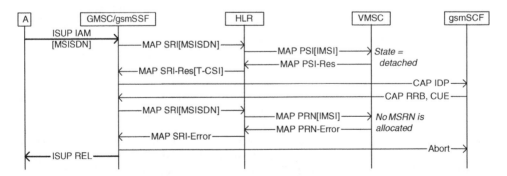

Figure 3.10   MT call establishment failure in GMSC

---

[15] When the subscriber is busy, the VLR will still allocate an MSRN for a call to that subscriber. The call set up failure due to the busy condition therefore occurs in the VMSC instead of the GMSC.

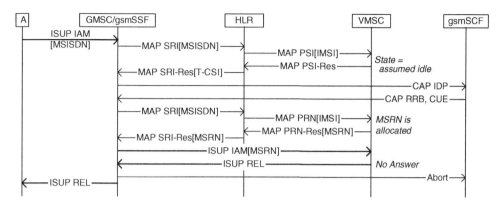

Figure 3.11   MT call establishment failure in VMSC

returns ISUP REL to the GMSC. The CAMEL phase 1 T-BCSM not having DPs associated with call establishment failure, the gsmSSF aborts the CAMEL dialogue. Figure 3.11 reflects the signal sequence for one of these conditions.

3.2.3 Mobile-forwarded Calls

GSM offers various forms of call forwarding. Call forwarding is specified in GSM TS 03.82 [40]. Call forwarding occurs when the network has determined that a mobile terminating call cannot be delivered to the called subscriber and an alternative destination for that call is available. This alternative destination may be a voicemail box. MT call handling in GSM is partly performed in GMSC and partly performed in VMSC. This two-fold handling of MT calls has resulted in the following grouping of call forwarding kinds:

- *Early call forwarding* – early call forwarding takes place in the GMSC. When the HLR determines that call forwarding should be applied to a call, then the HLR sends the FTN to the GMSC. The GMSC can now forward the call to the alternative destination, as indicated by the FTN.
- *Late call forwarding* – late call forwarding takes place in the VMSC. When the GMSC has received an MSRN from the HLR and has routed the call to the VMSC of the called subscriber, the VMSC may determine that call forwarding should be applied to this call. The VMSC will forward the call to the alternative destination, as indicated by the FTN that is present in the subscriber's profile in VLR. A prerequisite for call forwarding from the VLR is that the HLR has previously sent the corresponding FTNs to the VLR, typically during location update.

Early call forwarding has an advantage over late call forwarding. In the case of early call forwarding, no ISUP traffic link is established from the GMSC to the VMSC. Presuming that the GMSC resides in the HPLMN[16] and that the FTN is a destination in the HPLMN, the early call forwarding takes place entirely in the HPLMN. If late call forwarding occurs, an ISUP traffic link is already established from the GMSC to the VMSC. Again presuming that the FTN is a destination in the HPLMN, the late call forwarding includes the establishment of an ISUP traffic link from GMSC

---

[16] When basic optimal routing is applied, early call forwarding may occur in a PLMN other than the HPLMN.

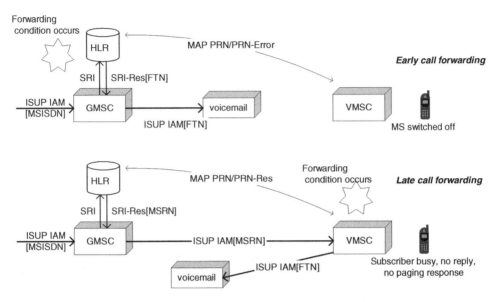

Figure 3.12   Early call forwarding vs late call forwarding

to VMSC in VPLMN as well as the establishment of an ISUP traffic link from VMSC in VPLMN to forwarding destination in HPLMN. Figure 3.12 depicts this distinction.

Call forwarding may be configured for all applicable basic services or for individual basic service groups. A subscriber may forward calls of different kinds to different destinations. A subscriber may configure the forwarding numbers on her terminal, depending on the terminal capabilities. Some terminals provide the possibility of setting different forwarding destinations for voice calls and for Fax and data calls.

GSM specifies four call forwarding categories, each of which may occur as early call forwarding, as late call forwarding or as both.

3.2.3.1 Call Forwarding – Unconditional

Call forwarding – unconditional (CFU) is the situation that the subscriber has defined in the HLR that all incoming calls shall be forwarded to a particular destination; this destination is defined by the FTN unconditional (FTN-U). The FTN-U is registered in the HLR and is sent to GMSC when the GMSC queries the HLR for terminating call handling. CFU is therefore always early call forwarding. When CFU is active for a CAMEL subscriber, the HLR returns the FTN-U together with the T-CSI to the GMSC. The GMSC invokes the CAMEL service before initiating the call forwarding. The CAMEL service will, however, not be aware that call forwarding will take place.[17] Hence, when the gsmSSF reports answer to the gsmSCF, this answer notification relates to the forwarded-to-destination, not to the called subscriber.

When the HLR sends FTN to the GMSC, it also sends O-CSI to the GMSC. The GMSC uses the O-CSI to invoke a CAMEL service for the forwarding leg. Figure 3.13 contains an example signal sequence diagram for this call case. The forwarding destination, i.e. the C-party, may in turn be a GSM subscriber belonging to the same network as the B-party. In that case, the GMSC would

---

[17] Interaction with call forwarding is introduced in CAMEL phase 2.

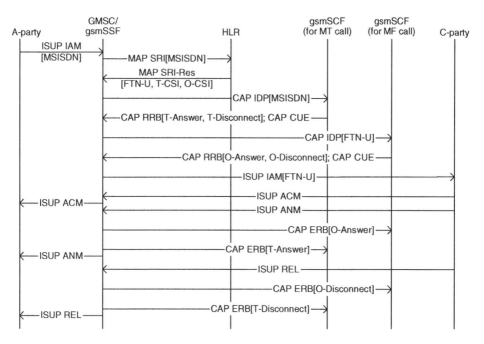

Figure 3.13 Signal sequence diagram for unconditional call forwarding

send MAP SRI to the HLR to obtain routing information for this call to the C-party. However, the signal sequence for the call to the C-party is not related to the signal sequence related to the MT call and the MF call for the B-party. If CFU is active in HLR whilst call forwarding not reachable is also active in HLR, then CFU takes precedence over call forwarding not reachable.

3.2.3.2 Call Forwarding – Not Reachable

Call forwarding not reachable (CFNRc) has three different sub categories:

(1) *Subscriber marked as 'detached' in HLR.* The subscriber has not performed a location update in the VLR for a configurable period; the VLR has therefore purged the subscriber from its internal register and has informed the HLR. The HLR marks the subscriber as detached. If a call arrives for this subscriber, then the HLR returns the FTN-not reachable (FTN-NRc) to the GMSC. Hence, this form of CFNRc is early call forwarding. The further call handling is similar to the CFU case.

(2) *Subscriber marked as 'detached' in VLR.* The subscriber has switched off the MS. The subscriber remains registered in the VLR, but the VLR marks the subscriber as 'detached'. The VLR, however, does not inform the HLR of this condition. As a result, when the HLR receives the MAP SRI for this subscriber, the HLR will not send the FTN-NRc to the GMSC. The HLR returns the T-CSI to the GMSC, resulting in CAMEL service invocation. When the GMSC sends the second interrogation to the HLR, the HLR sends MAP PRN to the VLR. However, since the subscriber is currently detached from VLR, the call cannot be completed and the VLR will not allocate an MSRN. The VLR returns MAP PRN error to the HLR; the HLR, in turn, sends the FTN-NRc to the GMSC. The GMSC now performs the call forwarding. This

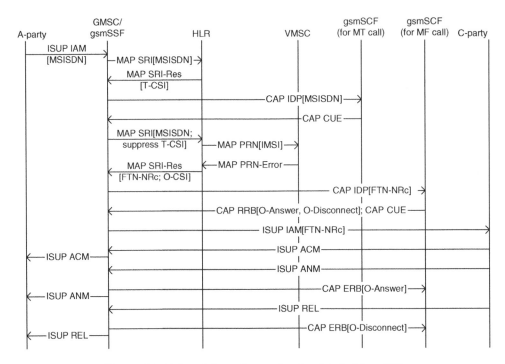

Figure 3.14   Early call forwarding – subscriber detached from VLR

forwarding case is also early call forwarding; see Figure 3.14 for an example of a sequence diagram.

(3) *No paging response.* When a call arrives for a subscriber and the call is delivered to the VMSC, then the VMSC will page the subscriber. If the VMSC does not get paging response, then the CFNRc condition occurs, provided that an FTN-NRc is registered in the VLR. The VMSC will now establish a call to the forwarded-to-destination. This forwarded call will be subject to a CAMEL service, provided that O-CSI is present in the VLR. This call forwarding is late call forwarding. The no paging response condition occurs when the subscriber is marked as 'Idle' in the VLR, but is temporarily out of radio coverage. If the subscriber is out of radio coverage for a certain duration, then the VLR marks the subscriber as 'Detached'. If the subscriber receives a call in that case, then the VLR will not allocate an MSRN and early call forwarding occurs in the GMSC.

3.2.3.3 Call Forwarding – Busy

If a subscriber busy condition occurs in the VMSC and the subscriber has an FTN-Busy (FTN-B) registered for the basic service of the incoming call, then Call Forwarding – Busy (CFB) occurs. CFB is always late call forwarding. The forwarded call will be subject to CAMEL, provided that the subscriber has O-CSI registered in the VLR. CFB has two sub categories:

(1) *Network-determined user busy* – when the mobile terminated call arrives at the VMSC and the subscriber is engaged in a CS call, then the VMSC determines that the subscriber is busy. The NDUB condition now occurs and the VMSC initiates call forwarding to the FTN-Busy

(FTN-B). If the called subscriber subscribes to Call Waiting (CW), then the VMSC may offer a second call to that subscriber.[18] In that case, there is no NDUB.
(2) *User-determined user busy* – when a subscriber receives a call and the call reaches the alerting phase, the subscriber may reject the incoming call ('push the call away'). The <u>user</u> determines that she is busy (not willing to answer the call), not <u>the network</u>; hence the term user-determined user busy. If the incoming call is a second call for a subscriber that has CW, then the called subscriber may also reject the call, i.e. apply UDUB.

Figure 3.15 shows an example signal sequence flow for UDUB, with CAMEL control of the forwarded call. It reveals that, when call forwarding occurs in the VMSC, the MT CAMEL service is not notified. Hence, when the MT CAMEL service receives the answer notification, it does not know that this answer is generated by the C-party rather than by the B-party. When the VMSC performs call forwarding, it may send a forwarding notification to the calling party. This notification,

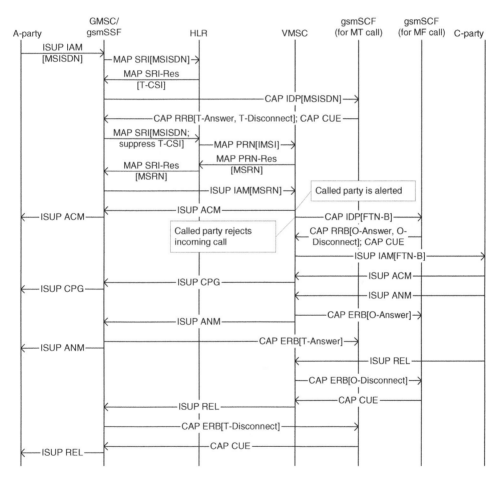

Figure 3.15   Call forwarding busy (UDUB)

<hr>

[18] If the second, incoming call uses a different bearer capability than the currently active call, then the VMSC will not offer the second call to the subscriber. In that case, NDUB occurs.

which is transported through ISUP, is subscription-dependent (see 'Ext-ForwOptions' in GSM TS 09.02 [54]) and is not detected by the gsmSSF in the GMSC.

3.2.3.4 Call Forwarding – No Reply

No reply occurs when a call that is offered to the subscriber is in the alerting phase, but the call is not answered within a pre-defined time. If the subscriber has an FTN-No Reply (FTN-NRy) registered for the basic service of the incoming call, then call forwarding – No Reply (CFNRy) occurs. The forwarded call will be subject to CAMEL, provided that the subscriber has O-CSI registered in the VLR. CFNRy is always late call forwarding.

When the no reply call forwarding service is provisioned in the HLR for a subscriber, a subscriber-specific no reply time in the range of 5–30 s may be set ('noReplyConditionTime'). This parameter is sent to the VLR during location update. When a call is offered to the subscriber, the serving MSC uses this value instead of the MSC default value.

3.2.3.5 SCP-induced Early Call Forwarding

A CAMEL service that is controlling an MT call in the GMSC may 'induce' call forwarding.[19] This results in the following further distinction for call forwarding: *GSM-based call forwarding* and *SCP-based call forwarding* (or 'service-based call forwarding').

If GSM-based forwarding is pending for a MT call, then a CAMEL service may effectively overwrite the GSM-based forwarding. The CAMEL service uses CAP CON to induce or overwrite call forwarding; see Figure 3.16.

Figure 3.16 shows an example whereby no GSM call forwarding is pending. The CAMEL service diverts the call to an alternative destination, i.e. applies call forwarding. Quintessential to SCP-based call forwarding is that the destination address contained in CAP CON ('destination routing address') differs from the called party number in CAP IDP. Otherwise, CAP CON results in call routing to the destination subscriber.

When the CAMEL service induces call forwarding, it provides the following information to the GMSC in CAP CON:

- *Original called party ID* – this information element contains the original number, i.e. the called party number from CAP IDP. If the call has already undergone one or more call forwardings prior to arriving at the GMSC, then this information element may already be present in ISUP

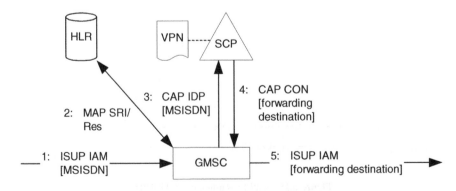

Figure 3.16   SCP-based unconditional call forwarding

---

[19] SCP-induced call forwarding is also referred to as call diversion.

IAM and included in CAP IDP. In that case, the CAMEL service does not need to provide this information element to the GMSC.

- *Redirecting Party ID* – this information element contains the MSISDN of the subscriber on whose behalf the call forwarding is performed. In other words, it is set to the called party number from CAP IDP. If the call has already undergone one or more call forwardings prior to arriving at the GMSC, then this information element may already be present in ISUP IAM and included in CAP IDP. In that case, the CAMEL service still includes this element in CAP CON and sets it to the called party number from CAP IDP.
- *Redirection Information* – this information element contains background information related to the call forwarding event. It includes:
  - *redirecting indicator* – this element indicates whether redirection information may be presented to the called party;
  - *original redirection reason* – this element indicates the redirection reason related to the first redirection of this call;
  - *redirection counter* – this element indicates the number of forwardings that have taken place; when the SCP induces forwarding, it increases the counter by one; when this counter has the maximum value, no further forwarding should take place; the maximum value may be 5, for example;
  - *redirecting reason* – this element indicates the reason of the current call forwarding; the CAMEL service sets this element to value that reflects the reason for the service to forward the call.

The encoding of these elements is defined in ITU-T Q.763 [137].

Since the T-BCSM of CAMEL phase 1 does not contain DPs for call establishment failure events, it is not possible for a CAMEL phase 1 service to induce any of the conditional call forwardings. CAMEL phase 2 is used for that. A CAMEL service may suppress GSM call forwarding by setting the redirection counter to its maximum value; see Figure 3.17. In that case, when a call establishment failure occurs, the GMSC or VMSC will not initiate call forwarding, but will release the call. A CAMEL phase 1 service will, however, not be notified of this release, as explained above.

When a MT call CAMEL service induces call forwarding in the GMSC, an MF call CAMEL service may be invoked for the forwarding leg. Prerequisite for the invocation of this CAMEL service is that the HLR has provided O-CSI in MAP SRI-Res and that CAP CON (from the MT call CAMEL service) contains the information element 'O-CSI applicable'. The Rationale is that, when the SCP induces call forwarding, it is aware of the destination of the call and may determine that no CAMEL service is required for the forwarding leg. For GSM-based call forwarding, on the

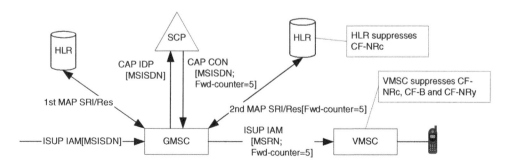

Figure 3.17   Forwarding suppression by CAMEL service

other hand, the MT call CAMEL service has no control over the applicability of O-CSI for the forwarding leg.

3.2.3.6 Optimal Routing of Late Call Forwarding

When late call forwarding occurs, the VMSC may apply optimal routing of late call forwarding (ORLCF). ORLCF has specific interworking with CAMEL phase 2 and is described in Chapter 4.

3.2.4 Any-time Interrogation

ATI is a feature that enables a CAMEL service to obtain subscriber information through the HLR. See Figure 3.18 for an architectural overview of Any time interrogation (ATI). Figure 3.19 depicts the associated sequence flow.

The SCP may use ATI to request one or both of *location information* and *subscriber state*. The SCP sends MAP ATI to the HLR associated with the subscriber whose information is required. The subscriber is identified with IMSI or with MSISDN in the argument of MAP ATI. When the HLR receives MAP ATI, it sends MAP PSI to the VLR where the subscriber is registered.

For addressing the HLR, i.e. sending MAP ATI to the HLR, various methods may be used:

(1) The SCP uses an SPC to identify the HLR; this method is fairly rigid, as the SCP would need to know the HLR address for the subscriber, possibly derived from IMSI or MSISDN.
(2) The SCP uses GT of the HLR; this method is more flexible since the operator may modify the SS7 configuration without impacting the SCP. However, the SCP would need to know the GT of the HLR, based on IMSI or MSISDN.

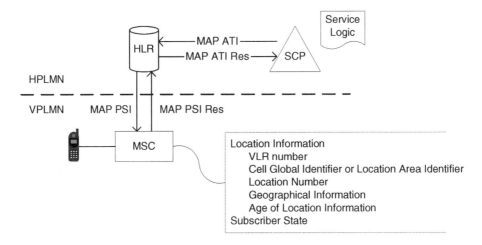

Figure 3.18   Architecture for any-time interrogation

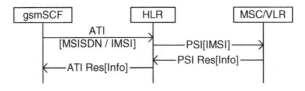

Figure 3.19   Signal sequence for any time interrogation

(3) The SCP uses IMSI or MSISDN as GT. This method is the most flexible. Subscribers may be allocated to various HLRs, e.g. grouped per IMSI range or per MSISDN range. An STP in the SS7 network derives the HLR address from the IMSI or MSISDN. In this way, the SCP does not need to know the HLR address. When the SCP sends MAP ATI, it indicates in the request whether IMSI or MSISDN is used to address the HLR; that indication is needed by the STP to derive the address of the HLR.[20]

ATI may be used within the context of a call or outside the context of a call. That is to say, the functional entity using MAP ATI may or may not be busy handling a call and the subscriber whose information is requested may or may not be engaged in a call (establishment). Example cases of ATI include:

Within the context of a call
- *Real distance charging (RDC)* – RDC entails that the charge for an MO call is based on the location of the calling and called subscriber. If the called subscriber is a GSM subscriber of the same network, then the SCP that is performing real-time charging for the calling party uses ATI to obtain the called party's location information.
- *Home Zone* – a VPN that is controlling a call to a GSM subscriber may apply ATI to check whether the called party is in her home zone. VPN would use ATI after the call was answered, to ensure that the VLR is updated with the current location. The reason is that the location information that is reported in CAP IDP for MT call handling was retrieved from VLR *before* the called subscriber is paged.

Outside the context of a call
- *Location server*: a location server may request the subscriber's location & state periodically or when requested by an application.
- *Call completion*: a call completion service may poll the subscriber's state, to establish a call when a subscriber has become idle.

Regarding to usage of ATI within the context of a call for the subscriber whose information is requested, the following applies:

- *Mobile originated call* – when an MO call is established, there is radio contact with the calling subscriber and hence the calling subscriber's current location is reported to the gsmSCF. Hence, the gsmSCF need not use ATI for the calling subscriber, but possibly for the called subscriber.
- *Mobile terminated call* – when an MT call is established, the called subscriber's VLR-stored location information may be reported to the gsmSCF. As explained earlier, the reporting of location information to the gsmSCF at this point depends on HLR setting. If the MT call CAMEL service for a subscriber always needs the location information, then the HLR should use MAP PSI during MT call handling. Otherwise, if the MT call CAMEL service needs the location occasionally, then the HLR may be configured not to use MAP PSI when handling an MT call for this subscriber. The MT CAMEL service may decide to use ATI during MT call establishment when needed (Figure 3.20). Using ATI during MT call handling results in additional signalling, compared with having the HLR use MAP PSI when processing MAP SRI; see bold arrows in Figure 3.20.
- *Mobile forwarded call* – the use of ATI when handling a mobile forwarded call would not be useful. When call forwarding takes place, there is no traffic path established with the MS of the subscriber. Hence, the subscriber's location within the serving MSC would normally not be relevant.

---

[20] When MNP is applied, a MNP signalling relay function (MNP-SRF) is required to route ATI to the correct HLR. See Chapter 6 for MNP.

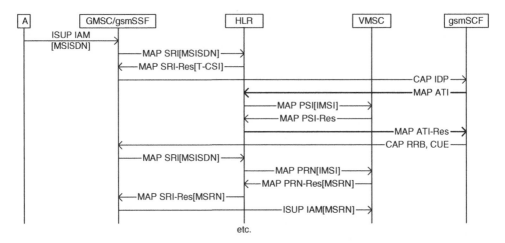

Figure 3.20   Using ATI for MT call handling

The SCP may use ATI for subscribers belonging to the same PLMN or for subscribers belonging to another PLMN in the same country or in another country. If SCP and HLR do not belong to the same network, then an agreement between the two involved operators is needed. The subscriber whose information is requested with MAP ATI does not need to be a CAMEL subscriber, i.e. she does not need to have O-CSI or T-CSI in HLR; a CAMEL service using MAP ATI may not have any knowledge about the subscriber whose information is required.

It is uncommon for a VLR to support MAP v3 but not support MAP PSI. Neither is it common for a VLR to screen on IMSI when receiving MAP PSI. The MAP PSI that may be sent to VLR may be used for two purposes:

(1) as part of the any-time interrogation procedure, i.e. as a result of MAP ATI from gsmSCF;
(2) as part of mobile terminating call handling, as a result of MAP SRI from GMSC, as explained above.

The VLR does not distinguish these cases for MAP PSI; it responds to MAP PSI in a uniform manner.

3.2.4.1 Location Information

The location information related to a subscriber is retrieved from the VLR; the subscriber is not paged during the Any Time Interrogation procedure. The subscriber may have changed location from one cell to another cell within the same location area. Such a change of location does not result in an update of the location information in VLR. The location that is returned to the HLR (and to the SCP) may therefore not be the current location of the subscriber. Location information is a set of data elements describing the subscriber's location. Refer to GSM TS 03.78 [38] for a list of the various elements that may be reported to SCP with MAP ATI-Res.

3.2.4.2 Subscriber State

The state of the subscriber may take one of the following values:

- *Assumed idle* – the subscriber is marked as Idle in the VLR. The term 'assumed' implies that the VLR deduces from its internal information that the subscriber is idle, and therefore capable of receiving a call for example. However, the subscriber could have entered a spot without GSM radio coverage. Such a change of position is not directly reported to VLR. If a call arrived for this subscriber whilst in this spot without coverage, the MSC/VLR would attempt to offer the call to the subscriber. The MSC/VLR would, however, not receive a paging response from the subscriber and the call establishment would fail.

- *Camel busy* – the subscriber is engaged in a call. Whilst the subscriber is marked *Busy*, she may have the capability to answer another incoming call. Answering a second call requires that the subscriber has call waiting supplementary service and that she has not more than one call active. The *CAMEL busy* indication does not, however, indicate whether the conditions for answering a second call are fulfilled.
- *Network determined not reachable* – the subscriber is marked in the VLR as not reachable. This response may be returned in the following cases: (1) the subscriber is explicitly or implicitly detached from the VLR; or (2) the VLR had not received a periodic location update from the subscriber for a defined duration, but had not yet purged the subscriber.
- *Not provided from VLR* – this response may be returned by the HLR in the following cases:
  - the HLR does not have a VLR address associated with this subscriber, i.e. the subscriber is currently not registered in any VLR;
  - the HLR has a VLR address associated with this subscriber, but the VLR does not support MAP v3; the HLR cannot send MAP PSI to that VLR;
  - the sending of MAP PSI to VLR has failed.

3.2.4.3 Location and Presence Server

One example of an application that uses ATI is a location and presence server. This entity is in fact not a gsmSCF and the application that sends MAP ATI is not a CAMEL service (Figure 3.21). The location and presence server may apply 'polling' for the subscriber information; it sends MAP ATI to the HLR on a regular interval, e.g. every minute. This method results in considerable load on the network (HLR, MSC, signalling network) and should be used for small number of subscribers only. The CAMEL phase 3 method mobility management (see Chapter 5) offers an improved method to keep track of the location of a subscriber.

3.3 Subscription Data

CAMEL phase 1 makes use of CAMEL subscription information (CSI) (Table 3.7).

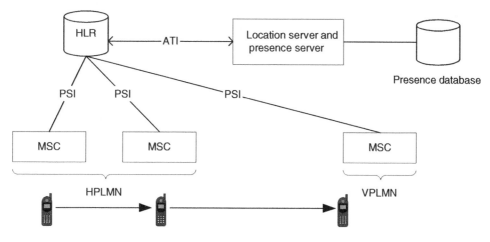

Figure 3.21 Using ATI for presence information

Table 3.7 CAMEL phase 1 subscription data

Subscription element	Description	Used in which entity
O-CSI	Originating CSI	MSC/VLR, GMSC
T-CSI	Terminating CSI	GMSC

Each subscriber may have an O-CSI and/or T-CSI in her HLR profile; different subscribers may have different O-CSI or T-CSI. O-CSI and T-CSI contain data elements that define how the gsmSSF shall invoke a CAMEL service.

3.3.1 Originating CSI and Terminating CSI

O-CSI and T-CSI contain the following elements.

3.3.1.1 gsmSCF Address

The gsmSCF Address identifies the location of the SCP where the CAMEL service resides. This address is in the format of a GT and complies with the number format rules as defined in ITU-T E.164 [128]. The following number formats are defined:

(1) international public telecommunication number for geographic areas;
(2) international public telecommunication number for global services;
(3) international public telecommunication number for networks.

The gsmSCF is defined in accordance with the format *International public telecommunication number for global services*. A GT may, as its name implies, be defined as a globally unique address. As a result, the gsmSCF address may be used from any visited PLMN, to address a unique CAMEL service in the HPLMN of the served subscriber. The HLR of the PLMN that a subscriber belongs to may send the same O-CSI of a subscriber to any visited PLMN that the subscriber roams to (provided that all conditions for the sending of O-CSI to a VPLMN are fulfilled).

As described in Chapter 2, the GT does not identify a particular SCP node. The use of GT depends on the network from where the CAMEL service is invoked:

(1) If the MSC or GMSC invoking the CAMEL service resides in the HPLMN of the served subscriber, then the CAMEL service invocation request may be sent directly to a STP in that HPLMN. The STP may derive the SPC of an SCP associated with this GT.
(2) If the MSC or GMSC invoking the CAMEL service resides in a PLMN which is not the HPLMN of the served subscriber, then the GT for this service is used to send the CAMEL service invocation request to a gateway STP of the PLMN, from where the service invocation request is forwarded to the HPLMN of the served subscriber.

In both cases, the serving MSC or GMSC does not need to know the SPC of the SCP (Figure 3.22). When an operator deploys IN *in the HPLMN only*, then the IN service invocations are always done from within the HPLMN. In that case, the network may be configured to use SPC for service invocation, instead of GT. However, due to the international nature of CAMEL, the gsmSCF address shall always be defined as GT.

The T-CSI is sent to GMSC in HPLMN only (if basic optimal routing is not used). Therefore, triggering of MT call CAMEL service occurs purely in the HPLMN, between GMSC and SCP. Hence, the gsmSCF Address in T-CSI *could* be configured in SPC format. However, the convention is that gsmSCF address for T-CSI is also in GT format. The use of GT is also needed for dynamic SCP load sharing.

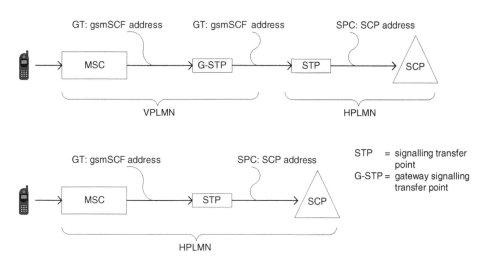

Figure 3.22   Using gsmSCF address to invoke a CAMEL service

3.3.1.2 Service Key

The service key (SK) identifies the CAMEL service in the gsmSCF. The serving MSC or GMSC places the SK from O-CSI or T-CSI into CAP IDP. The SCP platform uses the SK to select the corresponding CAMEL service. A gsmSCF may host several CAMEL services. Since the O-CSI is specific for a subscriber, different subscribers may subscribe to different CAMEL services. When a subscriber has an O-CSI and a T-CSI, then these CSIs may contain different SK values. If a subscriber has O-CSI and T-CSI for a CAMEL service like VPN, then O-CSI and T-CSI typically have different SK values, but may have the same gsmSCF Address.

The service key has an integer value, in the range $0-2,147,483,647$ ($2^{31} - 1$). An operator may use any SK value within this range. An operator should aim to use unique SK values for all service that are hosted in the network. Unique SK values facilitate network monitoring and post-processing systems. Large SK values should be avoided where possible. The reason is that, according to the BER applied in GSM, large integer values take up more data transmission capacity than small integer values.

3.3.1.3 Trigger Detection Point

The TDP identifies the point in the BCSM from where the CAMEL service is invoked. For CAMEL phase 1 (and phase 2), service triggering takes place at a defined point in the BCSM. These DPs are:

- mobile originated calls (O-BCSM) – DP collected info;
  Mobile terminated calls (T-BCSM) – DP terminating attempt authorized.[21]

A CAMEL phase 1 O-CSI or T-CSI with any other TDP value will be rejected by MSC or GMSC.

3.3.1.4 Default Call Handling

The default call handling (DCH) parameter in O-CSI and T-CSI is used to define gsmSSF behaviour in the case of signalling failure between gsmSSF and gsmSCF. One of the following behaviours for the gsmSSF may be set in O-CSI or T-CSI: *continue* and *release*.

---

[21] The ASN.1 definition for this DP is spelled with a 'z'. Textual descriptions use both 's' and 'z' spellings.

- *Continue* – the gsmSSF instructs the (G)MSC to continue call handling, without CAMEL control. This value may be used for VPN subscribers. Failure to invoke a VPN service may have the effect that a subscriber cannot benefit from, for example, number translation service, but it will normally not result in monetary loss for subscriber or operator.
- *Release* – the gsmSSF instructs the (G)MSC to release the call. This value may be used for pre-paid service, for example.

Default call handling may occur in the (G)MSC under the following conditions:

- *CAP IDP failure* – the gsmSSF has sent CAP IDP towards the gsmSCF, but receives an error indication from the SCP. The error may be 'MissingCustomerRecord', indicating that the served subscriber is not known in the SCP.
- *Tssf timeout at TDP handling* – the gsmSSF has sent CAP IDP towards the gsmSCF, but does not receive a response within the operation time of CAP IDP. This situation may be caused by a signalling error in the SS7 network or by overload in the SCP; see also the section on Tssf.
- *Tssf timeout at EDP handling* – the gsmSSF has sent CAP ERB towards the gsmSCF, but does not receive a response within the operation time of CAP ERB. This situation may have the same cause as a failure at TDP handling.[22] In CAMEL phase 1, the only EDP that may be armed in *interrupt mode*, resulting in the gsmSSF waiting for response from the gsmSCF, is the disconnect DP. Strictly speaking, applying DCH at DP disconnect has no effect. However, as will be explained below, DCH also relates to CDR generation.
- *Dialogue abortion* – a service processing failure in the SCP may have the effect that the SCP needs to abort an ongoing CAMEL dialogue. The gsmSSF applies DCH in that case.

Some call handling procedures in GSM TS 03.78 [38], such as CAMEL_OCH_MSC_DISC1 and CAMEL_MT_GMSC_DISC1, do not reflect the DCH in the case of Tssf timeout. The procedure definitions in GSM TS 09.78 [56] specify, however, the use of DCH in combination with both CAP IDP and CAP ERB.

Default call handling has the further effect that an indication is included in the CDR for this call. The following is extracted from GSM TS 12.05 [57]:

```
DefaultCallHandling ::= ENUMERATED {
    continueCall (0),
    releaseCall (1),
    ...}
```

When DCH has occurred, the CDR that is created for the call includes this default call handling parameter. The presence of this parameter indicates to post-processing systems that these CDRs need special processing. For example, if DCH occurs for a pre-paid subscriber and DCH in O-CSI has the value *continue*, then the post-processing system may use the CDR to calculate the call charge, verify the amount that was already deducted from the subscriber's account for this call and deduct any further outstanding debt for this call. For this reason, it is important that DCH also be applied when the signalling error occurs at DP disconnect.

---

[22] If an SCP is overloaded, it may ignore any incoming service request until the overload situation is resolved. Incoming operations related to an active service may, however, still be honoured during the overload situation. Overload handling in SCP is an implementation option.

3.4 Basic Call State Model

CAMEL phase 1 call control makes use of two state models: the originating call basic call state model (O-BCSM) and the terminating call basic call state model (T-BCSM). The BCSMs define the process in the MSC or GMSC, for the processing of a call. The O-BCSM is for the processing of MO and MF calls; the T-BCSM is for the processing of MT calls. It is only in GSM R96 that the call handling processes are strictly defined, primarily in GSM TS 03.18 [31]. The introduction of CAMEL in GSM R96 necessitated the definition of the BCSMs. The BCSMs contain defined points in the state model where the MSC may interact with the gsmSSF. The gsmSSF may, in turn, interact with the gsmSCF at these points. For each call that is established in the MSC, the corresponding BCSM is instantiated. That is to say, an MSC processing an MO or MF call instantiates an O-BCSM; a GMSC processing an MT call instantiates a T-BCSM. Although the BCSMs are defined as a result of the introduction of CAMEL, the instantiation of a BCSM in the MSC does not imply that a CAMEL service is invoked for that call. In principle, an O-BCSM instance is invoked for every MO or MF call and a T-BCSM instance is invoked for every MT call. If no CAMEL service needs to be invoked for the call, then the call processing continues through the BCSM according to GSM call processing rules, without CAMEL control. The call handling process in the MSC may not be aware of the invocation of a CAMEL service. This is reflected in the following strict functional separation:

- GSM TS 03.18 [31] specifies the basic call handling processes; these processes may call CAMEL procedures at defined points in the call handling processing;
- GSM TS 03.78 [38] contains the CAMEL procedures; the CAMEL procedures may, when required, send an internal signal to the gsmSSF.

3.4.1 Originating Basic Call State Model

Figure 3.23 depicts the CAMEL phase 1 O-BCSM. The DPs are the points in the call where interaction with the CAMEL service may take place. When the call is established, the BCSM is started in the state O_Null; the CAMEL service is started from DP Collected Info.

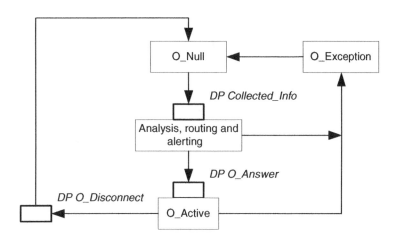

Figure 3.23 Basic call state model for MO and MF calls. Reproduced from GSM TS 03.78 v 5.8.0 Figure 7.2/1, by permission of ETSI

3.4.2 Terminating Basic Call State Model

Figure 3.24 depicts the CAMEL phase 1 T-BCSM. The T-BCSM is started in the state T_Null and the CAMEL service is started in DP terminating attempt authorized. The O-BCSM and T-BCSM for CAMEL phase 1 are derived from the BCSMs that are defined in ETSI CS1 (ETS 300 374-1 [158]) and ITU-T CS1 (Q.1214 [143]). The following sections explain some of the concepts related to the BCSMs.

3.4.3 Detection Points

A DP constitutes the occurrence of a call-related event. The event is often related to the reception of an ISUP message or DTAP message in the MSC. When such an event occurs, the MSC informs the internal gsmSSF, which decides on the action to be taken. The gsmSSF may report the event to the gsmSCF. The CAMEL phase 1 O-BCSM and T-BCSM contain the following DPs:

3.4.3.1 O-BCSM

DP Collected_Info
When an MO call or MF call is established, the O-BCSM is started at this DP. If O-CSI is available for the subscriber, then the MSC hands call control over to the gsmSSF, which invokes the CAMEL service. The call processing is suspended in this DP, until the gsmSCF has responded with a call continuation instruction.

DP O_Answer
When the MSC receives ISUP ANM or ISUP CON for the call, the O-BCSM transits to DP O_Answer. The gsmSSF process for this call may inform the gsmSCF about the answer event.

DP O_Disconnect
When the call is cleared, the O-BCSM transits to DP O_Disconnect. The gsmSSF process for this call may inform the gsmSCF about the disconnect event. The disconnect event may either be initiated by the calling party or by the called party; an indication is included in the notification to the gsmSCF.

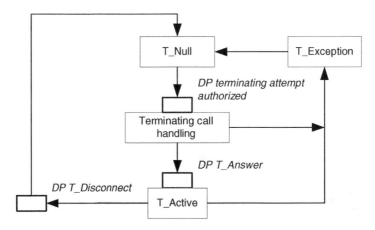

Figure 3.24 Basic call state model for MT calls. Reproduced from GSM TS 03.78 v5.8.0 figure 7.3/1, by permission of ETSI

3.4.3.2 T-BCSM

DP Terminating_Attempt_Authorized

When a GMSC is processing a MT call and has received T-CSI from HLR, the T-BCSM is started at this DP. The GMSC hands call control over to the gsmSSF, which invokes the CAMEL service. Call processing continues when the gsmSCF has sent a call continuation instruction.

DP T_Answer

When the MSC receives ISUP ANM or ISUP CON for the call, the T-BCSM transits to DP T_Answer. The gsmSSF may inform the gsmSCF about this event.

DP T_Disconnect

When the call is terminated by calling or called party, the T-BCSM transits to DP T_Disconnect. The gsmSSF may inform the gsmSCF about this event.

When a CAMEL service is started for a call, the gsmSCF maintains a mirror image of the respective BCSM instance. For example, when an MO call is set up, the CAMEL service receives the CAP v1 IDP operation in DP Collected_Info. The gsmSCF instantiates a CAMEL phase 1 O-BCSM in the DP. If the CAMEL service wants to know when the call is answered, it *arms* DP O_Answer in the O-BCSM instance in the gsmSSF. When the answer event occurs, the gsmSSF sends a notification to the gsmSCF, which in turn transits to DP O_Answer. The same method applies to DP O_Disconnect.

The interaction between the gsmSSF and the gsmSCF, when the BCSM has made a state transition to a particular DP, depends on the arming of that DP. See also Chapter 2 for generic BCSM principles.

3.4.3.3 TDP

When a DP is statically armed as TDP, the BCSM, when transiting to that DP, invokes a CAMEL service. The presence of O-CSI in the MSC constitutes the static arming of DP Collected_Info and the availability of T-CSI in the GMSC constitutes the static arming of DP Terminating_Attempt_Authorized. The other DPs for CAMEL phase 1 may not be armed as TDP.

3.4.3.4 EDP

When a CAMEL service is invoked, it may dynamically arm other DPs in the BCSM for this call. A CAMEL service for an MO or MF call may arm DP O_Answer and DP O_Disconnect; a CAMEL service for an MT call may arm DP T_Answer and DP T_Disconnect. When the CAMEL service arms DP O_Disconnect or DP T_Disconnect, it specifies the call leg for which the arming applies. See Chapter 2 for the arming modes EDP-N and EDP-R.

For CAMEL phase 1, the O_Answer and T_Answer DP may be armed as EDP-N only, not as EDP-R. The DPs O_Disconnect and T_Disconnect may be armed as EDP-N or EDP-R. The ability to arm DP disconnect on leg 2 as EDP-R does not imply the ability of the gsmSCF to generate a follow-on call. In other words, the gsmSCF is not allowed to respond to a disconnect indication on leg 2 by sending CAP CON, with the purpose of connecting the calling subscriber to another destination. Arming DP disconnect as EDP-R has the following purposes:

- having DP disconnect armed as EDP-R is required for the gsmSCF to have the capability to send CAP Release Call (RC) during the active phase of the call;
- arming DP disconnect as EDP-R enables the gsmSCF to send CAP RC to the MSC at the end of the call. Using CAP RC at the end of the call enables the gsmSCF to supply an ISUP release code. The ISUP release code is used in the ISUP signalling in the backwards direction, towards

the calling subscriber. By providing the ISUP release code, the CAMEL service may affect call termination behaviour in preceding exchanges.

3.4.4 Points in Call

The PIC for the CAMEL phase 1 BCSMs are 'O_Null', 'Analyse, Routing and Alerting' and 'O_Active' (for the O-BCSM) and 'T_Null', 'Terminating Call Handling' and 'T_Active' (for the T-BCSM). The BCSMs also depict O_Exception and T_Exception as PIC. However, these are not PIC in the strict sense of the word; O_Exception and T_Exception merely reflect an action that is taken by the gsmSSF. A brief description of the PICs follows.

3.4.4.1 O-BCSM

O_Null
The MSC starts the process of establishing an MO or MF call, which includes the performance of subscription checks and the instantiation of an O-BCSM. One may argue that O_Null is not a *true* PIC, since the call handling process in the MSC is not in a monitoring state, but is executing tasks, in preparation of transiting to DP Collected Info.

When the call is terminated, the BCSM transits back to O_Null and once all resources in the MSC for this call are released, the O-BCSM instance is released as well.

Analyse, routing and alerting
After the gsmSCF has instructed the gsmSSF to continue call establishment, the MSC enters the analyse, routing and alerting PIC. The MSC performs all tasks that are needed prior to sending ISUP IAM towards the destination of the call. The PIC also includes the waiting for call alerting message (ISUP ACM) and the waiting for call answer message (ISUP ANM). Whilst the call handling process in MSC is waiting for these call events, the gsmSSF finite state machine (FSM) is in the monitoring state.

O_Active
The call is in the active state. Both the call handling process in the MSC and the gsmSSF FSM are in the monitoring state, waiting for the call to terminate.

3.4.4.2 T-BCSM

T_Null
The GMSC starts the process of establishing an MT call. This includes the contacting of the HLR and the receipt of the T-CSI from HLR. T_Null is more genuinely a PIC than O_Null. The T_Null includes the sending of MAP SRI to HLR and (optionally) the sending of MAP PSI from HLR to VLR. Therefore, the call handling process in GMSC is waiting for a response from HLR before it can transit to DP terminating attempt authorized.

Terminating call handling
The GMSC takes action to connect the call to the called subscriber. This includes the sending of the second MAP SRI to HLR and the sending of MAP PRN from HLR to VLR. It further includes the sending of ISUP IAM to VLR and the receiving of ISUP ACM and ISUP ANM from VLR. The gsmSSF FSM is in the monitoring state, waiting for the call to be answered.

T_Active
This PIC may functionally be compared with O_Active.

3.4.5 BCSM State Transitions

During normal call processes, the call handling process in (G)MSC transits from one PIC in the BCSM to the next PIC in the BCSM. The normal sequence for an MO call would be, for example, O_Null → analyse, routing and alerting → O_Active → O_Null. When transiting from one PIC to the next, the BCSM passes through a DP. CAMEL phase 1 BCSMs do not include DPs for call establishment failure cases, such as busy or no answer. When such event occurs during PIC analyse, routing and alerting or during PIC terminating call handling, the BCSM transits to the exception PIC and from there to the null state. In this case, the gsmSSF aborts the CAMEL dialogue. Likewise, when the calling subscriber abandons the call before the call is answered, the gsmSSF transits to the exception PIC and aborts the CAMEL dialogue. CAMEL dialogue abortion is often regarded as a fault indication. However, these occurrences of dialogue abort are caused by limitations in the BCSM and are not erroneous.

Not all calls follow the transitions that are defined for the BCSM. The SCP may at various places in the BCSM induce a call release. These state transitions are referred to as 'transitions beyond basic call'. See Figure 3.25 for the transitions beyond basic call in the O-BCSM.

Transition (1) may occur when the CAMEL service determines that the subscriber is not allowed to establish this call, e.g. due to outgoing call restrictions. The gsmSCF sends CAP RC as a result of which the O-BCSM transits to the O_Null state. Transition (2) may take place when the CAMEL service applies SCP-based call duration control. When the CAMEL service determines during the call that the subscriber has reached the permissible credit limit, the SCP sends CAP RC and the O-BCSM transits to the O_Null state. The O-BCSM does not pass through DP O_Disconnect in this case.

A transition beyond basic call from PIC analysis, routing and alerting to PIC O_Null (3) is not very likely, as it would imply that the CAMEL service disallows the call after it has instructed the MSC to continue the call. A transition beyond basic call from DP O_Answer to O_Null is not possible, since DP O_Answer cannot be armed as EDP-R. As a result, when the O-BCSM transits to DP O_Answer, it reports the occurrence of this DP and immediately transits to PIC O_Active.

3.4.6 gsmSSF Process

The gsmSSF process acts as a relay between MSC and gsmSCF. The gsmSSF may report call-related events to the SCP. The gsmSSF may also suspend the call handling process in the MSC, in order to enable the CAMEL service to influence the call handling process. The gsmSSF contains a FSM that reflects the current state of the CAMEL dialogue with the gsmSCF; see Figure 3.26.

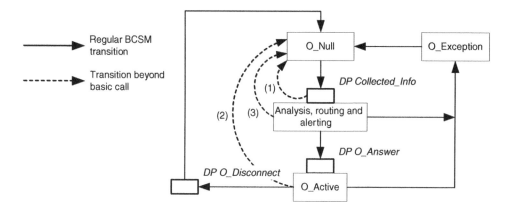

Figure 3.25   Transition beyond basic call in O-BCSM

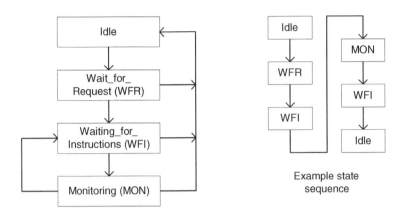

Figure 3.26   gsmSSF FSM

- *Idle* – this state of the FSM indicates that no gsmSSF process is currently active for the call. This condition may exist at the beginning of the call establishment, for example, when the MSC has not yet started a gsmSSF process for the call, or at the end of the call, when the gsmSSF process is released.
- *Wait_for_Request (WFR)* – this state applies when the MSC has invoked an instance of a gsmSSF process. The gsmSSF is now waiting for an indication from the MSC about the specific DP that was met, i.e. collected info or terminating attempt authorized.
- *Waiting_for_Instructions (WFI)* – the gsmSSF has contacted the gsmSCF and has suspended the call handling process. The call handling process remains suspended until the gsmSSF has received an instruction from the gsmSCF regarding continuation of the call handling. This FSM state may occur in the following cases: the gsmSSF has reported the occurrence of a TDP to the gsmSCF; and the gsmSSF has reported the occurrence of an EDP-R to the gsmSCF. For CAMEL phase 1, the only DP that may be armed as EDP-R is DP disconnect.
- *Monitoring (MON)* – This state applies when call handling in the MSC is in progress and the gsmSSF has at least one DP armed for reporting. Put differently, the gsmSSF is expecting to receive an indication from the MSC about the occurrence of a DP. If call establishment fails, the gsmSSF will receive an exception indication from MSC. The Monitoring state may exist during the PICs analyse, routing and alerting, O_Active, terminating call handling and T_Active.

Figure 3.26 shows an example state sequence for a successful call. The transition from Idle to WFR occurs at the beginning of the call. When the gsmSSF has sent CAP IDP to gsmSCF, the FSM transits to WFI; the transition to MON occurs when the gsmSSF has received CAP CUE or CAP CON. The occurrence of the answer event does not lead to a state transition, since answer cannot be reported as EDP-R. The transition from MON to WFI occurs when disconnect is reported as EDP-R. The gsmSCF responds with CAP CUE, to instruct the gsmSSF to continue call handling, as a result of which the FSM transits to idle.

3.4.7 Tssf Timer

The Tssf timer is a timer that is embedded within the gsmSSF process. Its function is to govern the response from the gsmSCF when the gsmSSF FSM is in the state WFI. This is to prevent the MSC keeping resources allocated when the gsmSCF would not respond, e.g. in the case of an SS7 communication problem. If the gsmSSF does not receive a response from gsmSCF within a pre-defined time, Tssf expires. The gsmSSF takes the following actions in that case:

- apply default call handling (DCH);
- place an indication in the CDR about the DCH event;
- close the CAMEL dialogue;
- terminate the gsmSSF process.

Tssf is started when the gsmSSF FSM transits from MON to WFI. When the gsmSSF FSM transits from WFI to MON, Tssf is stopped. In the case of FSM transition to Idle, the gsmSSF process is terminated, including Tssf. The Tssf has a value between 1 and 20 s; its exact value is operator-specific. Too short a value may lead to frequent Tssf expiry and call release; too long a value may keep resources allocated unnecessarily long in the case of SS7 communication failure.

The usage of Tssf is enhanced in CAMEL phase 2, as a result of the introduction of user interaction and the reset timer operation; see Chapter 4.

3.5 CAMEL Application Part

The present section describes the operations that are contained in the CAMEL application part v1 (CAP v1). CAP operations are described on two levels in the CAMEL specifications:

- *functional level* – GSM TS 03.78 [38] (for CAMEL phases 1 and 2) and 3GPP TS 23.078 [83] (for CAMEL phases 3 and 4) describe Information Flow (IF). Each IF may carry zero, one or more Information Element(s) (IE). The IF description includes usage of the various IEs and the conditions of the presence of an IE.
- *Syntax level* – GSM TS 09.78 [56] (for CAMEL phases 1 and 2) and 3GPP TS 29.078 [106] (for CAMEL phases 3 and 4)[23] describe the syntax of the CAP operations, including the parameters and errors that apply for the operation.

The Appendix lists the CAP operations.

3.5.1 Initial DP

The gsmSSF sends CAP IDP to the gsmSCF to start a CAMEL service. The sending of CAP IDP results from successful TDP processing during call establishment. The argument of CAP IDP contains a number of parameters that are used by the CAMEL service for its service processing. These parameters are filled with information from various sources:

- *subscriber-specific* – the parameter is obtained from the subscription data. Examples are service key, which is obtained from the CSI, calling party number for an MO call and called party number for an MT call.
- *MSC-generated* – the parameter is determined by the MSC or by the gsmSSF. Examples are call reference number and MSC address.
- *Call-specific* – the parameter is obtained from the signalling flow in the access network. For an MO call, the access network is the Radio Access Network. For an MF call and an MT call, the access network is ISUP.[24]

GSM TS 03.78 [38] specifies which parameters, also referred to as IE, will be present in CAP IDP for the different call cases (MO, MF, MT).

---

[23] For CAMEL control of IMS also 3GPP TS 29.278 [111].

[24] For a mobile-to-mobile call case, the VMSC and GMSC may reside in the same switch. In that case, MSC-internal ISUP is used between the VMSC and GMSC.

3.5.2 Request Report BCSM

The request report BCSM (RRB) operation is used by the gsmSCF to arm one or more DPs in the BCSM. When the gsmSCF arms a DP, it may specify a *call leg* for which the DP arming applies. For the disconnect event, the associated leg may be leg 1 (the calling party) or leg 2 (the called party); for the answer event, the associated leg may be omitted or may be set to *leg 2*. CAP RRB may be used to arm a DP in one of the follow modes:[25]

- *Notify and continue* – when the event associated with the DP occurs, the gsmSSF notifies the gsmSCF and continues call handling. If after notifying the gsmSCF there are no DPs armed in the BCSM, then the gsmSSF FSM transits to the idle state and the CAMEL dialogue is released.
- *Interrupt* – when the event occurs, the gsmSSF notifies the gsmSCF, suspends call handling and waits for instructions from the gsmSCF.
- *Transparent* – when the event occurs, the gsmSSF continues call handling without notifying the gsmSCF.

When a BCSM is instantiated, the DPs in that BCSM instance are not armed, i.e. are *transparent*. That implies that, if the CAMEL service is not interested in being notified about answer event, it does not need to use CAP RRB in transparent mode to disarm DP answer. The CAMEL service may simply leave DP answer unarmed.

Practically, a CAMEL phase 1 service does not need to use transparent arming. The reason is that the use of CAP RRB in CAMEL phase 1 is restricted to the gsmSSF FSM state WFI. The WFI state WFI occurs only at DP collected info and at DP disconnect. At DP collected info, the CAMEL service may decide not to arm a particular event; at DP disconnect, there is no need to disarm any event, since the only DP that may still be armed at that point in the call will be automatically disarmed by gsmSSF.

3.5.3 Event Report BCSM

The event report BCSM (ERB) operation is used in combination with CAP RRB. It is used to report the occurrence of an event, in the case that the event was previously armed. The mode in which the event is reported (EDP-N, EDP-R) depends on the arming state of the event. See also the description of request report BCSM (Section 3.5.2).

3.5.4 Continue

The CUE operation is used by the gsmSCF to instruct the gsmSSF to continue call handling. The CAMEL phase 1 BCSM has two places where CAP CUE may be used:

(1) *In response to CAP IDP* – the gsmSSF continues call handling with the available call-related information. That implies that an MO call is routed to the destination that is entered by the calling party.[26]
(2) *In response to DP disconnect when disconnect is reported in request mode* – CAP CUE has the effect that the gsmSSF continues call clearing. That implies that the ISUP REL that caused the disconnect event is propagated towards the calling subscriber.

---

[25] In CAP ERB, *notify and continue* mode is referred to as *continue* mode and *interrupt* mode is referred to as *request* mode.
[26] The MSC may still apply number translation during the analysis, routing and alerting PIC.

3.5.5 Connect

The Connect (CON) operation may be used in response to CAP IDP. CAP CON enables the CAMEL service to define or modify specific parameters in the call flow. When the gsmSSF receives a call parameter in CAP CON, it replaces the available parameter in ISUP by the parameter that is received in CAP CON. The following call-related parameters may be provided with CAP CON. All parameters in CAP CON, except destination routing address are optional.

- *Destination routing address* – this parameter defines the destination of the call. If CAP CON is used to modify a call-related parameter such as CPC, but not to change the destination of the call, then the CAMEL service shall use a destination routing address that is equal to the called party BCD number (MO call) or called party number (MF, MT call) in CAP IDP. Whereas Called Party BCD Number is coded in accordance with GSM TS 04.08 [49], destination routing address is coded in accordance with ITU-T Q.763 [137].
- *Calling Party's Category (CPC)* – the CPC serves as an indication of the type of calling party, such as *normal subscriber, operator* or *payphone*. A subscriber's CPC is provisioned in HLR and sent to VLR during registration. It may be used to indicate the subscriber's language preference within the HPLMN. CPC values are defined in ITU-T Q.763 [137]. Non-standard CPC values should normally be used in the HPLMN only.
- *Generic number (GN)* – GN is, as it name implies, a generic place holder for a number to be used in ISUP call signalling. The frame of a GN contains a *number qualifier*, which indicates what kind of number is contained in this parameter. ITU-T Q.763 [137] specifies generic numbers such as additional calling party number, additional called number, etc. CAMEL allows the use of GN in CAP CON only to define or modify the additional calling party number. The additional calling party number is often referred to as 'GN6', since it is contained in GN with number qualifier equal to 6. It would have been useful if CAMEL did not restrict the use of GN to GN6 only.[27]
- *O-CSI applicable* – a CAMEL service may use this parameter when it creates an outgoing call from the GMSC, when handling an MT call. When this parameter is present in this call case, the GMSC shall use O-CSI, if available, for the outgoing call leg. For CAMEL phase 1, the SCP may create an outgoing call from the GMSC only as a direct response to CAP IDP. From CAMEL phase 2 onwards, the outgoing call may also be created in response to call establishment failure or after a disconnect event.
- *Suppression of announcements (SoA)* – this parameter may be used when the CAMEL service is controlling an MT call in the GMSC. Its use is to suppress announcements in GMSC or VMSC resulting from call establishment failure. This option would be used when the CAMEL service wants to use service-specific announcements (CAMEL phase 2 onwards) and hence suppress generic, network-generated announcements; see Figure 3.27. When SoA is present in CAP CON, it may suppress announcements for both call establishment failure in GMSC and call establishment failure in VMSC. GMSC stores SoA internally and includes SoA in the second MAP SRI. HLR includes SoA in MAP PRN to VLR, which also stores SoA. If GMSC receives a negative result from HLR in response to the second MAP SRI, then it uses the stored SoA to suppress the announcement that it may otherwise generate. If a call is routed to VMSC, but call establishment fails in VMSC, then the VMSC uses the stored SoA to suppress the announcement that it may otherwise generate. A CAMEL phase 2 service may use SoA for call hunting. The service uses a list of alternative destination addresses to connect an incoming call. Presuming that these alternative destination addresses are MSISDNs of the same PLMN, the serving GMSC may send the MAP SRI for the call attempts to these destinations and include SoA in the respective

---

[27] Some vendors' MSC accept GN with number qualifiers other than 6. However, such deviation should be done only within the operator's HPLMN.

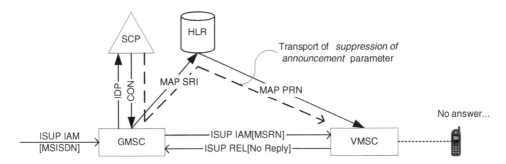

Figure 3.27   Transport of suppression of announcements indicator

MAP SRIs. This requires, however, that the GMSC supports the transport of SoA over internal ISUP, which is not specified by CAMEL.

- *Original called party ID* – if the CAMEL service forwards the call, then it includes this parameter in CAP CON. If the call was already forwarded prior to arriving at the GMSC, then this parameter should already be present in the ISUP signalling. In that case, the CAMEL service should not overwrite this parameter.
- *Redirecting party ID* – the CAMEL service uses this parameter when applying call forwarding, even when the call was already forwarded prior to arriving at the GMSC. It should contain the MSISDN of the served subscriber, i.e. the subscriber on whose behalf the call is forwarded. It may, however, also contain a VPN number or other number.
- *Redirection information* – this parameter may be used when applying call forwarding; it contains a number of variables that define the forwarding case. The following variables are included:
  - redirecting indicator, indicating the type of forwarding, such as call diversion or call rerouting;
  - original redirection reason, reflecting the reason for the first call forwarding for the call;
  - redirection counter, indicating the number of forwardings that have taken place; when the SCP applies forwarding, it increases the value of this variable by 1;
  - redirecting reason, indicating the forwarding reason, such as unconditional, busy or no reply.

A gsmSSF may copy original called party ID, redirecting party ID and redirection information, when received in CAP CON, transparently to ISUP IAM for the outgoing call, without checking whether these parameters contain correct information related to the specific forwarding case. The CAMEL service therefore should take care to apply sensible information in these parameters.

3.5.5.1 Setting the calling party number

CAMEL does not have the capability to modify the calling party number (CgPN) for a call. The rationale is that the CgPN should always be a true indication of the calling party for the call and no service should modify this number. CgPN may also be used by systems like lawful intercept (see GSM TS 02.33 [8]). The CgPN is set by the originating network and is sent to the called party. There exist call cases where a CAMEL service needs to change the CgPN; see Chapter 8 for example call cases. Some operators use a designated generic number (in CAP CON) to modify the CgPN. Operators should take care to use such a method only in the HPLMN.

3.5.6 Release Call

The gsmSCF may use CAP release call (RC) to disallow call establishment or to tear down an ongoing call. CAP RC carries a mandatory parameter, 'cause', which may be used in the ISUP signalling to the calling party; see Figure 3.28.

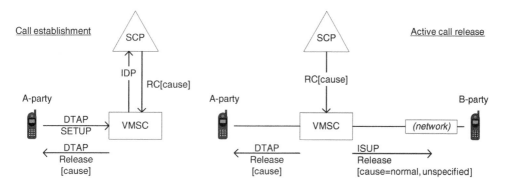

Figure 3.28   Usage of *cause* in CAP release call

If CAP RC is used during call establishment, then the cause is used for call clearing in a backwards direction. The MSC may use the cause to select an announcement and to send the cause code over DTAP or ISUP towards the calling subscriber. If CAP RC is used during an active call, then the MSC clears the call in both directions. The cause that is used towards the called party is not determined by the cause in CAP RC, but is set by the MSC, most likely 'Normal, unspecified'.

3.5.7 Activity Test

The activity test (AT) is used for testing the CAMEL dialogue between the gsmSCF and the gsmSSF. The SCP may send CAP AT at regular intervals to the gsmSSF, e.g. every 15 min. The only function of CAP AT is to verify the existence of the CAMEL dialogue. When the gsmSSF receives CAP AT, it returns an empty RESULT to the gsmSCF. If the gsmSCF does not receive an operation RESULT within the operation time for CAP AT, e.g. 5 s, then the gsmSCF terminates the CAMEL service. CAP AT is normally sent by the SCP platform, not by the CAMEL service. The arrival of CAP AT in the gsmSSF has no impact on any call handling process or on the BCSM. The sending of CAP AT is not dependent on the phase of the call or on the gsmSSF FSM state.

3.6 Service Examples

The current section describes a number of CAMEL phase 1 service examples.

3.6.1 Virtual Private Network

One of the principles of the VPN is that GSM subscribers belonging to a group use a customized dialling plan; the dialling plan emulates a private network, such as a PABX. One VPN member may call a member of the same VPN group by dialling that person's extension number; normally a 3- to 5-digit number. The VPN service translates the destination number into the public directory number of that VPN member; see Figure 3.29.

In the example in Figure 3.29, A-party and B-party belong to a particular VPN group. The VPN groups are defined in the SCP of the HPLMN operator. A-party may call the B-party by dialling '4523', the short number associated with the B-party. The A-party has O-CSI in the VLR, hence VPN service is invoked. VPN recognizes the called party number (CdPN) '4523' as an extension of the same group and therefore translates this short code to the public directory number of that subscriber, '+49 172 249 4589'. In addition, the SCP supplies an additional calling party number (A-CgPN), '3200', to the serving MSC of the A-party. Therefore, the B-party receives '3200' on her display instead of A-party's regular calling party number (CgPN) '+49 173 245 3211'.

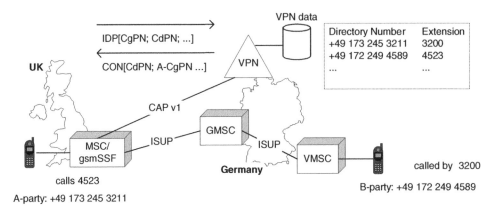

Figure 3.29   CAMEL phase 1 international VPN service

The VPN service may arm DP O-Answer, in order to receive a notification of successful call establishment; VPN may require this notification to generate a service CDR (a CDR generated by the SCP).

If the CdPN does not belong to the same VPN group, then VPN may decide to send CAP CUE, to continue the call without modifying the destination. Typically, destination numbers starting with '0' are not VPN extensions. It may occur that a short number, e.g. 8844, is at the same time a VPN extension of the VPN group of the calling party and a service code belonging to the serving PLMN. In the serving PLMN, 8844 may be a directory service available to both home subscribers and inbound roaming subscribers. If the VPN subscriber dials 8844, VPN may connect the subscriber to the indicated VPN member. To access the local directory service, the VPN subscriber may have to dial 08844. The 0 indicates to VPN that local access is required; VPN translates the CdPN to 8844, resulting in access to the directory service. The 0 serves as a 'break-out code'; details of such break-out codes are vendor-specific.

3.6.2 Pre-paid Route Home

Although the CAMEL phase 1 toolkit does not include on-line charging capability, CAMEL phase 1 is often used for pre-paid. A CAMEL phase 1 pre-paid service uses the 'route home' principle; see Figure 3.30.

The principle of using CAMEL phase 1 for international pre-paid is that the CAMEL service that is triggered from the VPLMN is used to force-route the call to a SSP in the HPLMN of the served subscriber. The SSP is an MSC with integrated SSF. The SSF may be compared with a gsmSSF, with the following differences:

- the SSF uses CS1 or a vendor-specific derivative of CS1;
- the SSF does not require a subscription element for service triggering; service triggering may also be number-based or trunk-based.

The CAMEL phase 1 service that is triggered from the VPLMN stores the information that is received in CAP IDP. When the SSP in the HPLMN re-triggers IN for this call, the service logic in SCP uses the stored information from CAP IDP to treat the call and to connect the call to the required destination. The destination number that is contained in CAP CON may have the following format:

Destination subscriber number $= <$ SSP routing address $><$ SCP Id $><$ Correlation Id $>$

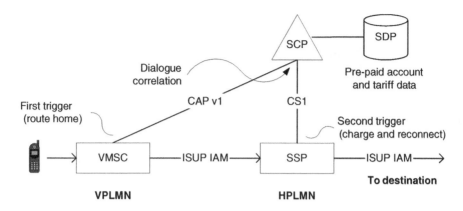

Figure 3.30 CAMEL phase 1 for international pre-paid

The *SSP routing address* contains digits that are used for routing the call to the SSP in the HPLMN. The CAMEL phase 1 route home mechanism is used only when the served pre-paid subscriber is in a PLMN other than the HPLMN. Therefore, the SSP routing address needs to contain country code, national destination code and subscriber number.

The *SCP Id* indicates the SCP in which the call data from CAP IDP is stored. The SCP Id is needed when the pre-paid service runs on more than one SCP. The SSPs in the HPLMN contain a table to map the SCP Id to SCP address. When this SCP address is in GT format, it should be the GT associated with this specific SCP.

The *Correlation Id* is used as an index into the file with stored CAP IDP data. The SCP may have several CAMEL phase 1 pre-paid calls in establishment. The correlation Id is needed during the establishment phase of the call only. As soon as the SCP has performed the correlation, the correlation Id may be returned to the pool of correlation Id's of that SCP. The SSP copies the entire called party number from ISUP IAM into the CS1 IDP. Hence, the SCP can retrieve the correlation Id.

An example structure of the destination subscriber number for the CAMEL phase 1 route home pre-paid service is the following:

SSP routing address	9 digits
SCP Id	2 digits
Correlation Id	4 digits
Total	**15 digits**

Once correlation has taken place, the SCP may apply on-line charging using the capabilities of CS1 or other protocol that is used between SSP and SCP. The CAMEL phase 1 service does not need to remain active after it has sent CAP CON. Figure 3.31 shows an example signal sequence for the CAMEL phase 1 route home service; signalling is reflected until the sending of ISUP IAM from SSP. Not all ISUP signalling is reflected. The ACH operation is described in Chapter 4.

The calling subscriber does not need to be aware that *route home* is applied for the call. However, dilemmas associated with the route home mechanism include:

- Additional service triggering and CAP/CS1 signalling; service triggering takes place from two nodes in the core network, leading to increased core network and signalling network load.
- Additional ISUP trunk usage. The call is always routed through an SSP in the HPLMN of the served subscriber, instead of directly to the destination. If the called party is a subscriber of a

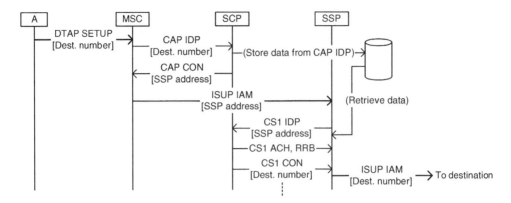

Figure 3.31   Example signal sequence flow for CAMEL phase 1 pre-paid service

network in another country than the HPLMN of the served subscriber, then call routing is sub-optimal. This dilemma manifests itself especially when the calling party calls a local destination; see Figure 3.32. In this call case, tromboning occurs.[28]

The operator has to decide how to charge the calling subscriber for CAMEL phase 1 roaming calls. Examples include:

- the call charge should reflect the actual ISUP usage;
- the call charge should be based on the distance between calling party and destination network, disregarding a possible tromboning connection in between;
- a flat fee is applied for international calling.

3.6.3 Short Number Dialling with CLI Guarantee

The route home mechanism used for CAMEL phase 1 pre-paid can also be used for 'CLI guarantee' (Figure 3.33). When establishing an international call, the calling party number (CgPN) may get

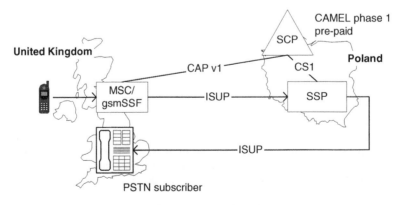

Figure 3.32   Tromboning between VPLMN and HPLMN with route home service

---

[28] The term 'tromboning' is derived from the shape of the brass instrument. An ISUP connection from one network to another and then back to the originating network resembles the shape of a trombone.

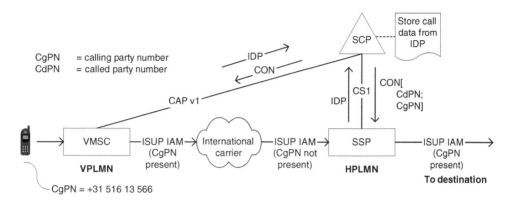

Figure 3.33   Number translation with CLI guarantee

lost in the transmission through the international carrier, due to characteristics of the carrier. The presence of the CgPN is, however, important when accessing a voicemail box; the CgPN is used to select the user's voicemail without the need for authentication.

When the CAMEL phase 1 service is triggered, it stores (some of) the parameters from CAP IDP, specifically the called party number (CdPN) and the calling party number (CgPN). The CAMEL service sends CAP CON, to route the call to an SSP in the HPLMN. The SSP applies retriggering, based on the called party number present in ISUP IAM. The retriggering uses CS1 or a proprietary protocol. When correlation has taken place in the SCP, the SCP sends CS1 CON, including a translated CdPN and the original CgPN.

The translated CdPN may be, for example, the SS7 address of the voice mailbox of the subscriber. The inclusion of the original CgPN in CS1 CON ensures that the CgPN is present in the ISUP signalling towards the destination.

Using CAP for retriggering would not be practical for the CLI guarantee, since CAP does not have the capability to modify the CgPN.

4

CAMEL Phase 2

CAMEL phase 2 is an extension of CAMEL phase 1. All functionality that is available in CAMEL phase 1 is also available in CAMEL phase 2. The following capabilities are added in CAMEL phase 2:

Call control-related
 (1) On-line charging
 (2) Call forwarding notifications
 (3) Follow-on calls
 (4) User interaction
 (5) Equal access
 (6) Enhancement to call control

Non-call control-related
 (7) Supplementary service invocation notification
 (8) Short forwarded-to-numbers
 (9) Conditional triggering
(10) USSD control

These features are described in detailed in Section 4.3.

4.1 Introduction

The first six listed features are directly related to call control, i.e. to CAP. These features may be regarded as enhancements to CAP. The call control enhancements may be used within a CAMEL relationship between gsmSCF and MSC/gsmSSF. In order to utilize the call control capabilities of CAMEL phase 2, a different protocol version is required, compared with CAMEL phase 1. In CAMEL phase 1, the protocol used between gsmSCF and MSC/gsmSSF is CAP v1; in CAMEL phase 2, CAP v2 is used. The distinction between CAP v2 and CAP v1 is made through the application context (AC), as further explained in Chapter 2.

CAP v2 is a superset of CAP v1. This means that it encompasses all the capability of CAP v1. However, even though CAP v2 is a superset of CAP v1, CAP v1 and CAP v2 are distinctively different protocols, invoked with different trigger data. An MSC/gsmSSF may request the establishment of a CAMEL phase 1 relationship or a CAMEL phase 2 relationship (or CAMEL 3 or 4). The gsmSCF may respond with the requested protocol only. If the MSC/gsmSSF requests the establishment of a CAMEL phase 1 relationship, then the gsmSCF is not entitled to respond with a CAMEL phase 2 relationship 'using the CAMEL phase 1 capability only'. This is a common mistake.

Practically, however, a mismatch between the requested CAMEL relationship (by gsmSSF) and available CAMEL services in the SCP should not occur. The CAMEL subscription data which is used to invoke a CAMEL service is administered by the same operator as the one that offers the CAMEL service in the SCP. Hence, the operator shall ensure that the available CAMEL services in the SCP correspond with the CAMEL subscription data.

As an aggregate result, there is no need for 'application context negotiation' between gsmSSF and gsmSCF. A gsmSCF shall not respond to a gsmSSF with a different (downgraded) AC than was requested by the gsmSSF. Note that GSM TS 09.78 [56] R96 and GSM TS 09.78 [56] R97 state that AC negotiation may be used. This, however, is not the case.

For an MSC to support CAMEL phase 2, it should support all CAMEL phase 2 capabilities. That includes not only the CAMEL phase 2 call control, but also non-call-related functionality, such as supplementary service invocation notification. An MSC/VLR that supports CAMEL phase 2 should support CAMEL phase 1 as well. That allows an HLR to decide which CAMEL data to send to such MSC/VLR. As an example, an MSC/VLR may support CAMEL phase 1 and CAMEL phase 2; a particular HLR may have a mix of CAMEL phase 1 subscribers (e.g. 'Route Home' pre-paid) and CAMEL phase 2 subscribers (e.g. VPN). Both CAMEL phase 1 and CAMEL phase 2 subscribers may register in that MSC/VLR. Therefore, when an operator upgrades its MSCs from CAMEL phase 1-capable to CAMEL phase 1 + CAMEL phase 2-capable, CAMEL phase 1 services may continue to be used in that MSC.

As described in Chapter 2, when a subscriber registers in an MSC (in HPLMN or in VPLMN), the serving MSC/VLR reports its CAMEL capability to the HLR. The MSC/VLR may support CAMEL phase 1 and CAMEL phase 2.[1] The HLR will then respond by sending GSM subscription data to the VLR, including CAMEL data, if available for the served subscriber. It is at this moment of registration that the HLR decides which CAMEL data to send to the VLR: CAMEL phase 1 data, CAMEL phase 2 data or no CAMEL data. Depending on the subscriber data in HLR, the HLR may also decide to disallow registration or to allow restricted registration. This process of subscription data transfer from HLR to VLR is described in detail in Chapter 2.

Once the HLR has provided the VLR with GSM subscription data, including CAMEL data, the subscriber is registered in that VLR as a CAMEL subscriber. The CAMEL services that may now be invoked by the subscriber are identified in the CAMEL subscription data, e.g. O-CSI. If the subscriber is a CAMEL phase 2 subscriber, then that is identified by the CAMEL capability handling (CCH) parameter in the O-CSI. That means that, for MO calls and MF calls for this subscriber, the MSC invokes a <u>CAMEL phase 2</u> service.

Similar to the O-CSI, of which a CAMEL phase 1 and a CAMEL phase 2 version may exist, there may be a CAMEL phase 1 and a CAMEL phase 2 version of T-CSI. Hence, the GMSC may invoke a CAMEL phase 1 or a CAMEL phase 2 service when handling a mobile terminating call. The *CAMEL Capability negotiation* that takes place between GMSC and HLR at the moment of MT call establishment is comparable to the CAMEL capability negotiation between VLR and HLR at the time of registration in VLR. The GMSC reports its CAMEL capabilities to the HLR[2] and the HLR will act by sending appropriate CAMEL data to the GMSC.

When the GMSC has received the CAMEL data (i.e. T-CSI), it will invoke a CAMEL service for the MT call. The CAMEL service that will now be invoked by the GMSC is identified in T-CSI. If the subscriber is a CAMEL phase 2 subscriber, then the CCH in T-CSI has the value *CAMEL2*. MT calls for that subscriber result in invocation of a <u>CAMEL phase 2</u> service.

A GMSC that supports CAMEL phase 2 also supports CAMEL phase 1; the HLR may therefore send CAMEL phase 1 subscription data to the GMSC for one subscriber and CAMEL phase 2 data for another subscriber. In addition to O-CSI, the HLR may send SS-CSI to the VLR. The HLR

---

[1] The CAMEL support in an MSC may be differentiated per IMSI range.

[2] Unlike the differentiation in CAMEL capability per IMSI range, as applied in VLR, GMSCs typically do not differentiate the reported CAMEL capability per MSISDN range.

may send a mix of CAMEL data to the VLR for one subscriber, e.g. CAMEL Phase 1 O-CSI and SS-CSI. SS-CSI does not come in different CAMEL versions, as is the case with O-CSI and T-CSI (and some other CSIs that are introduced in CAMEL phase 3). In other words, SS-CSI contains CAMEL trigger information, but no CCH parameter. SS-CSI may be sent to a VLR that supports CAMEL phase 2 or higher.

A third type of CAMEL phase 2 subscription data that may be sent to a CAMEL phase 2 MSC/VLR is the TIF-CSI. TIF-CSI, the translation information flag CSI, is used in combination with call deflection, to allow for call deflection to *Short Numbers*. The sending of TIF-CSI from HLR to VLR is specified in CAMEL phase 2 in GSM R98, but not in CAMEL phase 2 in GSM R97. As is the case for SS-CSI, TIF-CSI does not come in different CAMEL versions, i.e. TIF-CSI contains no CCH.

The conditional triggering feature may be used to limit triggering of CAMEL services to selected call cases. Trigger conditions may be included in O-CSI and T-CSI; the trigger conditions are checked when call establishment occurs.

Unstructured supplementary service data (USSD) control, finally, is an extension to the GSM USSD mechanism. USSD is a data communication protocol, between MS, MSC, VLR and HLR. CAMEL defines a mechanism to extend the USSD communication protocol between HLR and an external service handler, the gsmSCF. This extension allows a CAMEL service to communicate with an MS, by means of USSD signalling.

The remainder of the chapter provides a detailed description of each CAMEL phase 2 feature.

4.2 Architecture

Figure 4.1 depicts the network architecture for CAMEL phase 2. This network architecture reflects only the entities and protocols that are relevant for CAMEL.

4.2.1 Functional Entities

CAMEL phase 2 introduces a wide range of functionalities. Each functionality relates to a specific entity in the GSM network or may relate to a multitude of entities in the GSM network. The present section specifies the CAMEL phase 2 functionality for each involved entity in the GSM network. It also introduces new entities for which CAMEL functionality is defined.

4.2.1.1 HLR

In CAMEL phase 2, the HLR may contain more CAMEL trigger data than in CAMEL phase 1. The trigger data that may be provisioned for a CAMEL phase 2 subscriber in the HLR includes: O-CSI, T-CSI, SS-CSI, TIF-CSI and U-CSI. In addition, UG-CSI may be defined as a generic CAMEL trigger element. The CAMEL phase 2 HLR may function as a USSD signalling relay, using the contents of U-CSI or UG-CSI.

TIF-CSI is another data element that may be used internally in the HLR. It is used for the registration of short forwarded-to-numbers in HLR and for call deflection.

4.2.1.2 gsmSCF

A CAMEL phase 2 gsmSCF may respond to a CAMEL phase 2 relationship establishment request from a (G)MSC/gsmSSF over and above a CAMEL phase 1 relationship. In addition, there are a few more entities with which the CAMEL phase 2 gsmSCF may establish a CAMEL phase 2 relationship. These entities are the assisting gsmSSF and the intelligent peripheral. A CAMEL relationship with such an entity is for the purpose of in-band user interaction.

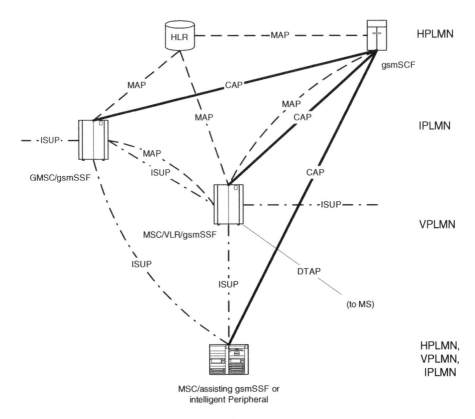

Figure 4.1   CAMEL phase 2 network architecture

Furthermore, the CAMEL phase 2 gsmSCF may establish USSD relationships with an MS and may function as end-point for USSD services requested by the MS. Finally, the CAMEL phase 2 gsmSCF may receive supplementary service invocation notifications from the MSC.

4.2.1.3 MSC/gsmSSF

The CAMEL phase 2 MSC may send supplementary service invocation notifications to the gsmSCF (using SS-CSI), and the CAMEL phase 2 MSC may establish a CAMEL phase 2 relationship with the gsmSCF (using O-CSI). The CAMEL phase 2 O-CSI may contain trigger criteria. If the MSC supports call deflection, then the MSC may also use TIF-CSI for short deflected-to-numbers.

4.2.1.4 GMSC/gsmSSF

The CAMEL phase 2 GMSC may establish a CAMEL phase 2 relationship with the gsmSCF (using T-CSI). Although the CAMEL phase 2 T-CSI may contain trigger criteria, these criteria are not sent to GMSC, but are checked in the HLR. Hence, conditional triggering for MT calls does not affect the GMSC.

On the other hand, since the GMSC may use O-CSI for MF calls, the triggering criteria in O-CSI may be used in the CAMEL phase 2 GMSC.

4.2.1.5 gsmSRF

An MSC or GMSC may contain a gsm specialized resource function (gsmSRF). The gsmSRF provides the capability of user interaction with a subscriber. When a gsmSCF has a CAMEL control relationship with the MSC or GMSC, the gsmSCF may use the CAMEL relationship to instruct the gsmSSF in that MSC or GMSC to connect the call to the gsmSRF. The gsmSCF may then instruct the gsmSSF to apply user interaction. User interaction may have the following forms:

- playing of pre-recorded announcements or tones to the user;
- announcing text messages to the user by means of text-to-speech conversion;[3]
- collecting in-band information (DTMF digits) from the user.

4.2.1.6 MSC/assisting gsmSSF

When a gsmSCF has a CAMEL control relationship with an MSC or GMSC, the gsmSCF may instruct the gsmSSF in that MSC or GMSC to establish a temporary speech connection (i.e. ISUP connection) with an MSC/assisting gsmSSF. An MSC/assisting gsmSSF is an MSC with integrated gsmSSF that can offer user interaction capability to other MSCs in the network or to MSCs outside the network. The user interaction runs in-band between the MSC/assisting gsmSSF and the user, via the MSC/gsmSSF from where the temporary connection was established.

The assisting gsmSSF may establish a CAMEL control relationship with the gsmSCF, to receive instructions for the user interaction. The MSC/assisting gsmSSF contains a gsmSRF for the user interaction. An MSC/assisting gsmSSF may at the same time be a (G)MSC/gsmSSF.

4.2.1.7 Intelligent Peripheral

The Intelligent Peripheral (IP) may offer the same user interaction capability as an MSC/assisting gsmSSF. A difference is, however, that an IP is a stand-alone node, dedicated to offering user interaction. The IP may establish a CAMEL control relationship with the gsmSCF. The IP may be regarded as a stand-alone gsmSRF. For that reason, and also to avoid confusion with the more common expansion of IP (internet protocol), the intelligent peripheral is sometimes referred to as a specialized resource point (SRP).

An IP may also act autonomously, without control by the gsmSCF. Such an IP does not need to support CAP v2. When such an autonomous IP is used, the gsmSCF will not know when an announcement is finished; the release by an IP is not reported to the gsmSCF. An autonomous IP may be used for voice-menu driven pre-paid voucher refill. Another example usage of autonomous IP is the ring-back tone (RBT) service. RBT entails a calling subscriber being connected to an IP during the alerting phase of the call; the IP plays a personalized ringback tone to the calling party.

4.2.2 Information Flows

The protocols for which CAMEL phase 2 functionality is defined are CAP and MAP. Table 4.1 provides an overview of the interfaces involved in CAMEL phase 2.

4.2.2.1 VLR–HLR

The MAP operations that may be used between VLR and HLR, as far as is relevant for CAMEL, are:

- update location
- restore data
- Insert subscriber data

---

[3] This is an optional functionality in the gsmSRF.

Table 4.1 CAMEL phase 2 interfaces

Interface	Protocol	Purpose
VLR – HLR	MAP v3	Transport of CAMEL trigger data from HLR to VLR
		Terminating call handling
		Retrieval of subscriber information
GMSC – HLR	MAP v3	Terminating call handling
gsmSCF – HLR	MAP v1, MAP v2, MAP v3	Any-time interrogation
		CAMEL control of unstructured supplementary service data
MSC – GMSC	MAP v3, MAP v4	Optimal routing of late call forwarding
MSC – gsmSCF	MAP v3	Supplementary service invocation notification
MSC/gsmSSF – gsmSCF	CAP v2	CAMEL control of mobile originated calls
		CAMEL control of mobile forwarded calls (late forwarded calls)
GMSC/gsmSSF – gsmSCF	CAP v2	CAMEL control of mobile terminated calls
		CAMEL control of mobile forwarded calls – early forwarded calls
		CAMEL control of mobile forwarded calls – late forwarded calls in combination with optimal routing
MSC/assisting gsmSSF – gsmSCF	CAP v2	CAMEL control of user interaction through assisting MSC
IP – gsmSCF	CAP v2	CAMEL control of user interaction through intelligent peripheral
(G)MSC – assisting MSC	ISUP	Temporary connection with assisting MSC, for user interaction purposes
(G)MSC – IP	ISUP	Temporary connection with intelligent peripheral, for user interaction purposes

- Delete subscriber data
- Provide subscriber info
- Provide roaming number

4.2.2.2 GMSC–HLR

The MAP operations that may be used between GMSC and HLR, as far as is relevant for CAMEL, are:

- send routing information.

4.2.2.3 gsmSCF–HLR

The MAP interface between gsmSCF and HLR is specific for CAMEL; it serves the following purposes: (1) any time interrogation; and (2) CAMEL control of unstructured supplementary service data. The ATI feature in CAMEL phase 2 is identical to the ATI feature in CAMEL phase 1. The USSD control signalling between gsmSCF and HLR is new in CAMEL phase 2.

4.2.2.4 VMSC–GMSC

The interface between the VMSC and the GMSC, for the purpose of optimal routing of late call forwarding (ORLCF) is conceptually the same as for CAMEL phase 1. In CAMEL phase 1, the MAP procedure used over this interface is MAP resume call handling (RCH) v3. From CAMEL phase 2 onwards, this interface may also use MAP RCH v4.

4.2.2.5 MSC–gsmSCF

The MAP interface between MSC and gsmSCF is specific for CAMEL; it is used for the supplementary service invocation notification (SSIN) procedure. The MAP procedure used over this interface is MAP SS invocation notification.

4.2.2.6 MSC/gsmSSF–gsmSCF

This interface serves the same purpose as in CAMEL phase 1. A difference is that, for CAMEL phase 2, the CAP v2 protocol is used for this interface.

4.2.2.7 GMSC/gsmSSF–gsmSCF

This interface serves the same purpose as in CAMEL phase 1. A difference is that, for CAMEL phase 2, the CAP v2 protocol is used for this interface.

4.2.2.8 MSC/assisting gsmSSF–gsmSCF

This interface is used for user interaction through a temporary connection with an MSC/assisting gsmSSF. The MSC/assisting gsmSSF may be used when the serving MSC does not have the required user interaction capability. The MSC/assisting gsmSSF uses CAP V2 towards the gsmSCF. The CAP V2 capability for this interface is a subset (and slight expansion) of the CAP V2 capability that is available for call control through the MSC/gsmSSF or GMSC/gsmSSF.

4.2.2.9 IP/gsmSCF–gsmSCF

This interface is used for user interaction through a temporary connection with an intelligent peripheral (IP). It has the same characteristics as the CAP v2 interface with the assisting gsmSSF.[4]

4.2.2.10 (G)MSC–assisting MSC

The interface between (G)MSC and assisting MSC is, unlike the above interfaces, a *traffic* interface. An ISUP connection may be established between an MSC or GMSC and an assisting MSC. The assisting MSC contains an assisting gsmSSF, which uses a CAP v2 interface with the gsmSCF for the purpose of user interaction control.

The ISUP protocol between (G)MSC and assisting MSC contains some IN-specific functionality, which may be used by CAMEL services but also by CS1 (or CS2 etc.) services.

4.2.2.11 (G)MSC–IP

The interface between (G)MSC and intelligent peripheral is the same as the interface between (G)MSC and assisting MSC. The intelligent peripheral contains a gsmSRF, which uses a CAP v2 interface with the gsmSCF.

---

[4] The CAP connect to resource (CTR) operation is not used in the interface between the IP and the gsmSCF.

4.3 Feature Description

The present section describes every main feature of CAMEL phase 2.

4.3.1 On-line Charging Control

On-line charging control allows the gsmSCF to monitor and control the duration of a call. 'On-line' refers, in this context, to the fact that the control of the call is done *as the call proceeds*. This is a clear difference from 'off-line' charging control; see the following comparison:

- *On-line charging* – the CAMEL service determines the charging rate of the call when the call is established. Call establishment will be allowed only when the served subscriber has sufficient credit in her account for this call. If call establishment is allowed, then the CAMEL service monitors the call and deducts money from the subscriber's account as the call proceeds. If the subscriber's credit falls below a defined threshold during the call, the CAMEL service may terminate the call.
- *Off line charging* – the charging rate of the call is determined afterwards, through CDR processing. The MSC that is serving the subscriber generates a CDR containing all relevant call details, such as called party number, time, location, subscriber identification, call duration, etc. This data is used to determine the cost of the call and to charge the subscriber.

A common mistake is to associate on-line charging strictly with pre-paid and off-line charging strictly with post-paid. However, on-line charging may be used for both pre-paid and post-paid subscribers.

- *Pre-paid subscribers* – a pre-paid subscriber has to obtain call credit prior to setting up outgoing calls or receiving incoming calls (when roaming). The on-line charging system monitors the calls of this subscriber and reduces the remaining credit, in accordance with the call rate and call duration. When the credit has reached a minimum level,[5] the call is terminated. The subscriber has to replenish her account before she can set up or receive new calls.
- *Post-paid subscribers* – a post-paid subscriber pays for incoming and outgoing calls afterwards and so will build up a debt. At regular intervals, the debt has to be settled with the operator. The on-line charging system monitors the calls of this subscriber. The on-line call monitoring may be used for features like:
  - spending control – the subscriber may call for a maximum amount per month;
  - credit control – the operator wants to have tight control over the available credit for subscribers; in this context, 'credit' refers to the amount of money the subscriber may still spend on calls, etc.;
  - call monitoring – within a company, phone usage may need to be monitored in real time;
  - the on-line charging system may generate call records, reflecting the network usage by the operators' subscribers.

The CAMEL on-line charging mechanism may be used for both types of subscribers.

4.3.1.1 Credit Reservation

CAMEL does not specify how an on-line charging system shall behave. However, one principle commonly used in on-line charging is the 'credit reservation' (Figure 4.2).

---

[5] In commercial pre-paid installation, the minimum amount may be negative. This may depend on local regulations.

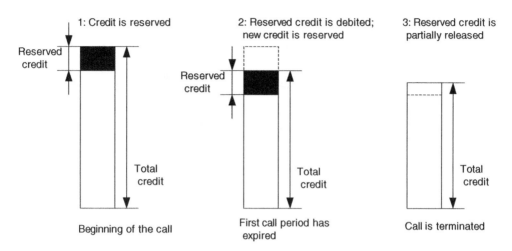

Figure 4.2   Credit reservation during a call

Step 1

When the call is established, the on-line charging service determines the rate of the call and reserves an appropriate amount of credit. The on-line charging service grants the user call time in accordance with the reserved credit. The reserved credit cannot be claimed by another charging process.

Step 2

When the granted call time is consumed, the charging service deducts the reserved credit from the user's credit balance and reserves an additional amount of credit. The charging service grants the user additional call time in accordance with the reserved credit. Step 2 may be repeated a number of times, depending on the call duration and available credit.

Step 3

The call is terminated by the calling or called party. The charging service deducts a part of the last reserved credit from the user's credit balance. This part corresponds to the actual call time of the user. The remaining part of the reserved credit is released, i.e. is unreserved. The released credit may now be used for other charging services.

As an operator's option, the subscriber may be charged in multiples of full call periods, e.g. multiples of 30 s. In that case, the actual call time is rounded up to the nearest multiple of 30 s call periods.

The on-line charging mechanism may be used for the different call cases: mobile-originated calls (Figure 4.3); mobile-terminated calls (Figure 4.4); and mobile-forwarded calls (Figures 4.5 and 4.6).

For an MO call, the CAMEL charging service is invoked from the MSC/gsmSSF. The MSC/gsmSSF may be located in the subscriber's HPLMN or in another PLMN.

The service data point (SDP) is described in ETSI European Norm (EN) 301 140-1 ('Capability Set 2', CS2) [159]. An SDP may contain call rating tables or subscriber account information. ETSI EN 301 140-1 [159] defines the interface between the gsmSCF and the SDP. The use of the SDP is, however, not specified by CAMEL. The Rating function may also reside internally in the SCP; this is an operator/vendor's option. The charging calculation (based on information received from the MSC/gsmSSF) may be done internally in the SCP or may be performed by an external entity such as an SDP. In the latter case, the SCP passes the relevant information received from the MSC/gsmSSF on to the SDP, which will then calculate the rate of the call.

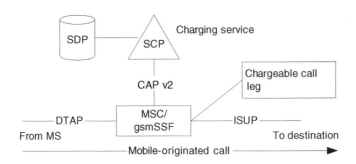

Figure 4.3   MO call charging

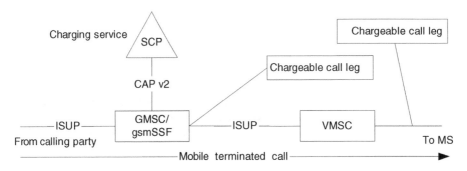

Figure 4.4   MT call charging

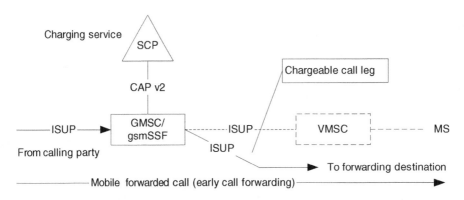

Figure 4.5   Mobile forwarded call (early forwarding) charging

The interface that will be used between the SCP and an external entity such as SDP is not specified by CAMEL either. According to CS2, an SS7-based protocol using the transaction capabilities (TC) services may be used.

All information that is required to determine the charge of the call is reported by gsmSSF to the SCP at the time of service invocation. This information may include parameters such as:

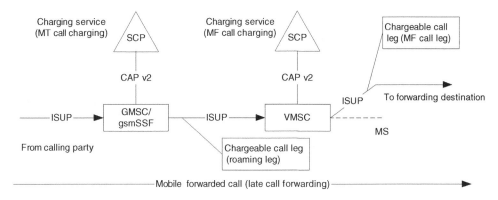

Figure 4.6   Mobile forwarded call (late forwarding) charging

- static information – location of the calling subscriber, access type (GSM/UMTS), time and time zone, etc.;[6]
- call-related information – destination, call type (speech, data, Fax, etc.);
- identification of the calling subscriber – MSISDN and IMSI.

The charge of the MO call also includes the costs of the usage of the radio access network.

For an MT call, the CAMEL charging service is invoked from the GMSC/gsmSSF. In most cases, the GMSC is located in the HPLMN of the served subscriber. As a result, the MT call charging is done within the HPLMN of the served subscriber, even if the subscriber is currently roaming in a different country. Refer to Section 4.8.1 for cases where the GMSC is located in a network other than the HPLMN.

The MT call CAMEL charging service receives information such as:

- static information – location of the called subscriber,[7] access type, time and time zone,[8] etc.
- call related information – calling subscriber, call type;
- identification of the called subscriber – MSISDN and IMSI.

The MT call charging may be based on the distance of the roaming leg, i.e. the call leg between the GMSC and the VMSC; the charging rules are operator-specific. The further away the called subscriber is from her HPLMN, the higher the charge will be for receiving a call. When a subscriber is in the HPLMN, then MT calls may be free of charge. The call up to the GMSC is normally paid for by the calling subscriber. The charging of the roaming leg between GMSC and VMSC does not normally affect the charging of the call leg up to the GMSC.[9]

The cost of the roaming leg between GMSC and VMSC may also differ per roaming partner, i.e. it is not solely dependent on the distance. An operator may use different interconnect carriers for traffic connections with different roaming partners. Also, the operator may have different accounting

---

[6] The reported time is the time of the serving MSC; if the MSC service area spans several time zones, then the charging service may need to use the location of the subscriber (e.g. cell Id) as an indication of the local time for the subscriber.

[7] As described in Chapter 1, the location of the called subscriber is fetched from the VLR, during the MT call handling process.

[8] This is the time of the GMSC.

[9] One exception is distance-based charging, whereby the charge of an outgoing call depends on the location of the called party. Such a charging mechanism may be achieved, for example, with any-time interrogation.

agreements with different roaming partners. Furthermore, a multi-country operator may apply a preferential rate for the roaming leg when the called party resides in one of its subsidiary/partner networks.

Early call forwarding is the call forwarding that occurs in the GMSC. Call forwarding is extensively described in Chapter 3. When early call forwarding occurs, a CAMEL service may be invoked from the GMSC/gsmSSF on behalf of the forwarding party. The charge of the forwarded call is mainly based on the destination of the forwarded call leg. If the destination of the forwarded call is in the HPLMN, such as voicemail, then the early-forwarded call is normally free of charge.

In the case of early call forwarding, there will not be a roaming leg established between the GMSC and VMSC. Hence, there will not be MT call charging (unless collect call charging applies). The feature 'Call forwarding notification' allows the SCP to drop the charge for the MT call in the case of early call forwarding. Call forwarding does normally not affect the call charge up to the GMSC, i.e. the calling party pays the same price for the call, even when the call is forwarded.

In the case of late call forwarding, the roaming leg to the VMSC is already established when the forwarding occurs. From the VMSC, a forwarding leg will be established. The CAMEL service that is controlling the MT call will not know that forwarding is taking place in the VMSC.[10] As an aggregate result, the charging of the roaming leg between GMSC and VMSC will continue.

The forwarded leg in the VMSC may be subject to a CAMEL charging service. There will now be two charging services active for this call: one MT call charging service from the GMSC; and one MF call charging service from the VMSC. Hence, the called subscriber, on whose behalf the forwarding is taking place, pays both the roaming leg and the forwarding leg.

If the late call forwarding occurs when the called subscriber is in the HPLMN, then the MT CAMEL charging service may have been suppressed during call establishment; that is, the HLR may have suppressed the sending of T-CSI to the GMSC for that MT call. In that case, only the MF call CAMEL charging service will be invoked. If the forwarded call is to a destination in the HPLMN, then the forwarding CAMEL service will not normally charge for this forwarded call. As a result, in the case of late call forwarding in the HPLMN, both the terminating part of the call and the forwarding part of the call will be free of charge for the called subscriber. [11]

4.3.1.2 Control of Call Duration

The main mechanism for CAMEL call duration control is call duration monitoring in the gsmSSF. When a call is established and the gsmSCF has gained control over the call, the gsmSCF may instruct the gsmSSF to monitor the duration of the call and to send a call duration report to the gsmSCF after a pre-defined period of time. The call monitoring in the gsmSSF starts as soon as the call has reached the active state, i.e. the called party has answered the call.

The call is divided into call periods. After every call period, the gsmSSF sends a report to the gsmSCF, informing the gsmSCF about the elapsed duration of the call. The gsmSCF may send a new request for call duration monitoring to the gsmSSF. This sequence of (1) request for call duration monitoring and (2) generation of charging report continues until one of the call parties terminates the call or the available credit has reached a minimum level. At such a moment, the gsmSSF or gsmSCF will terminate the call.

---

[10] When call forwarding occurs in the VMSC, the VMSC may send an ISUP progress message in the backwards direction. This progress message is, however, not picked up by the gsmSSF in the GMSC and is not reported to the MT call CAMEL service. This dilemma may be solved with ORLCF.

[11] Some operators charge late call forwarding from the HPLMN even when the destination of the forwarded call is in the HPLMN.

The CAP operations that are used for this mechanism are:

- *apply charging (ACH)* – this is the instruction from the gsmSCF to the gsmSSF to start or continue monitoring the call duration;
- *apply charging report (ACR)* – this is the report that is sent from gsmSSF to gsmSCF at the end of a call period or when the call is released. In addition, when call set up failure occurs, such as called party busy or no answer, the gsmSSF also sends a charging report (if previously requested).

Figure 4.7 gives a graphical example representation of this process. The figure reflects only those CAP operations that are relevant for call duration control.

The SCP sends the first CAP ACH before gsmSSF has processed the call answer event. There are two methods of sending the first CAP ACH.

(1) *In response to IDP* – when the gsmSSF receives CAP ACH, it is ready to start timing call duration. When call answer is detected, the actual timing in gsmSSF starts. The answer event (DP answer) may be armed, but does not need to be armed in interrupt mode. In the case that call establishment fails, the call period that was expected to start (at call answer) is regarded as having started and stopped. The gsmSSF sends an empty charging report.
(2) *At call answer* – the SCP needs to arm the answer event in interrupt mode. When the answer event is reported, the SCP sends CAP ACH to the gsmSSF, followed by CAP CUE. The gsmSSF now starts call monitoring directly when it receives CAP ACH. The call answer message (ISUP ANM) is propagated in the backwards direction (e.g. DTAP CONNECT message in the case of an MO call) after the gsmSSF has received CAP CUE. This is relevant for CAMEL control of advice of charge, as will be seen in a next section. When call establishment fails, no charging report is sent to the SCP. The reason is that the gsmSSF has not yet received CAP ACH in that case.

Use of method (1) or method (2) does not affect the on-line charging. A gsmSSF is capable of handling either method. The CAMEL service may decide which method to use. A CAMEL phase 3 service may have a specific reason to defer the sending of the first CAP ACH until the answer is detected; see Chapter 5.

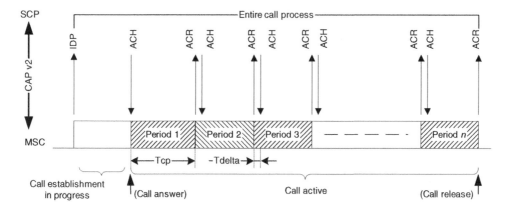

Figure 4.7 On-line call charging sequence (example)

The first CAP ACH *may* also be sent in the period between IDP and call answer. The gsmSSF will in that case wait for call answer, as in method (1). However, this method is not very useful and may lead to unexpected service behaviour, as it is not known when call answer will occur.

Defining the optimum call period duration requires careful consideration by service logic design. Short call periods give the operator tighter control over the call, but result in more signalling between gsmSSF and gsmSCF. The call period duration also depends on the rate of the call. For a local call (low rate), the call period may be longer than for a long distance call (high rate).

To prevent charging inaccuracy due to delay between the sending of a CAP ACR by gsmSSF and the receiving of the subsequent CAP ACH by gsmSSF, a delta timer (T_{delta}) is used. The gsmSSF starts the delta timer when it sends CAP ACR. When it receives CAP ACH, the gsmSSF subtracts the delta time from the next call period.

A CAMEL pre-paid service may, as a vendor/operator option, use longer call period durations for subscriber with available credit above a certain value. Alternatively, a pre-paid system may combine CDR based charging with on-line CAMEL charging, as follows:

- When available credit is in excess of e.g. $50.00, keep O-CSI and T-CSI de-activated in the HLR. Calls will be charged off-line, by means of CDR processing.
- When available credit falls below e.g. $50.00, activate O-CSI and T-CSI. Calls will be charged on-line through CAMEL. The CDRs for the calls that were charged through CAMEL contain an indication that CAMEL was applied for these calls. Hence, these calls will not be charged off-line by the billing system.

Accumulated Call Duration Report
During an ongoing call that is subject to call duration control, gsmSSF and gsmSCF exchange charging operations. The gsmSCF sends CAP ACH, granting a defined call duration, and the gsmSSF responds, when the granted call time has lapsed, with CAP ACR. Whereas the CAP ACH contains the permissible call duration for the next call period, the resulting CAP ACR contains the *accumulated* call duration (Figure 4.8). Call periods of 300s are used. The maximum call duration that can be reported is 24 hours.

Forced Call Release
When at the beginning of a call or during a call the gsmSCF has determined that the call may continue for a *partial call period* only, the gsmSCF may instruct the gsmSSF to release the call when the call period expires. This mechanism is knows as 'forced call release'. The reasons for applying forced call release include, but are not limited to:

- the subscriber is a pre-paid subscriber and her credit has reached a minimum level;
- the subscriber is a post-paid subscriber and she has built up the maximum amount of debt.

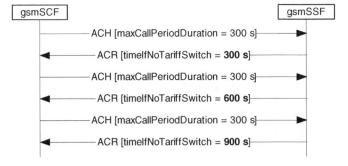

Figure 4.8 Cumulative charging reports

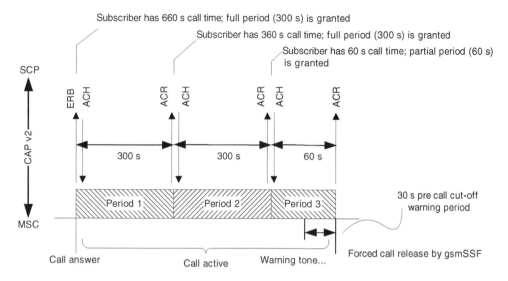

Figure 4.9   Forced call release

See Figure 4.9 for a graphical representation. When the final call period is complete, the gsmSSF still sends a charging report to the gsmSCF, so the gsmSCF can calculate the final amount for the call.

The forced call release mechanism serves the following purposes:

- *safety* – the gsmSSF will release autonomously; should a signalling error occur between gsmSSF and gsmSCF, then this will not affect the call release;
- *reduced signalling* – when the final call period is complete, there is no additional instruction required from gsmSCF to release the call;
- *accuracy* – call release occurs exactly at the moment that the final call period is complete.

Warning Tone
A feature that is tightly coupled with the forced call release is the warning tone. When the gsmSCF instructs the gsmSSF to force-release the call when the call period is complete, the gsmSCF may include an indication that a warning tone should be played 30 s prior to call cut-off (Figure 4.9). The warning tone should be played to the served subscriber only:

- *MO call* – the calling subscriber is the served CAMEL subscriber; the calling party pays for the call; therefore, the warning tone is played to the calling party.
- *MT call* – the called subscriber is the served CAMEL subscriber; the called party pays for the call (for an MT call, this is the roaming leg between GMSC and VMSC);[12] therefore, the warning tone is played to the called party.
- *MF call* – the CAMEL service should not use a warning tone during an MF call, since the served CAMEL subscriber is not part of an MF call.

CAMEL does not specify whether the speech connection between the calling and called party remains active during the playing of the warning tone. Two implementations are possible (vendor-specific). (Figure 4.10). Since the warning tone has a total duration of 1 s, no problem is expected.[13]

---

[12] In the case of *collect call*, the called party may also pay for the call up to the GMSC.
[13] In CAMEL phase 4, a flexible warning tone is introduced which may have a longer duration.

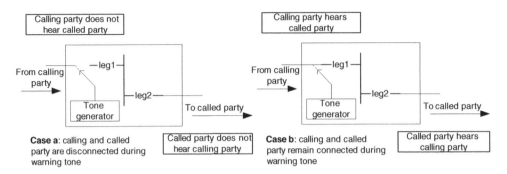

Figure 4.10   Handling of CAMEL warning tone in MSC

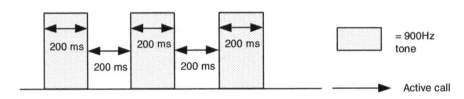

Figure 4.11   CAMEL warning tone

The structure of the warning tone is depicted in Figure 4.11. CAMEL does not specify the signal level of the warning tone. That is left to MSC vendors.

The CAMEL service ensures that the final call period, preceding the forced call release, has a minimum length of 30 s. Otherwise, the warning tone cannot be played 30 s prior to call release.

The warning tone can be used only in combination with forced call release. A CAMEL service may use the CAP release call (RC) operation to terminate a call instead of using forced call release. One example is the case where the CAMEL service wants to include free-format data in the MSC-generated CDR just prior to releasing the call. The forced call release leads to immediate termination of the CAMEL dialogue between gsmSSF and gsmSCF. As a result, the gsmSCF no longer has the possibility to include free-format data in the MSC-generated CDR. To overcome that dilemma, the CAMEL service would, after having received the charging report related to the final call period, use CAP furnish charging information (FCI), followed by CAP release call. The effect is, however, that no warning tone can be played. This dilemma is resolved in CAMEL phase 3, where the warning tone is decoupled from the forced call release in CAP ACH.

It cannot be guaranteed that a served subscriber will *always* receive the warning tone prior to forced call release. When on-line call charging is ongoing and a new call period starts, the CAMEL service may deduce that there is sufficient credit available for at least two full call periods. Hence, the SCP would not use the forced call release for that call period. Whilst the call period is ongoing, the served subscriber may send an SMS. The sending of the SMS may be charged on-line and may result in real-time credit deduction. By the time that the current call period call is complete, the SCP deduces that there is not sufficient credit anymore to start another call period. The SCP may release the call at that moment, but the user does not receive a warning tone. In other cases where multiple service instances access the same user credit, similar situations may occur.

Missing Apply Charging Operation
When the MSC/gsmSSF has sent the CAP ACR during an ongoing call, the call continues and the gsmSCF sends a subsequent CAP ACH. If, however, this CAP ACH does not arrive as a result of a communication failure between gsmSCF and gsmSSF, then the call at the MSC/gsmSSF will

continue.[14] Although the gsmSSF does expect a new CAP ACH, it takes no action if this new CAP ACH does not arrive. The result may be that the call continues without call duration control. In addition, the CAMEL service logic would 'hang'; the service logic remains active, but will not receive the CAP ACR. The next CAP activity test (AT) operation that is sent by the SCP to the gsmSSF will, however, fail. The SCP may now terminate the CAMEL service. A pre-paid service will typically not charge for a call period if the SCP does not receive a CAP ACR for that call period. The credit reserved for that call period will in that case be released, i.e. it becomes available again for other calls. The precise effect for the gsmSSF depends on the arming of the disconnect DP.

DP Disconnect Armed as EDP-N

The call continues until the calling or called party disconnects from the call. At the end of the call, the gsmSSF reports the disconnect event and sends the call information report(s), if requested. If the communication error condition still exists, then the reporting of the disconnect event and the sending of the call information report(s) may fail.

When the gsmSCF detects the communication error for this call (with activity test), it may include an indication in the service CDR that the call will be rated off-line.

DP Disconnect Armed as EDP-R

When the gsmSSF has reported the disconnect event and has sent the call information report(s), it transits to the waiting for instructions state. If the communication error condition still exists, then the gsmSSF will not receive any instruction from the gsmSCF. As a result, the Tssf timer in the gsmSSF expires and the gsmSSF applies default call handling (DCH). The application of DCH by the gsmSSF may result in a marking in the MSC-generated CDR for this call (see Figure 4.12 for a graphical explanation of this).

The DCH marking in the MSC-generated CDR may be used by the CDR post-processing system to filter out these CDRs and forward them to a billing system. The CDR may be correlated with the CDR that is generated by the gsmSCF. In this way, the charge levied for the call can be compared with the actual call duration.[15]

This post-call reconciliation is not standardized by CAMEL; it is a vendor/operator's option. CAMEL phase 3 specifies a mechanism to have the gsmSSF terminate the call when a following CAP ACH does not arrive at the gsmSSF within a pre-defined duration.

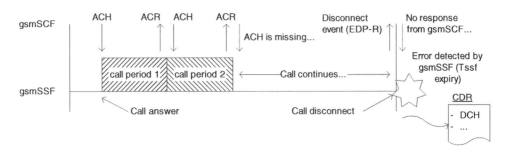

Figure 4.12 Default call handling at call release

---

[14] GSM TS 02.78 for R97 specifies in Section 9.4 that: 'If the report is not confirmed by the CSE within a specified time, the IPLMN/VPLMN shall release the call.' That functionality is, however, not implemented in GSM TS 03.78 for R97.

[15] To facilitate this post-call comparison, the gsmSCF may place a charge indicator in the CDR by means of the FCI operation, at the beginning of the call.

4.3.1.3 Call Information Reports

An on-line charging service may request various information elements from a call. The gsmSSF may be instructed to collect certain time stamps and measure specific durations. The following two CAP operations are used here:

- call information request (CIRq) – CIRq is used by the gsmSCF to ask the gsmSSF for the call information report;
- call information report (CIRp) – CIRp is used by the gsmSSF to send the requested call information report to the gsmSCF.

Whereas call duration control (using CAP ACH and CAP ACR) applies to an outgoing call connection (leg 2 in the BCSM), the call information reports may be requested per call leg, i.e. the incoming leg ('leg 1') and the outgoing leg ('leg 2'). Per required call information report, CAP CIRq and CAP CIRp are needed. The requested information elements are reported to the gsmSCF at the end of the call leg for which the specific information is requested. Information elements may be requested from the gsmSSF (Table 4.2).

A CAMEL service may initiate a follow-on call and request a call information report for each successive leg 2. Each successive outgoing call connection generates its own call information report and will send that report to the SCP when the respective call connection is released. The information reported for the respective outgoing call legs may differ for each call leg.

The call information report is generated by gsmSSF for both successful and unsuccessful call connection. The request for call information report, if required, will always be sent in direct response to IDP. The reason is that the timing for the call attempt elapsed time starts at IDP. For a follow-on call, the request for the call information report is sent just prior to the CAP CON that leads to the follow-on call.

Call Attempt Elapsed Time
For leg 2, this is the duration between the CAP CUE or CAP CON that leads to the outgoing call set up and the answer of that call. For leg 1, this variable always has value 0. The rationale is that, when a call is established, leg 1 is considered to be active for the entire call duration.

Some operators charge a calling party not only for an *active* call connection, but also for a call attempt. The call attempt elapsed time may be used to fulfil such a requirement. However, the call attempt elapsed time is reported to the SCP only at the end of the call connection, so it may be difficult for a pre-paid charging service to reserve an appropriate amount of credit for the call attempt elapsed time. Accurate charging, including the call attempt elapsed time, can be finalized only at the end of the call.[16]

Table 4.2 Call information report

Information element	Description
Call attempt elapsed time	This data element indicates the time it takes for the call to be connected
Call stop time	The call stop time is a time stamp indicating the exact moment that the indicated leg is released
Call connected elapsed time	This data element indicates the duration of the call leg in active state, as from answer onwards
Release cause	This data element indicates the cause of the release of the indicated leg

---

[16] As will be discussed later, a similar dilemma exists for calls that cross a tariff zone.

Call Stop Time

Leg 1 and leg 2 may be released at different moments. A practical use case is a call whereby a connection to a particular destination is established. When the called party releases the call, the CAMEL service creates a follow-on connection to a different destination. In that case, the initial leg 2 is released while leg 1 remains active. Hence, these legs will have different call stop times.

Call Connected Elapsed Time

This is the actual active duration of a call leg. For leg 1, the call duration is the total duration from IDP until call release.

Release Cause

This variable indicates the cause of the release of the call leg. Different legs in the call may have different release cause. Refer to Figure 4.13 for a graphical representation of the call information reports of an example call.

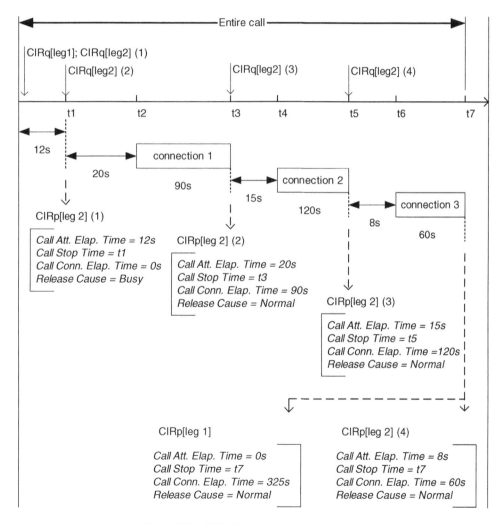

Figure 4.13   Call information requests and reports

The example call in Figure 4.13 consists of the following connections:

- a failed call attempt to destination 1; the called party has induced busy ('UDUB') after 12 s;
- the CAMEL service induces a follow-on call to a second destination; the call is answered after 20 s and is released by the called party after 90 s;
- the CAMEL service induces a second follow-on call to a third destination; the call is answered after 15 s and is released by the called party after 120 s;
- the CAMEL service induces a third follow-on call to a fourth destination; the call is answered after 8 s and is released by the called party after 60 s.

The total call duration reported in the call information report on leg 1 is 325 s.

4.3.1.4 Free-format Charging Information

Any call that is established in an MSC results in the generation of a CDR. Each call type results in one or more specific CDRs. Refer to Chapter 7 for description of the CDR generation process. The CDR that is generated in the MSC contains certain CAMEL specific information fields. These fields facilitate the processing of these CDRs by a billing gateway or similar entities. The contents of these CAMEL specific information fields are specified in the relevant GSM specifications. Hence, the service logic has no influence on these fields.

To enable a CAMEL service to place service-specific data in the CDR, 'free format data' is introduced. A CAMEL phase 2 service has the possibility of placing free format data in the CDR. As its name implies, the contents of the free format data are not defined by the GSM specifications; the contents are defined by the CAMEL service. The free format data parameter is therefore a placeholder for service data.

Free format data is carried in the FCI CAP operation. See Figure 4.14 for the use of FCI in a mobile originated call. The CAMEL service may provide free format data per call leg. If a call consists of several successive outgoing call legs, then each outgoing call leg may have its own set of free format data associated with it.

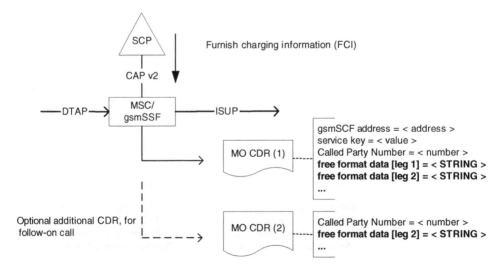

Figure 4.14 Placing of free format data in CDR

The free format data has a maximum length of 40 octets. A CAMEL service may provide free format data for a call leg at any moment during the call. Each time the CAMEL service sends free format data to a particular leg in the call, the free format data overwrites any free format that was previously provided by the CAMEL service for that leg. The free format data for a particular leg is retained internally in the gsmSSF until the call leg is released. At that moment, the free format data is placed in the CDR associated with that leg. If the MSC generates intermediate CDRs, e.g. for a long-duration call, then the intermediate CDR will not contain free format data, since the call leg is not yet released. See figure 4.15; the CAMEL service ensures that successive FCIs contain the accumulated free format data for the respective call leg.

When a CS1 service uses FCI, the free format data that is carried in FCI is normally immediately placed in the CDR. A CS1 service may use FCI for event charging. Successive intermediate CDRs for the call may contain free format data, each set of free format data reflecting the charge for an event.

Examples of the usage of FCI in CAMEL include:

- *Charge indication* – when a call is controlled by a VPN service, the VPN service may place an indication in the CDR that this call qualifies for a reduced rate. An example of a call that qualifies for a reduced rate is a call that is established in a user's home zone. The operator's billing system shall, on detection of this particular information in the CDR, apply a reduced rate for the call. This free format data may be placed in the CDR at the beginning of the call, when the VPN service has ascertained whether the call qualifies for the reduced rate.
- *Forced call release indication* – a pre-paid CAMEL service may indicate in the CDR, by means of the free format data, that a call is released by the service due to insufficient credit. This free format data may be placed in the CDR when the last call period commences. If the Tcp timer expires, the gsmSSF writes the free format data to the CDR. If the call is released by the calling

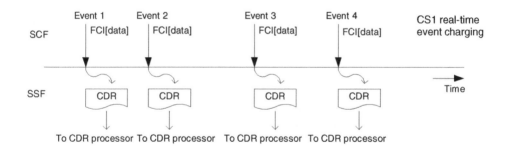

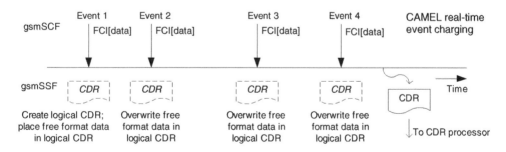

Figure 4.15 CDR creation in CS1 vs CDR creation in CAMEL

or called party during the last call period, then the CAMEL service gains control over the call (through EDP-R reporting) and may overwrite the free format data in gsmSSF by an empty string (FCI contains a minimum of one octet free-format data). This has the implicit effect that the free-format data is removed from the CDR.

The free-format data mechanism may also be used during user interaction. When the establishment of an outgoing leg is preceded by user interaction, then the free format data that is received during the user interaction is considered to be associated with this outgoing leg. Hence, an FCI that is subsequently received for the outgoing leg will overwrite the free format data that was received during the user interaction. In CAMEL phase 3, the capability of the free format data mechanism is slightly enhanced.

4.3.1.5 CAMEL Control of Advice-of-charge

Advice of charge (AoC) is the GSM supplementary service that enables an operator to place charge advice information (CAI) on the display of the MS of a subscriber. The CAI advises the user about the cost of a call. There are two AoC categories: (1) advice of charge – information (AoC-I); and (2) advice of charge – charge (AoC-C). AoC-I category is for information purposes; the information is only placed on the MS display. If the MS or the network is not capable of providing the charge advice to the user, then that has no effect on the user's ability to establish or receive calls.

AoC-C category, on the other hand, is for MS-based charging purposes. The MS needs to receive the charge advice in order to charge the user of the MS. If the MS or the network is not capable of providing the charge advice to the user, then the user cannot use the MS for establishing or receiving chargeable calls.[17]

When a GSM user subscribes to AoC, the serving MSC generates the CAI elements. For an MO call, the serving MSC is the MSC from where the call originates. For an MT call, the serving MSC is the MSC where the call terminates. CAI elements are sent to the MS at call answer, in a FACILITY message over DTAP. The MS converts the CAI elements to information text strings. The conversion from CAI elements to the information text strings is based on operator settings on the SIM card.

The charge advice information that may be provided to a user is also known as 'e-parameters'. The e-parameters in Table 4.3 may be sent to a user. See also GSM TS 02.24 [4]. Refer to Figure 4.16 for a functional overview.

For MO calls, the AoC information is sent to the MS together with the Answer indication from the MSC. This enables the MS to start timing the chargeable call duration immediately. For MT

Table 4.3 Overview of e-parameters for advice of charge

Element	Description
e1	Units per interval
e2	Time interval [s]
e3	Scaling factor
e4	Unit increment
e5	Units per data interval
e6	Segments/data interval
e7	Initial/time interval [s]

---

[17] This restriction does not apply to emergency calls.

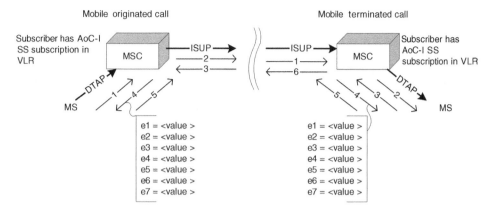

Figure 4.16 Transfer of advice of charge information. MO call: (1) MS sends DTAP SETUP to MSC, to set up a call; (2) MSC sends ISUP IAM to the destination, to establish call connection < more messages exchange takes place >; (3) MSC receives ISUP ANM from the destination, as indication that the call has been answered; (4) MSC sends DTAP CONNECT to MS, to indicate that the call is answered; the MSC also sends the FACILITY information to the MS, containing the AoC information; (5) MS sends DTAP FACILITY result to MSC, to acknowledge reception of the AoC information. MT call (1) MSC receives ISUP IAM from GMSC < more messages exchange takes place >; (2) MSC sends DTAP SETUP to MS to inform the subscriber of the incoming call; (3) MSC receives DTAP CONNECT from MS, as an indication that the call has been answered; (4) MSC sends DTAP FACILITY to MS, containing the AoC information; (5) MS sends DTAP FACILITY result to MSC, to acknowledge receipt of the AoC information; (6) MSC sends ISUP ANM to the GMSC, to indicate that the call has been answered.
Not all DTAP messages and ISUP messages, related to call set up, are reflected in Figure 4.16.

calls, the AoC information is sent to the MS in immediate response to the Answer indication from the MS.

CAMEL Control

For a call that is subject to CAMEL on-line charging, it is not the serving network that determines the cost of the call, but the on-line charging service in the SCP. In line with that paradigm, charge advice information for on-line charged calls should be generated by and provided by the SCP, rather than by the serving MSC. For that purpose, CAMEL phase 2 has the mechanism whereby the SCP can send e-parameters to the serving MSC, via the CAP interface (Figure 4.17).

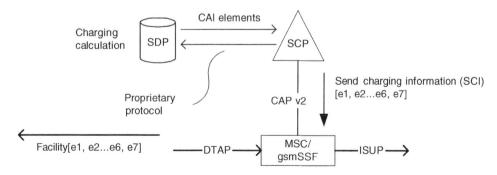

Figure 4.17 SCP-provided advice of charge information for MO call

CAMEL control of Advice of Charge (in CAMEL phase 2) may be used in MO calls only. The reason is that the CAI elements need to be provided to the serving MSC; the serving MSC sends the CAI elements to the MS over the radio network. For MO calls, the SCP has a CAMEL relationship with the serving MSC. For an MT call, however, the SCP has a CAMEL relationship with the GMSC, not with the serving visited MSC. The GMSC has no direct link with the radio network for the MT call.[18] ISUP does not offer the capability to transport the CAI elements from GMSC to VMSC. Hence, the SCP does not have the capability to send CAI elements to the MS for an MT call. CAMEL phase 3 offers the possibility for the serving MSC to establish a CAMEL relationship with the SCP for a terminating call. Therefore, CAMEL phase 3 allows CAMEL control of AoC for MT calls.

The CAI elements may be generated by the SDP or by the SCP. When the SCP sends CAP ACH to gsmSSF, in order to prepare the gsmSSF for monitoring the call duration, it may also send the CAI elements to gsmSSF. The sending of CAI elements to gsmSSF is done with the CAP send charging information (SCI) operation.[19] There are two moments in the call establishment process at which SCI may be sent:

(1) *In response to IDP* – when the gsmSSF receives SCI, it stores the CAI elements until the call answer is detected. When the call answer is detected, the MSC sends the CAI elements to the MS. In the case that call establishment fails, the MSC discards the stored CAI elements.
(2) *At call answer* – the SCP needs to arm the answer event in interrupt mode. When the call answer is detected, the gsmSSF informs the SCP and waits for instructions. The SCP responds with CAP SCI (and CAP ACH), followed by CAP CUE. The MSC now sends the CAI elements to the MS. In the case that call establishment fails, no CAI elements are received from gsmSCF.

In both methods, the CAI elements are provided to the MSC/gsmSSF before the MSC sends the answer indication to the MS. As discussed before, this is required in order to enable the MS to start timing the call immediately after answer. A CAMEL phase 3 service may have a specific reason to defer the sending of the first SCI until the answer is detected. Refer to Chapter 5.

The CAMEL service logic shall take care not to provide the first set of CAI elements after answer. In that case, the MSC may have sent CAI elements to the MSC already. The behaviour of the MS may in that case be unpredictable. As soon as the MSC receives CAI elements for a call from the SCP, it stops any internal process for the calculation of CAI elements for that call. CAI elements provided by the SCP overwrite any CAI elements already generated by the MSC.

Service designers should be aware that the advice of charge information remains, as its name implies, an *advice* of the charge of the call. The actual charge that is levied against the served subscriber is strictly not defined by the CAI elements. The following applies.

• *Off-line charging* – the charge for the call is calculated by analysing the relevant CDR from the MSC. The CDR may include AoC information, reflecting the CAI elements that are sent to the MS. This is information is, however, not necessarily used for calculating the charge of the call.
• *On-line charging* – the charge for the call is calculated by the CAMEL service. The CAI elements are used for information to the served subscriber only and may be placed in the CDR for off-line processing.

Required Support
For the SCP to be able to send the CAI elements to the MS, the following conditions need to be fulfilled:

---

[18] An MSC may be VMSC and GMSC at the same time. The GMSC entity and the VMSC entity in this node remain, however, different functional entities. Hence, even when GMSC and VMSC are co-located for a call, AoC cannot be applied to the MT call in CAMEL phase 2.

[19] The use of the SCI operation in CAP is fundamentally different from the use of the SCI operation in CS1.

(1) The MSC supports the AoC supplementary service. If the MSC does not support AoC and a subscriber with an AoC subscription attempts to register in that MSC, then the HLR will be informed about the lack of support of AoC in that MSC (through MAP ISD-Res). The HLR may decide, as a vendor's or operator's option, to disallow registration in that case.

(2) The user subscribes to the AoC supplementary service. If the user has no AoC subscription, then the MSC will not forward any SCP-provided CAI elements to the MS.

(3) The MS supports the AoC supplementary service. If the served subscriber has a subscription to AOC-C and the MSC fails to send the CAI elements to the MS due to the MS not supporting AoC, then the MSC will release the call and the SCP will be informed with an appropriate cause indication.

The SCP may not know whether the above conditions are fulfilled. Hence, when an SCP sends the SCI to an MSC, the MSC may not always be able to process this instruction. This has led to confusion in the field; the following two implementations have been found:

- *implementation 1* – when the MSC receives CAP SCI but cannot process this instruction, the MSC ignores the CAI elements from SCI and continues call processing;
- *implementation 2* – when the MSC receives CAP SCI but cannot process this instruction, the MSC returns an SCI Error indication to the SCP.

In cases where the above issue leads to unexpected service behaviour, a market adaptation is required. This dilemma is resolved in CAMEL phase 3. CAMEL phase 3 specifies that implementation 1 shall be followed. The MSC will not simply *discard* CAP SCI when it cannot process the CAI elements. The reason is that CAP SCI may also contain a tariff switch time, which should not be discarded.

4.3.1.6 Tariff Zones

An operator may apply multiple tariff zones in the call charging. A tariff zone is a period of time (during the day) for which a particular tariff applies for a call. A day may be divided into multiple tariff zones. The operator may, for example, have a high-rate zone and a low-rate zone, as in the following example.

Monday – Friday:	00:01–08:00	low rate
	08:01–18:00	high rate
	18:01–24:00	low rate
Weekend days:		low rate
Public holidays:		low rate

CAMEL phase 2 provides a mechanism to cater for the multiple time zones in the CAP operations that are used for on-line charging. This is reflected in Figure 4.18. In this example, the call is answered at 17:52:00. The SCP provides a call period duration of 300 s and charging starts at that moment. After 300 s call duration, at 17:57:00, the first call period is complete and the gsmSSF generates a charging report, indicating the elapsed time for the active part of this call, i.e. 300 s. The SCP allows for another 300 s call time. The next call period will cross the tariff zone border from 18:00. The SCP therefore includes a tariff switch (Tsw) in CAP ACH. There are 180 s between the start of the second call period and the tariff switch-over time; hence, Tsw is set

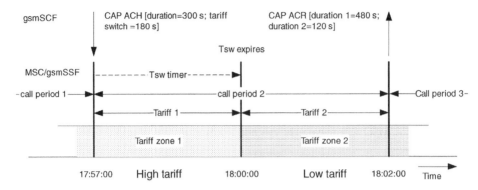

Figure 4.18   Multiple tariff zones

to 180 s. The gsmSSF starts the tariff switch timer when it has completed the processing of CAP ACH.

After 180 s into the second call period, the tariff switch occurs. The gsmSSF stores the current call duration and starts a second call duration timer. The second call period timer continues running until the allowed call duration of 300 s for this call period has elapsed. At that moment, a second charging report is generated. This second charging report contains the following reported values:

- *Time since last tariff switch* – this is the chargeable call part in the second tariff zone. In the current example, it has a value of 120 s (= 300 s call period duration − 180 s tariff switch time).
- *Tariff Switch Interval* – this is the chargeable call part in the first tariff zone. In the current example, it has a value of 480 s (= 300 s of the first call period duration and 180 s into the second call period).

The SCP may now apply a high tariff to the first 480 s of the call (tariff switch interval) and low tariff to the remaining 120 s of the call (time since last tariff switch). If the call continues, i.e. a third and subsequent call period is started, then this remaining call time is charged at the low rate. If no subsequent tariff switch takes place, the gsmSSF continues to report just the accumulated lapsed call time in the low tariff zone for each charging report in the remainder of this call. The value that is reported in time since last tariff switch will increase with the call duration for successive call periods, i.e. the reported values will be 120 s (600 s = 10 min total call time), 420 s (900 s = 15 min total call time), 720 s (1200 s = 20 min total call time), etc.

As indicated in a previous paragraph, the gsmSSF starts the tariff switch timer as soon as it has received this timer. Hence, the tariff switch-over moment is related to an absolute point in time. This is different from the call period timer, which starts when the call is answered. As a result, an on-line charging service may not know when the call is established how much call time will fall into the first tariff zone. In the above example, call establishment may have started at 17:51:00, i.e. the call setup time was 60 s. The call could, however, have been answered at 17:51:30 instead of 17:52:00, as suggested in the example. In that case, there would be an additional 30 s call time in high tariff zone.

A pre-paid service should therefore reserve enough credit at the beginning of the call to allow for the maximum call time falling into the high tariff zone. The call will in any case be correctly charged when the charging reports from gsmSSF are received.

Multiple Tariff Zones

Although it is not likely, there may be more than one tariff zone switch-over event during a call. Refer to Figure 4.19 for an example.

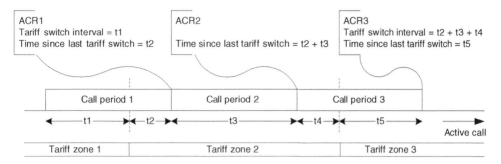

Figure 4.19 Charging reports for multiple tariff zones

- *First charging report* – in the call period related to the first charging report (ACR1), a tariff switch has occurred. Therefore, the call durations in tariff zone 1 (t1) and the call duration in tariff zone 2 (t2) are reported.
- *Second charging report* – for the second call period, no tariff switch occurred. Therefore, the charging report related to the second call period contains only the call duration in tariff zone 2 (t2 + t3).
- *Third charging report* – for the third call period, a tariff switch has occurred again. Therefore, the charging report related to the third call period contains the call duration in tariff zone 2 (t2 + t3 + t4) and the call duration in tariff zone 3 (t5). The information that is reported in this third charging report no longer contains the *complete charging information*; the call duration in tariff zone 1 (t1) is not included.

Therefore, for calls that span three or more tariff zones, the on-line charging system has to maintain this information internally as the call progresses.

Advice of Charge at Tariff Zone Change Over
When the charge for a call changes while the call is ongoing, the advice of charge information on the MS display should be updated accordingly. The advice of charge supplementary service allows a new set of CAI elements to be sent to the MS during the call. When the MSC enters a new tariff zone for an ongoing call, it generates a new set of CAI elements and sends these to the MS, with an intermediate FACILITY message.

When the gsmSCF is applying on-line call duration control for a call, the gsmSCF may at some point in the call provide a tariff switch time at the start of a new call period. During the call period that will now start, the call will switch over to a new tariff zone. In order to have the MSC send updated CAI elements to the MS at switch-over to the new tariff zone, the gsmSCF may provide the gsmSSF with a new set of CAI elements at the beginning of this call period. The gsmSSF will store the CAI elements and instruct the MSC to send these elements to the MS when the tariff switch occurs.

4.3.1.7 Multiple Credit Reservations

A previous section outlined that, for on-line charging, credit is reserved at the beginning of every call period. The credit is then deducted from the user's account at the end of that call period. The operator should take care to set the credit reservation value to an optimum value, taking the following considerations into account:

- small credit reservations give tight charging control, but lead to excessive CAP signalling and increased processing load on charging system and gsmSSF alike;

- large credit reservations put the operator at risk, in the case of signalling failure between gsmSSF and gsmSCF;
- large credit reservations block other charging processes from accessing the user's credit; this may have the result that the user is denied access to other services during the call and that may in turn lead to loss of revenue for the operator.

For certain call cases, it is even required that at least two processes access the credit simultaneously. One example is late call forwarding when roaming (Figure 4.20).

In the example sketched in Figure 4.20, the terminating call leg and the forwarding call leg are both subject to on-line charging. The respective on-line charging services for these call legs access the same credit for this subscriber, residing in the SDP that contains the credit of the served subscriber. These credit accesses occur in the following order:

- *MT call leg* – credit is reserved from the subscriber's account for the roaming leg between GMSC (in HPLMN) and VMSC (in VPLMN).
- *MF call leg* – when late call forwarding takes place in the VMSC in VPLMN, a second on-line charging service is invoked; this second charging service takes care of the on-line charging of the forwarding leg between VMSC in VPLMN and the voicemail system in HPLMN. Credit is reserved from the subscriber's account for the forwarding leg.

Should the on-line charging service for the MT call leg have reserved the entire available credit for this subscriber, then the on-line charging service for the MF call leg would fail to reserve credit, since all available credit is already reserved. The result is that the call fails and the operator loses revenue. Partial credit reservation, however, would allow for the forwarding leg to take place (Figure 4.21). Credit should be reserved in chunks of, for example, 1 min call time.

In the example in Figure 4.21, the two charging services may be reserving and consuming credit simultaneously. While the voicemail deposit is ongoing (during which there will be MT call leg charging and MF call charging), the subscriber may send and SMS. Owing to the partial credit reservation, the charging service that handles the SMS charging can reserve credit for the sending of the SMS.

4.3.2 Call Forwarding Notifications

When a GSM subscriber receives a call, the CAMEL-based on-line charging for this MT call takes place from the GMSC. The charging of the MT call may be needed: because the subscriber is

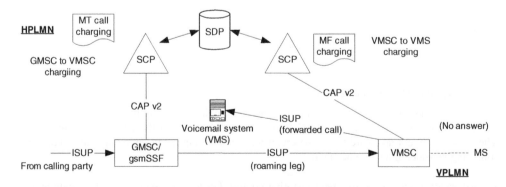

Figure 4.20 Multiple credit access

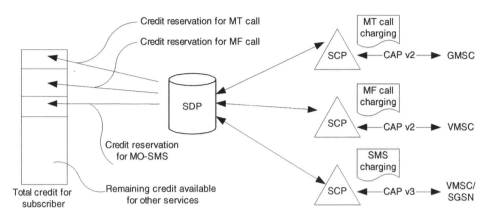

Figure 4.21   Partial credit reservation

currently registered in another country and hence must pay for the roaming leg between GMSC and VMSC; or because the operator's charging regime specifies that terminating calls are always chargeable, even when registered in the home network. Owing to the nature of IN service handling for MT calls, the CAMEL service is started before the GMSC has received the MSRN to route the call to the MSC where the subscriber is currently registered. When the MT call CAMEL service is started, it receives an indication of the location of the terminating subscriber, so it can apply charging for the roaming leg, based on the current location of the subscriber. For example, if the terminating subscriber is located in a distant country, then the roaming charge will be higher than when she is in a nearby country.

When the GMSC receives an FTN from the HLR instead of an MSRN, early call forwarding takes place. An example of such a call forwarding case is when the subscriber's MS is switched off. The HLR does not receive an MSRN from the VLR and will send the FTN not reachable (FTN-NRc) to the GMSC. The GMSC forwards the call to the destination indicated by the FTN. If this is a local destination, such as a voicemail box, then a local call is established from the GMSC, instead of a call to the MSC where the subscriber is currently registered.

When the call to the forwarding destination is established, an answer message is sent over ISUP from the forwarding destination towards the GMSC. This answer message is signalled to the MT CAMEL service, provided that the corresponding detection point in the T-BCSM (T_Answer) is armed. The gsmSSF for the terminating call control starts monitoring the call duration.

The above sequence of events poses a dilemma. If the called subscriber is roaming abroad, then the pre-paid charging service bases the call charge for the roaming leg on the location (country, network) where the subscriber is roaming. If forwarding takes place in the GMSC, then the roaming leg is replaced by a connection to a different destination, e.g. a voicemail box in the HPLMN. If the terminating charging service was not aware of this forwarding event, then incorrect charging would be applied.

4.3.2.1 Early Call Forwarding

Figures 4.22 and 4.23 depict the forwarding notification feature for early call forwarding. The call forwarding notification is for the purpose of *conditional call forwarding*. The reason is that the *unconditional call forwarding* is already indicated in the CAP initial DP operation, through the presence of the 'GSM forwarding pending' information element. When that information element is present, the CAMEL service may, as a service logic design option, decide to terminate its

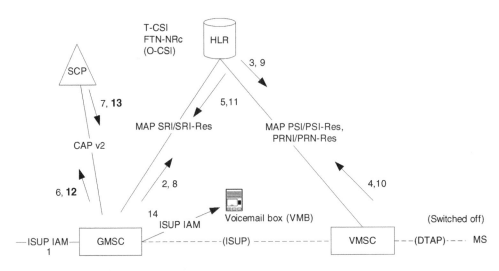

Figure 4.22   Call forwarding notification – early call forwarding

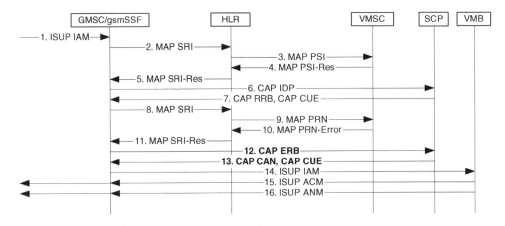

Figure 4.23   Call forwarding notification – early call forwarding; sequence diagram

processing. The GMSC will in that case initiate the call forwarding unconditionally, as instructed by HLR.

One other case where the CAP initial DP operation may contain the 'GSM forwarding pending' information element is the scenario in which the called subscriber is marked as 'detached' in the HLR. Normally, when a subscriber turns off her MS for short durations, the HLR is not informed. When an MS is switched off for a long duration, the VLR may signal the detached state to the HLR.[20,21] The conditional call forwarding cases for which the notification may be sent to the SCP include: call forwarding not reachable – MS switched off; and call forwarding not reachable – no paging response.

---

[20] The presence of the 'gsm forwarding pending' information element in CAP initial DP does not indicate the call forwarding reason. It may be *unconditional* or *not reachable*.

[21] When the presence of the forwarding pending flag in CAP IDP results from the fact that the MS is marked as 'detached' in HLR, it may occur that the GMSC also sends the conditional call forwarding notification to

4.3.2.2 Call Forwarding Not Reachable – MS Switched Off

This is the regular not reachable case. When the GMSC sends the second MAP SRI to the HLR, to obtain the MSRN, the HLR sends MAP PRN to the VMSC to request the MSRN. In the case that the MS is switched off, the VLR responds with MAP PRN-Error. The HLR returns the appropriate FTN to the GMSC.

The receipt of the FTN-NRc in the GMSC is interpreted by the GMSC as an indication that the call establishment to the B-subscriber has failed. Therefore, the GMSC informs the SCP of this call establishment failure by means of the EDP that is associated with the not reachable event, i.e. the T_Busy EDP, provided that this DP was previously armed by this CAMEL service. As discussed before, the not reachable event is reported by means of the busy DP; the event notification contains the cause. The cause indicates the actual event that led to the busy DP. The bold numbers in Figures 4.22 and 4.23 relate to the call forwarding notification.

The SCP will make the following distinction in the receiving of the T_Busy notification:

- T_Busy without the '*call forwarded*' information element included – the call establishment failed due to a busy or not reachable subscriber. No GSM call forwarding is active.
- T_Busy with the '*call forwarded*' information element included – the call establishment failed due to a busy or not reachable subscriber. GSM call forwarding is active.

In the second case, the CAMEL service may send CAP CUE to the GMSC to allow the GSM call forwarding to take place. Depending on service logic requirements, the service logic may precede the CAP CUE by CAP CAN to cancel all detection points and reports. As a result, the CAMEL service will be terminated. Alternatively, a CAMEL service may in this case want to remain active until call answer or later.

The alert reader will have noticed that the call forwarding notification for the busy case cannot occur in the GMSC for early call forwarding, since the busy call forwarding event occurs in the VMSC. Hence, in the early call forwarding notification, the cause code will indicate subscriber not reachable.

4.3.2.3 Call Forwarding Not Reachable – No Paging Response

If pre-paging is supported by GMSC, HLR and VMSC, then the called subscriber is paged in the VMSC when the VMSC receives MAP PRN. If the VMSC fails to receive paging response from the subscriber, even though the subscriber's MS may be marked 'attached', then the VMSC does not allocate an MSRN. Further processing of the terminating call case, including GSM call forwarding and call forwarding notification, is the same as for the not reachable – MS switched off case.

4.3.2.4 Late Call Forwarding with Optimal Routing

Figures 4.24 and 4.25 depict the forwarding notification feature for optimal routing at late call forwarding (ORLCF). The following call forwarding events may occur:

- *Subscriber busy* – the busy event may be NDUB or UDUB.
- *No reply* – the subscriber does not answer the call within a predefined time after the start of alerting.

---

the SCP, even though the SCP has already been informed through the forwarding pending flag in CAP IDP. This inconsistency is resolved in CAMEL phase 3.

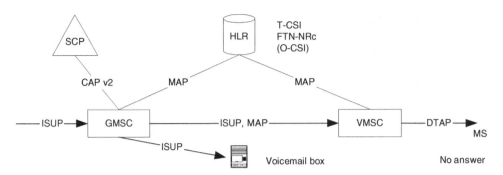

Figure 4.24   Call forwarding notification – ORLCF

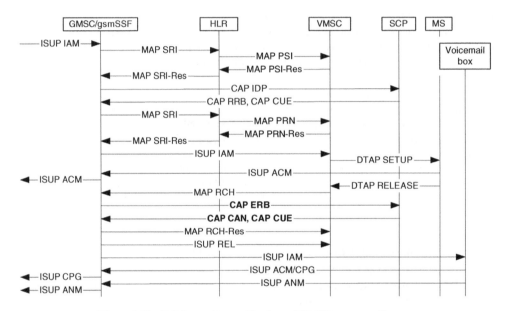

Figure 4.25   Call forwarding notification – ORLCF; sequence diagram

- *Subscriber not reachable* – the VMSC receives no paging response from the MS.
- *Call deflection* – the called subscriber deflects the incoming, alerting call.

The cause of late call forwarding is included in the MAP RCH and is used in the GMSC to select the corresponding DP.

In the ORLCF call forwarding notification, both the T_Busy EDP and the T_No_Answer EDP may be reported. The cause code for T_Busy may indicate not reachable or busy. Further behaviour of the CAMEL service is similar to the early call forwarding case. In both cases, the CAMEL service is informed that the roaming leg will not be established and that GSM forwarding will take place in the GMSC. For a typical terminating IN service, it may not be relevant whether the call forwarding notification is due to early call forwarding or ORLCF.

4.3.2.5 CAMEL Forwarding Leg Service

When the GMSC initiates the GSM call forwarding as a result of receiving the FTN from HLR (early call forwarding) or VLR (ORLCF), a CAMEL service may be invoked for this forwarded leg. This CAMEL service is invoked if the FTN that is received from HLR (MAP SRI-Res) or VLR (MAP RCH) is accompanied by an O-CSI.

4.3.2.6 CAMEL Service Induced Forwarding

When the call forwarding notification occurs, the SCP may act in various ways. Examples include:

- send CAP CUE to the GMSC – the GMSC will initiate the GSM call forwarding to the FTN that is received from HLR or from the VLR;
- send CAP RC – the GMSC will release the call;
- send CAP CON to the GMSC[22] – the GMSC will initiate call forwarding to the destination number that is contained in CAP CON; this destination address overrides the FTN received from HLR (in MAP SRI Res) or VLR (in MAP RCH).

In the case that the CAMEL service sends CAP CON, it can suppress the invocation of the CAMEL service for the forwarded leg. For example, if the CAMEL service for the terminating call forces call forwarding to a destination that carries no charge, then this CAMEL service may suppress the forwarding CAMEL service, since no charging for the forwarding leg is needed.

4.3.3 Follow-on Calls

A CAMEL service may create a follow-on call when an outgoing call leg is released. The creation of a follow-on call ensures that the outgoing call connection is replaced by another outgoing call connection. The CAMEL service is dependent on the release of the outgoing call leg in the network, i.e. the CAMEL service cannot force-release the outgoing call leg in order to create a follow-on call.[23] A follow-on call may be generated in the following call cases:

- the call establishment to the called party fails (busy, route select failure or no answer);
- the called party disconnects from an active call.

A follow-on call may be generated in all three call scenarios for which CAMEL service may be invoked, that is, MO calls, MF calls and MT calls. Examples of follow-on calls include:

- *Call diversion* – when call establishment fails, the CAMEL service tries a number of alternative destinations in succession. Call diversion may be used for MT call handling, whereby the called subscriber is called on her GSM phone first; if the call fails, the call diversion service connects the call to an alternative destination. An alternative destination may be a GSM MSISDN or a destination in another network such as the PSTN.
- *Operator-assisted calling* – when a subscriber tries to set up a call via a selected carrier and the call fails, the CAMEL service may reconnect the call via a different carrier.

---

[22] If CAP CON contains a destination routing address that is equal to the called party number that was reported in CAP IDP, then the CAP CON is treated as CAP CUE with modified call information.

[23] A CAMEL service may use release call (RC) to release a call, but RC results in the release of the entire call, and so cannot be used for follow-on calls. In addition, a CAMEL service can have the gsmSSF release the call when a call period expires. However, that method of call release also results in the release of the entire call, and so cannot be used for follow-on calls either.

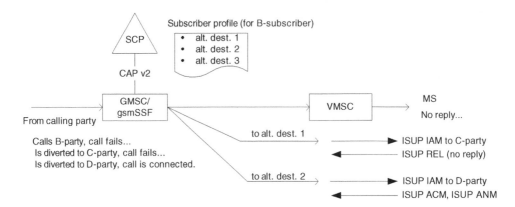

Figure 4.26 Example of follow-on call (call diversion)

Figure 4.26 gives a graphical representation of the follow-on calls. The ISUP REL from the called party or from one of the alternative destinations is not propagated backwards towards the calling party. Hence, whilst the call diversion is in progress, the calling party remains waiting for an answer. When all alternative call destinations fail, then the call diversion service allows the ISUP REL (from the last alternative destination) to be propagated in the backwards direction towards the calling subscriber.

To prevent unnecessarily long call setup time, a call diversion should normally define a short no reply timer for each outgoing call leg. As an example, the service may instruct the gsmSSF, for each outgoing call leg attempt, in the GMSC to generate a no answer event if the call is not answered within, for example, 15 s.

In a particular scenario, one or more of the alternative destinations may itself also be subject to a call diversion service. In such a case, the call diversion service of the B-subscriber may receive a no answer event once all alternative destinations associated with the first alternative destination associated with the B-subscriber are exhausted. Adequate setting of the no answer timer in the gsmSSF should prevent this situation.

To prevent such a situation, a call diversion may suppress further forwarding for each outgoing leg, by setting the forwarding counter in the ISUP IAM that results from the CAP CON, to the maximum value.

4.3.3.1 Interaction with Call Forwarding

There is a subtle difference in the creation of follow-on calls within an MO call CAMEL service and an MT call CAMEL service. This is depicted in Figures 4.27 and 4.28. When a follow-on call is created within an MO call, then one outgoing leg, for which call establishment failed or which was released by the remote party, is replaced by another outgoing leg. Effectively, there is after the follow-on call still one call connection between the calling party and the destination. Therefore, the SCP may leave the forwarding counter in the ISUP IAM at 0.

When a follow-on call is created within an MT call in the GMSC, then the original outgoing leg, which is the roaming leg to the called party, is replaced by an outgoing call leg to a different destination. Once the follow-on call is created, the GMSC is effectively no longer acting as a GMSC for MT call handling. Rather, it is acting as transit MSC, for the purpose of forwarding the call to its forwarding address. Therefore, the SCP should, within an MT call, increment the forwarding counter in the ISUP IAM by 1. In both call cases (MO and MT), however, the SCP may still decide to set the forwarding counter to the maximum value to prevent further forwarding for an individual follow-on call leg.

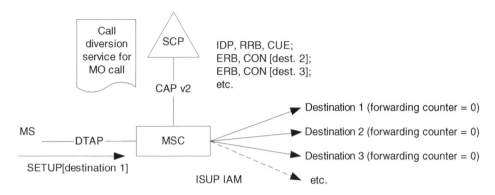

Figure 4.27   Call diversion for MO calls

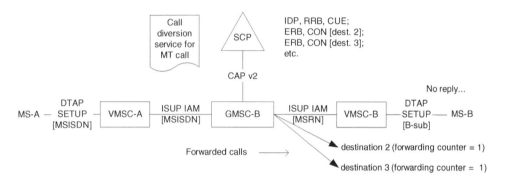

Figure 4.28   Call diversion for MT calls

4.3.3.2 Parameter Usage

When the SCP connects a call, it may provide call parameters to the gsmSSF. The parameters in CAP CON apply, however, only to the leg that is connected as a result of that CAP CON. If the establishment of that call leg fails or the call leg is disconnected by the remote user, then these parameters that were conveyed in CAP CON are no longer applicable.

If the SCP creates a follow-on call, then it needs to include again the required parameters in CAP CON. In the example in Figure 4.29, a call is initiated, but establishment fails. Parameters that the MSC/gsmSSF receives in CAP CON are used for one outgoing call leg only. If a follow-on call is created, then the parameters from the previous CAP CON no longer apply.

In Figure 4.29, 'call connect' refers to the CAP CUE or CAP CON in response to the IDP. This CAP CUE or CAP CON is used to connect the call to its destination. 'Call reconnect' refers to the CAP CON in response to the call establishment failure or call release. The following categorization of parameters in the outgoing ISUP IAM may be made:

- *Static parameters* – these are parameters which are provided by the incoming connection to the MSC (e.g. the MS) or are derived in the MSC/gsmSSF and which remain unchanged when a follow-on call is created. Examples are calling party number, user service information and user-to-user (UUS) information. Not all of these parameters need to be present, e.g. if UUS information is not provided by the MS in the DTAP SETUP message, then it will not be present in the resulting ISUP IAM. The first outgoing call leg and all subsequent follow-on call legs

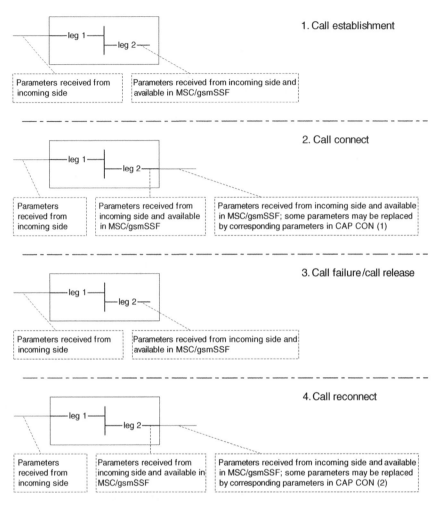

Figure 4.29 Usage of parameter in CAP CON for follow-on calls

within the call will have the same calling party number. If UUS information is provided by the MS, then all call legs will have the same UUS information, etc.

- *Dynamic parameters* – these are parameters which may be provided by the incoming connection or may be derived in the MSC/gsmSSF and which may be provided by the CAMEL service (if not yet present at call set up) or may be changed by the CAMEL service (if already present at call set up). An example is called party number; this parameter is always available at call set up. Other examples are calling party's category, additional calling party number and redirection information; these parameters may be present at MT or MF call set up. The first outgoing call leg and all subsequent follow-on calls legs within the call may have different permutations and values of the dynamic parameters.

Designers should consult the description of the Initial DP and the Connect information flows in GSM TS 03.78 [38]. These descriptions indicate which call parameters may be present at call set up and which parameters may be provided or changed by the SCP.

4.3.3.3 On-line Charging

When follow-on calls are created, each outgoing call leg may be subject to on-line charging. That means that the CAMEL service may use CAP operations ACH/ACR, CIRq/CIRp, FCI and SCI for each individual call leg.

Apply Charging

The CAMEL service may use CAP ACH when it creates a follow-on call. CAP ACH may be sent to the gsmSSF when the release of the previous call leg is reported to the SCP. Although the gsmSSF is at that point still suspended as a result of the previous call, the gsmSSF will accept CAP ACH and apply it for the follow-on call. Alternatively, the CAMEL service may send CAP ACH at call answer for the follow-on call.

Call Information Request

When a follow-on call is created, CAP CIRq may be used for leg 2 only. The reason is that leg 1 is active for the entire call, including follow-on calls. If CAP CIRq [leg 2] is used for a follow-on call, then it is sent to the gsmSSF when the release of the previous call leg is reported to the SCP.

Furnish Charging Information

Free format data may be provided for each individual outgoing call leg. Each set of free format data will be stored in the CDR associated with the (follow-on) call for which the free format data was sent. Since leg 1 is active for the entire call, including follow-on calls, any free format data that is sent for leg 1 overwrites the already available free format data for leg. FCI for the follow-on call shall be sent to the gsmSSF after the SCP has instructed the gsmSSF to create the follow-on leg. The reason is that, when call failure or call release is reported, the SCP should still have the possibility to include free format data in the CDR for the call leg that was just released. Such a requirement does not exist (and would not make sense) for CAP ACH and CAP CIRq.

Send Charging Information

For each (follow-on) call leg, the SCP may apply advice of charge, with the restriction that CAMEL control of AoC is possible only for MO calls. CAP SCI may be sent to the gsmSSF together with CAP ACH. The following shall be borne in mind when using SCI for follow-on calls. When a follow-on call is created when no answer has been received from a remote party yet, then no answer indication is sent to the MS and no AoC information is sent to the MS. Any AoC information stored in the gsmSSF, as a result of a previous SCI operation, is discarded when call establishment fails. For the follow-on call, the SCP may send new AoC information.

If a follow-on call is created when an answer has already been detected from a remote party, then answer indication is sent to the MS and AoC information is also sent to the MS. When the outgoing call is released, the CAMEL service should send an SCI to the gsmSSF; this SCI has the purpose of sending 'zero-rate' AoC information to the MS.

When the follow-on call is answered, the CAMEL service can send a new set of AoC information to the gsmSSF, which will forward it to the MS; this time, the AoC information contains values that relate to the charge of the follow-on call. Service Logic designers shall follow the rules and preconditions that are specified in GSM TS 03.78 [38] for the sending of SCI to the gsmSSF.

4.3.3.4 User Interaction

Typically, the call diversion instructions from the service will be preceded by user interaction. For details on user interaction, refer Section 4.3.4. Figure 4.30 gives a presentation of the user interaction for follow-on calls.

In Figure 4.30, the user interaction consists of an announcement to the calling user. Alternatively, the user interaction may include a request for user input. The user may be prompted to press * on the keypad to be connected to the voicemail box or to press # to be connected to an operator.

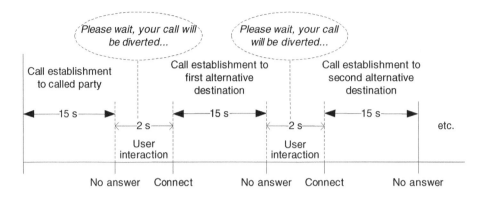

Figure 4.30   Follow-on calls in call diversion service

However, including a request for user input in the user interaction requires a *both-way through-connection* with the calling user. If the call diversion is performed in the originating MSC, then the user input may be done without charging the calling subscriber for the user interaction, i.e. the subscriber is charged only when a successful connection is established to the desired destination (operator or voicemail box).

If, on the other hand, the call diversion is performed in an MSC other than the originating MSC, then the request for user input has the effect that the calling subscriber is charged for the user interaction menu. The reason is that the opening up of the speech path from the calling subscriber requires an ISUP ANM to be sent to the originating MSC. The ISUP ANM starts the charging process in that MSC.

The user interaction that is applied before creation of a follow-on call may be subject to its own on-line charging. The gsmSCF uses the ACH-ACR mechanism for this purpose (Figure 4.31).

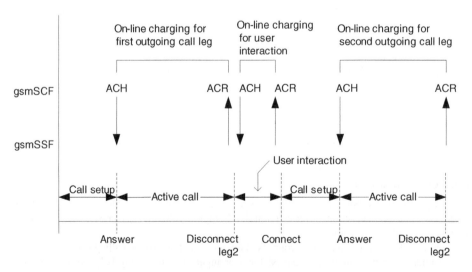

Figure 4.31   Charging for user-interaction for a follow-on call

4.3.3.5 No Reply Timer

When the creation of follow-on call is done within the call diversion service, the operator may want to use a shorter no reply timer than the network no reply timer. In the examples of the present section, the call diversion instructs the gsmSSF to generate a no answer event after 15 s. After 15 s, the outgoing call attempt will be aborted by the gsmSSF, i.e. an ISUP REL is sent towards the destination exchange. For the follow-on call, the service may again instruct the gsmSSF to generate a no answer event after 15 s. However, there is, with respect to the no reply timer, a practical limit to the number of times the call diversion service may create a follow-on call. (Figure 4.32).

In the example in Figure 4.32, the SCP is creating follow-on calls, whereby each call has a no reply timer ('T9') value of 20 s. 'T9' for the gsmSSF-internal no reply timer is not official terminology; however, this timer mimics the T9 timer in MSC. After 85 s, the MSC-internal no reply timer expires. No further follow-on calls are possible and the call will now be released.

4.3.3.6 CDRs

When follow-on calls are created, multiple CDRs may be generated in the MSC. In practice, operators often suppress the generation of a CDR when call establishment fails. Failed calls do not lead to billing, so CDRs for those calls may not be needed.[24] If multiple successful outgoing (follow-on) calls are established, then these calls will be reflected in respective CDRs. These CDRs, however, do not each have to contain the full set of data fields that is specified for CDRs. CDR contents for specific call cases are specified in GSM TS 12.05 [57].

Generation of CDRs in the MSC for various call scenarios is often done in a proprietary manner. An example of CDR generation for follow-on calls is given in. Table 4.4.

The two CDRs in Table 4.4 are linked to one another with the network call reference (NCR) and MSC address (MSCA). The NCR and MSCA are meant for correlating CDRs that are generated by different entities, such as MSC and SCP. NCR and MSCA may, as an implementation option, also be used to link CDRs for follow-on calls, as in the current example. In the case of a follow-on call, the CDRs that pertain to the individual outgoing legs are created in the same node (MSC). The CDR that relates to the follow-on call does not need to contain data elements that will not change during the call, such as IMSI, calling party number, etc.

4.3.4 User Interaction

User interaction (UI) is the method of having an IN service interact with the calling party,[25] during service logic processing. UI may be applied during call establishment, to advise the calling party

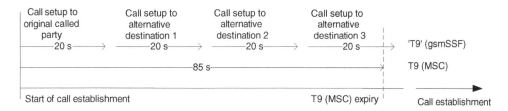

Figure 4.32   MSC no reply timer expiry

---

[24] An operator may, however, decide to also have the MSC generate CDRs for failed calls, for statistical purposes.
[25] User interaction with <u>called</u> party requires call party handling (CPH) functionality. CPH is supported in CAMEL phase 4.

Table 4.4   CDR creation for follow-on call

CDR type = MO call record	CDR type = MO call record
CDR for original outgoing call leg	*CDR for follow-on call*
Served IMSI	Network call reference
Served IMEI	MSC address
Served MSISDN	...
Calling number	Answer time
Called number	Release time
...	Call duration
Answer time	...
Release time	CAMEL call leg information
Call duration	CAMEL destination number
...	CAMEL modification
gsm-SCF address	Generic numbers
Service key	...
Network call reference	
MSC address	
...	

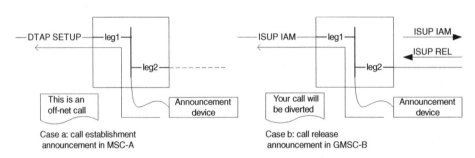

Figure 4.33   Announcement connections at call establishment and at call release

about the tariff of the call. When call diversion is applied, an IN service may play an announcement to inform the caller about the call diversion.

User interaction is not applied to an individual call leg, but to the call connection bridge (Figure 4.33). User interaction is done by means of connection to a gsmSRF or an intelligent peripheral (IP; also known as specialized resource point).

In Figure 4.33, case a, the connection to the gsmSRF in MSC-A is made before the ISUP IAM is sent out. The calling party has not received an answer message and hence will not be charged for the duration of the announcement. In case b, the connection to the gsmSRF in the GMSC is made after an ISUP REL is received from the called party. If an ISUP ANM had been sent to the calling party already for this call (e.g. follow-on call after answer), then the calling party will be charged for the duration of the announcement.

To play an announcement to a calling subscriber, a traffic connection needs to be established between the calling party and a gsmSRF. This traffic connection may be an MSC-internal connection or may be an external ISUP/TDM channel.[26] Different call connection configurations are possible for the connection to the gsmSRF, as will be explained further on.

---

[26] Other traffic connection protocols may also be used, e.g. for the connection of an external gsmSRF to an MSC, DSS1 may be used. This is not further specified by CAMEL

In CAMEL phase 2, user interaction may be applied during call establishment and at call release. An example of the former is the playing of an announcement at the beginning of the call. An example of the latter is the playing of a call diversion announcement. In the latter example, the user interaction is applied *in between* successive calls.

User interaction cannot be applied during an active call. The reason is that user interaction is in essence a traffic connection, i.e. ISUP/TDM channel, between the calling party and a user interaction device. The gsmSSF state model does not have the capability to monitor a user interaction connection and an outgoing call connection at the same time. User interaction during an active call is introduced in CAMEL phase 4. Two forms of user interaction exist: (1) announcement playing; and (2) information collection.

4.3.4.1 Announcement Playing

For the playing of an announcement, a 'one-way connection' or 'both-way connection' is needed between gsmSRF and the calling party. When a call is initiated in the GSM network, a backwards speech path is automatically opened. Such a speech path is needed during call set up for ring-back tone, call progress message, etc. Injecting an announcement in the speech path does not affect the call state.

The playing of an announcement is done with the play announcement (PA) CAP operation. The announcements that a CAMEL service may play to a calling party are pre-recorded, residing in the gsmSRF. The announcements are stored in a format that is applicable to the network in which the gsmSRF is located, e.g. a-law encoded PCM speech for ITU-T ISUP networks and μ-law encoded speech for ANSI ISUP networks. CAMEL allows the following types of announcements to be played:

- *tone* – the *tone* is a pre-recorded tone. The CAMEL service may indicate a specific tone to be played and the duration of the tone. The tone may be a single frequency tone or a dual frequency tone (e.g. one of the standard DTMF tones). *Note*: the playing of a tone with long duration should be avoided since such a tone may become distorted due to the speech-encoding techniques that are applied in the GSM radio access network.
- *message* – the *message* is a pre-recorded announcement.
- *text* – the *text* is a text string that will be converted to speech and will be played to the calling party. The conversion from text to speech is an optional capability. A gsmSSF will indicate to the CAMEL service (in CAP IDP) whether it supports this capability.
- *message string* – the *message string* is a concatenation of pre-recorded messages.
- *variable message* – the *variable message* is a pre-recorded message including one or more variable elements. The CAMEL service selects a variable message and provides the required parameters for that message. The following variable messages may be pronounced: integer, number, time, date and price. If the CAMEL service wants to pronounce a duration, then the service provides the identifier of the variable message and a parameter containing the duration. The gsmSRF then creates the following message: 'The duration of the call was' <HH> 'hours and' <MM> 'minutes'. The values for HH and MM are provided by the CAMEL service.

When a CAMEL service has ordered the playing of an announcement, it may request the gsmSSF to send a signal to the CAMEL service when the playing of the announcement is complete. The CAMEL service may, upon receiving this signal, continue call establishment. This signal from gsmSSF to the CAMEL service is the specialized resource report (SRR) CAP operation. SRR is linked to a specific PA operation instance. The method of *linking* CAP operations is part of the transaction capabilities (ITU-T Q.773 [141]). When a CAMEL service receives an SRR operation, the SRR operation contains an indication of the PA operation it relates to. The CAMEL service therefore knows <u>which</u> announcement is complete.

4.3.4.2 Information Collection

In CAMEL, 'information collection' entails the collection of DTMF digits from a user. A user may be requested to press digits on the MS to indicate a choice within a menu. The digits are reported to the CAMEL service. Information collection is done with the prompt and collect user information (PC) CAP operation.

DTMF digits entered on an MS are transported to the MSC as 'out-band' information, i.e. they are not included in the speech signal. The DTMF digits are sent to the MSC by signalling means. Transport of DTMF from MS to MSC is specified in GSM TS 03.14 [29] (semantics) and GSM TS 04.08 [49] (protocol).

Within the MSC, the DTMF tones may be transported in the following ways:

- *in-band information* – the DTMF signal that the MSC receives from the MS is converted to PCM-coded audio information and is injected (superimposed) into the speech channel. The speech channel is connected to the gsmSRF for information collection. Detection by the gsmSRF of a DTMF tone requires a key recognition device (KRD). The KRD analyses the PCM coded speech and informs the gsmSRF when a DTMF tone is detected.
- *out-band information* – the DTMF signal that the MSC receives from the MS is transported internally in the MSC as out-band information. The DTMF signals are converted to PCM coded audio information when a traffic connection to an external destination or to a gsmSRF is established. When BICC capability set 2 (BICC CS2) is used in a network, DTMF tones may be transported between MSCs as out-band information. Refer to 3GPP TS 29.205 [107] and ITU-T Q.1901 [144]. Transporting DTMF tones as out-band information is more efficient than transporting as in-band information.

When a CAMEL service uses CAP PC to collect information, it may precede the information collection by playing a tone or an announcement to the user. The tone or announcement that may be played follows the rules that apply to play announcement. When digits are detected, the gsmSRF generates a report, containing the collected digits. This report has the form of a RESULT component for the CAP PC operation. The PC Result is relayed through the gsmSSF to the gsmSCF.

The CAMEL service indicates to the gsmSSF the criteria for generating a prompt and collect RESULT. These criteria include:

- *Minimum number of digits* – this criterion indicates the minimum number of digits, including optional start digit(s) and end of reply digit(s), that should be entered before a result is generated.
- *Maximum number of digits* – this criterion indicates the maximum number of digits, including optional start digit(s) and end of reply digit(s), that should be entered before a result is generated.
- *End of reply digit* – this criterion indicates that a result should be generated when these digits are detected. The user may enter a variable length digit string, such as a bank number, and confirm the input by pressing #.
- *Cancel digit* – this criterion enables the user to cancel the input when a wrong key is pressed.
- *Start digit* – this criterion indicates that digit collection will start when a specific digit combination is detected.
- *First digit time-out* – this criterion defines how long the gsmSRF will wait for the first digit to be detected. If no digit is entered within the first digit time-out, a CAP PC-Error is generated.
- *Inter-digit time-out* – this criterion defines how long the gsmSRF will wait for subsequent digit(s) to be detected. If no subsequent digit is entered within the inter-digit time-out, a CAP PC-Result is generated.
- *Error treatment* – this parameter indicates the action the gsmSRF will take when an error condition occurs. An error condition occurs when no (valid) digits are entered within the duration defined by the first digit time-out criterion. The error treatment is normally set to 'report error

to the SCF'. Alternatively, the error treatment may indicate that the gsmSRF will repeat the prompt to the user or that the gsmSRF will take a help action, such as playing a predefined announcement.

- *Interruptable announcement indicator* – this parameter indicates whether the playing of the announcement or tone that precedes the digit collection may be interrupted by the first digit that is entered by the user. With this option, users may skip a message when they are acquainted with the menu.
- *Voice information* – this parameter indicates whether information collection will be done by means of speech recognition. A CAMEL service shall use this option only when the gsmSSF has indicated (in CAP IDP) that the gsmSRF supports this capability.
- *Voice back* – this parameter indicates whether the collected information will be pronounced to the user prior to generating a CAP PC-Result to the CAMEL service. A CAMEL service shall use this option only when the gsmSSF has indicated (in CAP IDP) that the gsmSRF supports this capability.

4.3.4.3 Charging

When a CAMEL service wants to obtain information from a user with user interaction, a 'both-way through connection' between MS and gsmSRF is required. The speech path in the direction from MS to network and the gsmSRF is not open before the MS has received an answer indication from the network. Hence, digits cannot be conveyed to the gsmSRF before answer. Therefore, the CAMEL service needs to ensure that an answer indication is sent to the calling MS.[27] The place in the network where the user interaction is applied determines what effect this 'emulated answer' has on the charging for the calling subscriber (Figures 4.34 and 4.35).

When user interaction is applied in the originating MSC, the CAMEL service may establish a both-way through connection with the MS without charging the calling subscriber. This is reflected in Figure 4.34. The *connection established* indication from the gsmSRF is forwarded to the MS, in the form of DTAP Connect, but bypasses the charging mechanisms in the MSC.

When the user interaction is applied in an MSC other than the originating MSC, the answer signal that is sent towards the MS, for the purpose of creating a both-way through connection, traverses the regular charging mechanisms of the originating MSC (Figure 4.35). As a result, the MSC creates a MOC CDR; charging commences at the beginning of the user interaction. If the calling subscriber is a pre-paid subscriber, then pre-paid charging is started as well.

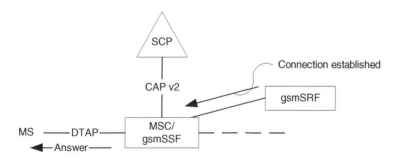

Figure 4.34   User interaction in MSC, no implicit impact on charging

---

[27] The collecting of DTMF digits from the MS as part of user interaction should not be mixed up with the DTMF mid-call event in CAMEL phase 4. Collecting DTMF digits with a mid-call event is not possible before the O-BCSM governing the calling party leg has transited to the active state.

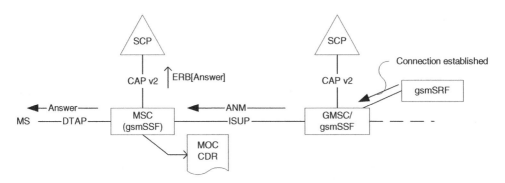

Figure 4.35   User interaction in GMSC with implicit impact on charging

The both-way through connection is requested with the service interaction indicators two (SII2) information element in the connect to resource (CTR) operation or the establish temporary connection (ETC) operation. See the following sections for more information on these CAP operations.

To prevent immediate charging of a calling subscriber, when that subscriber is diverted to a CAMEL user interaction menu, the following mechanism may be applied. The service connects the calling subscriber to a gsmSRF with one-way connection and instructs the playing of an announcement. The calling party will not be charged for this announcement. When the playing of the announcement is finished, the service re-establishes the connection to the gsmSRF, this time with both-way through connection. The service prompts the calling party to enter a digit on the keypad, to indicate a choice in the menu. The sending of the prompt at this point is optional. Charging of the calling subscriber now starts.

4.3.4.4 Cancellation of User Interaction

When a CAMEL service has initiated the playing of an announcement or the collection of information, it may be required at one point that this user interaction be stopped. This may be done with the CAP cancel (CAN) operation. A CAP PA operation or a CAP PC operation may be cancelled prior to the commencement of its execution or when its execution has already started.

If multiple CAP PA operations are sent to the gsmSSF in a single TC message, then there may be one or more CAP PA operations pending, i.e. waiting to be executed. A single one of these pending PA operations, or the PA operation that is currently executed, may be cancelled (Figure 4.36). It may occur that the execution of an announcement has progressed to a point where it cannot be cancelled. If a CAMEL service attempts to cancel such operation, then it will receive a CAP CAN-Error from the gsmSSF ('CancelFailed').

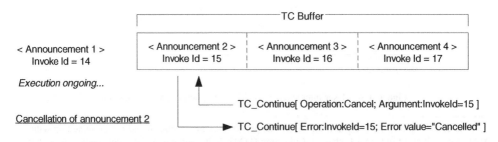

Figure 4.36   Cancellation of an announcement

4.3.4.5 User Interaction for Non-speech Calls

User interaction by means of temporary connection may not be used for all types of calls; refer to GSM TS 09.78 [56], section A.5. That section specifies that the transmission medium requirement in the ISUP IAM that results from CAP ETC shall be set to '3.1 kHz audio'. Applying user interaction to call types such as video telephony (Bearer Service 30, Unrestricted Digital Information, UDI) or for a SCUDIF call[28] (dual capability: tele service 11, speech, and bearer service 30, UDI) may require adaptation to the serving MSC. Care shall be taken in that case that the user interaction does not interfere with the data exchange with the connected (i.e. calling) party.

The architecture for user interaction may be divided into three main categories: (1) integrated gsmSRF and external gsmSRF; (2) assisting gsmSSF with internal gsmSRF or external gsmSRF; and (3) intelligent peripheral. These architectures will be discussed in detail in the following sections.

4.3.4.6 Integrated gsmSRF and External gsmSRF

An MSC or GMSC may have internal gsmSRF or external gsmSRF connected to it. This architecture is depicted in Figure 4.37. The gsmSCF may instruct the gsmSSF to establish a connection between the call connection bridge and an internal gsmSRF or an external gsmSRF. Connection to a gsmSRF is permissible only when the gsmSSF had indicated in CAP IDP that the MSC in which it (the gsmSSF) resides, has user interaction capabilities.

Connection to a gsmSRF is done by means of CAP CTR. The syntax of CAP CTR is as indicated in Figure 4.38. The parameter *resourceAddress* indicates whether the call connection bridge shall be connected to an internal gsmSRF or an external gsmSRF.

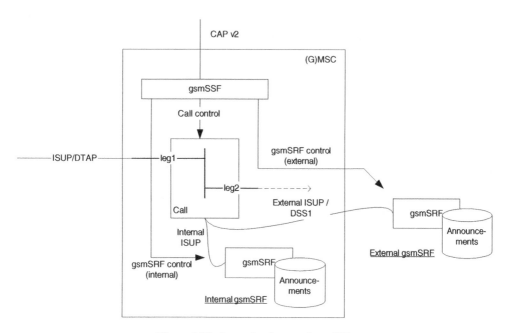

Figure 4.37 Internal and external gsmSRF

---

[28] Service change and UDI fallback; refer to 3GPP TS 23.172 [89].

```
ConnectToResourceArg ::= SEQUENCE {
    resourceAddress                    CHOICE {
            ipRoutingAddress                [0] IPRoutingAddress,
            none                            [3] NULL
            },
    extensions                         [4] SEQUENCE SIZE(1..numOfExtensions) OF
                                                 ExtensionField    OPTIONAL,
    serviceInteractionIndicatorsTwo [7] ServiceInteractionIndicatorsTwo
                                                               OPTIONAL,
    ...
}
```

Figure 4.38 Syntax of CAP CTR operation. Reproduced from GSM TS 09.78 v6.5.0, Section 6.3, by permission of ETSI

- *Internal gsmSRF (resourceAddress = none)* – a default internal gsmSRF will be used for the user interaction. The traffic connection between the call connection bridge and the gsmSRF is an internal ISUP protocol.
- *External gsmSRF (resourceAddress = IPRoutingAddress)* – the call connection bridge will be connected to an external gsmSRF. The traffic connection between the call connection bridge and the external gsmSRF may be ISUP, DSS1 or another suitable protocol. Connection to external gsmSRF is permissible only when this capability is explicitly indicated by gsmSSF in CAP IDP.

Using internal gsmSRF is the more common method. The reason is that using external gsmSRF requires that the gsmSRF control operations (such as CAP PA or CAP PC) are relayed between gsmSSF and external gsmSRF over the ISUP or DSS1 connection. Neither ISUP nor DSS1 has a predefined mechanism to convey these user interaction control operations. A proprietary control mechanism would be needed.

The SII2 parameter in CAP CTR contains an indication whether a both-way through connection is required with the calling party, in other words, whether an ISUP ANM or DTAP connect will be sent in the backwards direction. See the previous section for the usage of the both-ways through connection.

Within a CAMEL service logic processing, CAP CTR may be sent to the gsmSSF after an answer has been sent to the calling party already. An example is: user Interaction is applied when B-party disconnects – in that case, the calling party has received the answer indication already at call answer.

In that case, sending CAP CTR to the gsmSSF with a both-way through connection indicator will not lead to another answer indication being sent to the calling subscriber. Instead, a call progress indicator will be sent to the calling subscriber.

Once the gsmSSF has established the connection with the gsmSRF, the gsmSSF finite state machine (FSM) enters a 'relay state'. In this state, the gsmSSF will relay CAP operations between the gsmSCF and the gsmSRF. The CAP operations, CAP errors and CAP results that may be relayed are listed in Table 4.5.

A CAMEL service may at any time disconnect the gsmSRF from the call with the disconnect forward connection (DFC) CAP operation. When the gsmSSF has processed CAP DFC, its FSM transits back to a state in which it may receive instructions for connection of the call to the required destination.

Alternatively, when a CAMEL service sends CAP PC or CAP PA to a gsmSRF, it may request that the gsmSRF automatically disconnects itself from the call connection after the CAP PA or CAP

Table 4.5 CAP relay between gsmSCF and gsmSRF

gsmSCF → gsmSRF	gsmSRF → gsmSCF
Play announcement (PA)	Play announcement error (PA-Error)
Prompt and collect (PC)	Prompt and collect error, result (PC-Error, PC-Result)
Cancel (CAN)	Cancel error (CAN-Error)
	Specialized resource report (SRR)

Note: the relaying of CAP CAN from gsmSCF to gsmSRF applies only when the argument of CAP CAN contains an Invoke Id.

PC instruction has been completed and the CAP SRR has been issued. The receipt of the SRR then serves as an indication to the CAMEL service that the connection is terminated and that the gsmSSF is ready to receive further instructions. An example signal sequence for user interaction with integrated gsmSRF is provided in Figure 4.39.

Charging
When a CAMEL service has connected the calling subscriber to a gsmSRF, it may apply on-line charging. The principles of the call period, tariff switch, delta timer, forced call release, warning tone, advice-of-charge, etc. are all applicable to the on-line charging during user interaction. When the resource connection is released, on-line charging is stopped. The CAMEL service may provide new charging instructions for the connection of the call to the requested destination.

Free format data may be provided to the gsmSSF during user interaction. However, free format data that is provided during user interaction is associated with the call leg to which the user interaction is related. This implies, for example, that when free format data is provided during user interaction at call establishment and free format data is also provided when the call is active, then the latter free format data overwrites the former free format data. In the end, this user interaction is associated with this outgoing call leg.

4.3.4.7 Assisting gsmSSF with Internal gsmSRF or External gsmSRF

There are cases where a CAMEL service cannot use internal or associated gsmSRF for user interaction with the served subscriber. These cases include:

- The CAMEL service is controlling the call through a CAP dialogue with an MSC in a visited country. The gsmSRF in the MSC in the visited country does not have the announcements required for this CAMEL service.
- The MSC does not have an integrated or associated gsmSRF. For example, an operator may equip a limited number of MSCs in the network with a gsmSRF.
- The MSC has an integrated gsmSRF, but the gsmSRF does not have the capability required for this particular CAMEL service. For example, all MSCs in the operator's network have an integrated gsmSRF, but only a limited number of gsmSRFs are equipped with the 'generation of voice announcements from text' facility.

In those cases, the CAMEL service may establish a temporary connection. A temporary connection is a traffic (ISUP) connection between (1) the MSC where the CAMEL control is running and (2) an MSC with an 'assisting gsmSSF'. This is reflected in Figure 4.40.

Figure 4.40 suggests that the MSC/gsmSSF resides in the VPLMN. The temporary connection scenario may also be used when the MSC/gsmSSF resides in the HPLMN. Furthermore, CAMEL

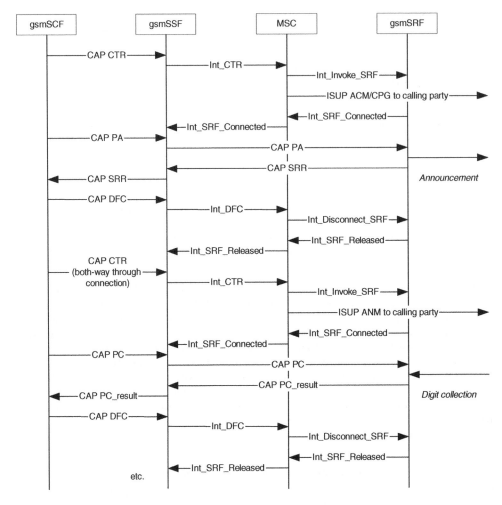

Figure 4.39 Example sequence diagram for user interaction

allows for a scenario whereby the assisting MSC resides in a network other than the HPLMN. Such scenario is, however, not very likely.[29]

The temporary connection is established with CAP ETC. The effect of sending CAP ETC to the gsmSSF is that the gsmSSF instructs the MSC to send ISUP IAM to the MSC which contains the assisting gsmSSF (assisting MSC). The address of the assisting MSC is contained in CAP ETC. This temporary connection may be an international connection.

On reception of the ISUP IAM, the assisting MSC invokes an instance of an assisting gsmSSF and forwards the ISUP connection to this assisting gsmSSF. The assisting MSC may be a regular MSC which may also function as a visited MSC or as GMSC. The assisting gsmSSF, on the other hand, has different behaviour compared with a regular gsmSSF.

---

[29] Practically, a multi-country operator may consider the use of an intelligent peripheral in the VPLMN. Roaming subscribers from different countries, belonging to the same operator group, may use that IP.

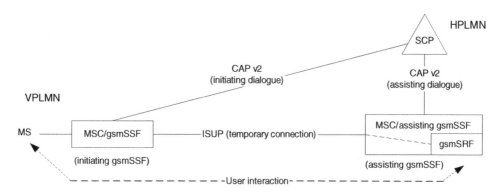

Figure 4.40   User interaction through temporary connection

The task of the assisting gsmSSF is to initiate an 'assisting CAP dialogue' with the gsmSCF. This assisting CAP dialogue may then be used by the gsmSCF to send user interaction instructions. The gsmSCF may instruct the assisting gsmSRF to connect the ISUP connection to an internal gsmSRF or to an associated gsmSRF. The user interaction capability that is available through the assisting dialogue follows the rules that apply for the non-assisting user interaction case, i.e. serving MSC with integrated gsmSRF.

Assisting gsmSSF CAP Dialogue
The assisting gsmSSF CAP dialogue is used for user interaction only. A dedicated application context (AC) is used for that dialogue. This AC contains a different set of CAP operations. The supported CAP operations in the assisting CAP dialogue are:

- *Activity test (AT)* – used for continuity check during an active assist dialogue.
- *Assist request instructions (ARI)* – used by assisting gsmSSF to start the assisting dialogue.
- *Cancel (CAN)* – used by gsmSCF to cancel a pending or ongoing PA or PC operation.
- *Connect to resource (CTR)* – used by gsmSCF to connect the incoming ISUP connection to internal or external gsmSRF.
- *Disconnect forward connection (DFC)* – used by gsmSCF to disconnect the incoming ISUP connection from internal or external gsmSRF. When this operation is executed by the assisting gsmSSF, the gsmSCF may send a subsequent CTR operation, to connection the incoming ISUP connection to another gsmSRF.
- *Play announcement (PA)* – used to instruct the playing of an announcement or tone.
- *Prompt and collect user information (PC)* – used to start the collection of digits from the calling subscriber.
- *Reset timer (RT)* – used to reset the Tssf application timer in assisting gsmSSF, when the assisting gsmSSF is waiting for instructions from gsmSCF.
- *Specialized resource report (SRR)* – used by gsmSRF to inform gsmSCF about the completion of the PA operation.

A typical signal sequence in the assisting gsmSSF dialogue is presented in Figure 4.41.

The assisting MSC always generates an ISUP ANM towards the initiating gsmSSF when the resource connection is established. This answer is needed for the following purposes:

- The ISUP ANM from the assisting MSC serves as an indication to the initiating gsmSSF that the resource connection is successful. The initiating gsmSSF may start the on-line charging process. Note that this ISUP ANM from the assisting MSC is not propagated to the calling party.

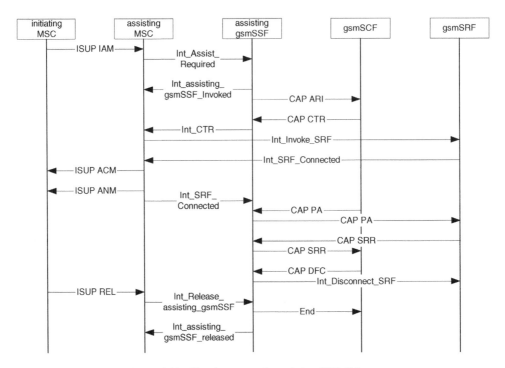

Figure 4.41   Signal sequence for assisting CAP dialogue

- The traffic connection between the initiating gsmSSF and the assisting MSC runs over ISUP and may be an international connection. The ISUP ANM from the assisting MSC is used by the initiating MSC to create a CDR. The CDR may be used by the operator in whose PLMN the initiating MSC resides, for accounting purposes.

The termination of the temporary connection is always initiated by the initiating MSC. When the user interaction through the assisting MSC is complete, the gsmSCF sends CAP DFC to the initiating gsmSSF. The initiating gsmSSF will then disconnect the temporary connection with the assisting MSC, which will in turn end the assisting CAP dialogue with the gsmSCF.

Dialogue Correlation
When the temporary connection is established, the gsmSCF needs to correlate the assisting CAP dialogue with the original CAP dialogue that was established when the CAMEL service started. The correlation of these dialogues is done by *SCF Id* and *correlation Id*.

- *SCF Id* – the SCF Id represents the address of the gsmSCF where the CAMEL service is running. When the gsmSCF sends CAP ETC to the gsmSSF, it includes the SCF Id in CAP ETC. The gsmSSF includes the SCF Id in the ISUP IAM that is sent to the assisting MSC. The assisting gsmSSF uses the SCF Id to derive the SS7 address of the gsmSCF. The assisting gsmSSF then uses this gsmSCF address for the sending of the CAP ARI operation. It is hereby assumed that the gsmSCF and the assisting gsmSSF are owned by the same operator. The assisting gsmSSF may have a list with gsmSCF addresses and may use the SCF Id as index into this list.
- *Correlation Id* – the correlation Id is an identifier of the *correlation process* that is currently taking place in the gsmSCF. The correlation Id is taken from a pool of correlation Id's in

the gsmSCF. The gsmSCF includes the correlation Id in CAP ETC; the gsmSSF includes the correlation Id in ISUP IAM that is sent to the assisting MSC. The assisting gsmSSF, in turn, includes the correlation Id in CAP ARI. When CAP ARI arrives at the gsmSCF, the gsmSCF uses the correlation Id to correlate this assisting CAP dialogue with an originating CAP dialogue. Once the correlation has been done, the correlation Id may be returned to the pool of correlation Id's.

This correlation mechanism is depicted in Figure 4.42.

The transportation of the SCF Id and the correlation Id between gsmSCF, initiating gsmSSF and assisting MSC may be done in two ways:

- *Method 1* – CAP ETC contains the parameter 'assisting SSP IP routing address', which is the address of the assisting MSC. SCF Id and correlation Id may be embedded in this parameter. SCF Id and correlation Id will form part of the address digits of the assisting SSP IP routing address. Assisting SSP IP routing address is a concatenation of assisting MSC address, SCF Id and correlation Id.
- *Method 2* – CAP ETC may contain dedicated parameters for the transport of SCF Id and correlation Id. The initiating gsmSSF includes SCF Id and correlation Id as separate parameters in ISUP IAM.

Method 1 has the advantage that no special ISUP requirements exist. The transportation of SCF Id and correlation Id is transparent from the ISUP signalling point of view. However, the number of permissible address digits in the called party number in ISUP IAM is restricted. ITU-T E.164 [128] recommends that no more than 15 digits are used.

An example structure of the assisting SSP IP routing address with embedded SCF Id and correlation Id is:

Assisting MSC address	9 digits
SCF Id	2 digits
Correlation Id	4 digits
Total	**15 digits**

Two digits for the SCF Id allow indexing into a table containing 100 gsmSCF addresses.[30] As a further refinement, one digit of the SCF Id may be used to select the AC version of the assisting

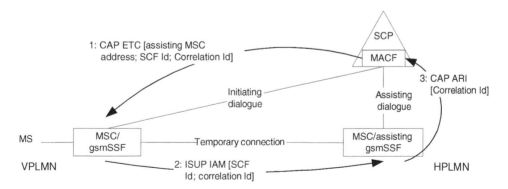

Figure 4.42 Dialogue correlation for temporary connection

---

[30] Presuming that digits 0–9 are used for address digits.

CAP dialogue. Such refinement may be needed if the assisting gsmSSF supports multiple versions of the CAP assist dialogue, e.g. CAP v2 assisting dialogue and CAP v3 assisting dialogue.

Four digits for the correlation Id allow for a maximum of 10 000 simultaneous temporary connection establishment processes. The correlation Id will be returned to the pool of correlation Id's as soon as correlation is done. Keeping the correlation Id occupied longer than needed may result in a situation where all correlation Id's are occupied.[31]

Nine digits remain for the assisting MSC address. The structure of the called party number in ISUP, when coded as an international public telecommunication number for geographic areas, is as follows:

Called party number = <CC> + <NDC> + <SN>
CC = country code (1–3 digits)
NDC = national destination code
SN = subscriber number

When the initiating gsmSSF is located in another country than the assisting MSC, then the called party number will contain the CC. If, for example, the CC is a two-digit code and the NDC for the GSM network is a two-digit code, then five digits remain for the SN. This should constitute an acceptable demand on the operator's numbering plan.

The operator should take care in structuring the assisting SSP IP routing address and bear the following in mind:

- If the number of digits for correlation Id or SCF Id is increased, then fewer digits will be available for the SN to address the assisting MSC; fewer digits for addressing the assisting MSC means a higher demand on the operator's numbering plan.
- A delimiter-digit may be required between the digits used for addressing the assisting MSC and the digits needed for SCF Id and correlation Id. The number of digits available for the SN will be decreased by one, increasing the demand on the operator's numbering plan by a factor of 10.
- If a variable length correlation Id is used, then a delimiter-digit may be needed between SCF Id and correlation Id, resulting in the above-described effect.

Method 2 has a number of advantages compared with method 1.

(1) The full length of the called party number in ISUP IAM may be used for addressing the assisting MSC.
(2) The full address of the gsmSCF may be conveyed to the assisting gsmSSF, instead of the SCF Id. In this way, the assisting gsmSSF does not need to retrieve the SCF Id from the called party number and index into a table.
(3) Correlation Id of arbitrary length may be used.

Method 2 requires that the ISUP connection between the initiating gsmSSF and the assisting MSC complies with ISUP 97 or later. If a CAMEL service cannot deduce from the address of the initiating MSC that ISUP 97 is guaranteed between the initiating MSC and the assisting MSC, then the CAMEL service should use method 1.

---

[31] The rules for releasing the correlation Id in the gsmSCF, when correlation is done, may be compared with the rules for releasing the MSRN, as soon as an MSC has correlated an incoming ISUP IAM with a terminating call process that was started when the VLR received MAP PRN.

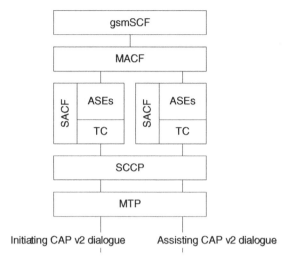

Figure 4.43 Dialogue correlation through MACF. Reproduced from GSM TS 09.78 v6.5.0, figure 2, by permission of ETSI

Multiple Association Control Function

CAMEL specifies the use of the multiple association control function (MACF) for the correlation of the two CAP dialogues, when a temporary connection is established. This is depicted in Figure 4.43.

The use of the MTP layer, SCCP layer and TC layer is described in Chapter 2.[32] A single association control function (SACF) is used per CAP dialogue (or MAP dialogue, where applicable). The MACF provides the link between the two CAP dialogues.

CAMEL formally specifies that the initiating CAP dialogue and the assisting CAP dialogue will use the same application context, e.g. both CAP v2 or both CAP v3. However, a gsmSCF may, if it has that capability, allow for a mix of CAP ACs for the initiating and assisting CAP dialogue.

A practical example is the case where an operator has a CAMEL phase 2 service in operation, which uses an assisting MSC for user interaction. The assisting MSC is from another vendor than the serving MSCs. When the operator upgrades the CAMEL service to CAMEL phase 3, the operator may want to leave the assisting MSC capability at CAMEL phase 2. As a result, the initiating CAP dialogue will be CAP v3 and the assisting CAP dialogue will be CAP v2.

4.3.4.8 Intelligent Peripheral

The (IP) is conceptually the same as an MSC with integrated assisting gsmSSF (assisting MSC). A functional difference is the following:

• The assisting MSC has an integrated or associated gsmSRF. When a temporary connection is established with an assisting MSC, the gsmSCF instructs the assisting gsmSSF to connect the ISUP connection to the internal or associated gsmSRF. The gsmSCF may disconnect the gsmSRF and connect the ISUP connection to another internal or associated gsmSRF.

---

[32] When SIGTRAN is used in a network, the CAP protocol uses the IP connectivity layer instead of the MTP layer. The MACF rules in the gsmSCF remain valid; the SCCP layer and upward are functionally not affected when SIGTRAN is used.

- The IP is a stand-alone gsmSRF. When a temporary connection is established with an IP, there is implicit connection to the resource functionality in the IP. Hence, the gsmSCF does not need to send CAP CTR before sending CAP PA or CAP PC.
- The IP uses a different CAP AC from the assisting gsmSSF. CAP CTR and CAP DFC are not required for an IP and are not included in the CAP protocol for the IP.
- The reset timer operation is not supported in the CAP AC for the IP; a CAMEL service cannot refresh the Tssf timer in the IP. The Tssf value in the IP should therefore be set to allow for the pre-recorded announcement with the longest duration, to prevent expiration of the Tssf timer.

Other than that, the sequence diagrams that are used for user interaction through an IP are the same as the sequence diagrams that are used for user interaction through an assisting MSC.

An IP may be used for special user interaction cases. Functionality such as speech recognition, text-to-speech conversion or voice-back may be supported in a dedicated IP, but may not be supported in the (assisting) MSC. When user interaction is done through an IP, the IP reports its supported capability in the IPSSPCapabilities parameter in CAP ARI.

4.3.4.9 Autonomous User Interaction

One special case of user interaction is the autonomous user interaction. Although this method of user interaction is not formally specified by CAMEL, it is applied in practice. The user interaction cases described so far have in common that the CAMEL service connects a calling subscriber to a gsmSRF and then instructs the gsmSRF to play announcements. The CAMEL service is thereby in control of the announcements played. Likewise, when a CAMEL service collects information (i.e. digits) from a calling subscriber, the CAMEL service has a process in which it can process the collected information.

There may be cases, however, where it is desirable to have an IP act autonomously. The IP plays announcements and collects digits from the user, without instructions from the CAMEL service. Examples include a voucher refill system, service registration and a personal ring-back tone. For that purpose, the CAMEL service may connect a calling subscriber to an associated gsmSRF or to an IP. The CAMEL service will not apply instructions such as play announcement or prompt and collect.

- *Associated gsmSRF* – when the CAMEL service uses CAP CTR to connect the calling subscriber to an autonomous gsmSRF, then the CAP signalling sequence is exactly in accordance with the CAMEL specifications. The only difference is that the CAMEL service does not use CAP PA or CAP PC. CAP CTR will in this case contain resourceAddress = ipRoutingAddress. As a result of not using CAP PA or CAP PC, the CAMEL service will not receive CAP SRR or CAP PC-result. Therefore, the CAMEL service has to use another stimulus to decide to terminate the resource connection. One such stimulus may be a CAMEL service-internal timer expiry or the reporting of an ISUP event.
- *Intelligent peripheral* – when the CAMEL service uses CAP ETC to connect the calling subscriber to an autonomous IP, the gsmSCF will not receive CAP ARI. When this method is applied, it should be ensured that the absence of CAP ARI does not lead to exception handling in the MACF in the gsmSCF. CAP ETC does not need to contain SCF Id and correlation Id, since there will not be CAP dialogue correlation.

Using ETC for the connection to an autonomous IP is the more common method. One reason is that the use of CTR for connection to an associated gsmSRF is not a mandatory CAMEL capability. Refer to BIT 0 (IPRoutingAddress) in parameter IPSSPCapabilities in GSM TS 09.78 [56].

One aspect that these methods have in common is the requirement that the CAMEL service include additional information in the traffic connection to the IP (or gsmSRF). This additional information is needed by the IP, e.g. to select the announcement that shall be played. Information that may be needed by the IP may include (the list is not exhaustive): calling party number; and original called party number. CAMEL does not specify which call parameters shall be present in the ISUP connection that results from CAP ETC. Refer to GSM TS 09.78 [56], section A.5 for mandatory parameters in this information flow. Therefore, when using CAP ETC for connection to an autonomous IP, the inclusion of information elements like the ones listed above, into the ISUP connection, shall be accomplished by proprietary means in the serving MSC.

To facilitate the connection to autonomous IP, CAMEL phase 4 in 3GPP Rel-6 introduces the capability for the CAMEL service to include the above-listed information elements in CAP ETC. The serving MSC shall then copy these information elements in the ISUP connection to the autonomous IP. When the connection to autonomous IP is used for mobile terminating call handling, the original called party number would typically contain the called party number of the called destination subscriber.

4.3.5 Equal Access

Equal access is an optional feature in CAMEL phase 2. It is specified for use in North American countries only. It is therefore also referred to as North American equal access (NAEA). NAEA enables a subscriber to select a preferred carrier for long distance calls.

Three ways exist for a subscriber to select a carrier: (1) subscribed; (2) manually selected; and (3) service-induced.

4.3.5.1 Subscribed

The subscriber profile in the HLR may contain the information element NAEA preferred carrier identification. This information element has the form of a three- or four-digit number string. The HLR may send this element to the VLR during location update. When the subscriber initiates mobile originating calls or mobile forwarded calls, the NAEA preferred carrier identification is used to select the preferred carrier. The HLR may also send this element to the GMSC during terminating call handling. The GMSC then uses the NAEA preferred carrier identification to select the preferred carrier for the roaming call leg between GMSC and VMSC. The structure and usage of the NAEA preferred carrier identification are described in ANSI T1.113.3 [160].

4.3.5.2 Manually Selected

When the subscriber initiates a MO call, she may include a carrier selection code in the called party BCD number. The serving MSC may verify that the subscriber's subscription profile allows the use of the requested carrier.

4.3.5.3 Service-induced

The CAMEL service may provide a carrier code for a call. The carrier code may be subscribed in a service subscription database associated with the CAMEL service. Figure 4.44 gives a graphical representation of these equal access selection methods for mobile originated calls.

For mobile forwarded calls, only the subscribed method and the service-induced method are available.[33] Figure 4.45 reflects the carrier selection for mobile terminated calls.

---

[33] If late call forwarding is the result of call deflection, then the subscriber may have entered a manually selected preferred carrier on the MS. That possibility is not further specified by CAMEL.

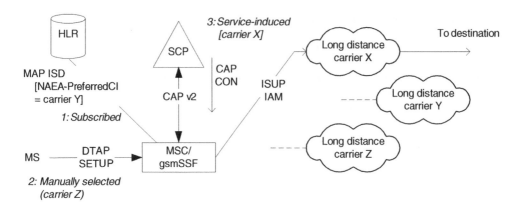

Figure 4.44   Carrier selection for mobile originated calls

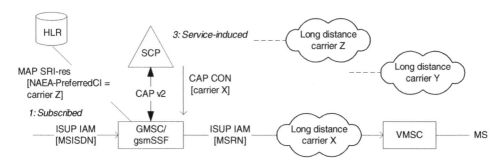

Figure 4.45   Carrier selection for mobile terminated calls

An HLR may, depending on its capability, store two NAEA preferred carrier identification values:

- one value which is included in MAP ISD and which is used for mobile originated calls and for mobile forwarded calls (late call forwarding); and
- one value which is included in MAP SRI-res and which is used for mobile terminated calls and mobile forwarded calls (early call forwarding).

However, such a capability is not explicitly specified by GSM.

4.3.5.4 CAMEL Information Transfer

The following carrier-selected information may be transferred by CAMEL operations.

- *Initial DP (IDP)* – carrier identification code: this code identifies the preferred carrier for the call; carrier selection info: this parameter indicates how the preferred carrier for this call is selected, e.g. subscribed or dialled.
- *Connect (CON)* – carrier identification code: see IDP; carrier selection info: see IDP; originating line info: this parameter contains an indication of the telephony service where the carrier was originally defined; charge number: this parameter contains a number string. This number identifies the party that shall be charged for the call. The charged party may differ from the calling party for an MO call.

- *Establish temporary connection (ETC)* – for ETC, the same carrier selection-related parameters are specified as for connect. The rationale is that ETC may be used for long distance temporary traffic connections.

Although CAP CTR may be used for connection to an external gsmSRF, the connection to an external gsmSRF will not run over a long distance carrier. For that reason, no carrier selection information is specified for CAP CTR.

4.3.6 Enhancement of Call Control

Over and above what is explained in previous sections, the following call control-related enhancements are introduced in CAP v2.

4.3.6.1 Reporting of Time and Time Zone

The initial DP operation contains the time and time zone. This is the time related to the MSC where the gsmSSF is located. This may not have to correspond to the time that applies to the served subscriber, e.g. for MT call control, the time of the GMSC is reported, but the CAMEL service receives no indication of the local time of the served subscriber. The following elements are reported: year, month, day, hour, minute, second and time zone. The time zone is reported as relative time zone, related to GMT, in multiples of quarter of an hour. The relative time zone may therefore have a value between $-48 (= \text{GMT} - 12 \text{ h})$ and $+48 (= \text{GMT} + 12 \text{ h})$.

4.3.6.2 Forwarding Pending Indicator

The forwarding pending indicator may be included in CAP IDP for MT call control. This indicator is present when the T-CSI in the SRI-res is accompanied by an FTN. The FTN may in this case relate to unconditional forwarding or to forwarding when not reachable. In the latter case, the sending of the FTN is due the fact that the subscriber is marked as detached in the HLR.

4.3.6.3 Enhancement to Call Reference Mechanism

The call reference mechanism that is included in CAMEL phase 1 has no proper support for forwarded calls. In CAMEL phase 2 for mobile forwarded calls the *GMSC address* is reported, in addition to the *MSC address*. Refer to Chapter 7 for a full description of the call reference number mechanism.

4.3.6.4 Alerting Pattern Control

A CAMEL service may influence the *alerting pattern* that will be used to alert the called subscriber. The alerting pattern may be used in MT calls only (Figures 4.46 and 4.47).

In the example in Figures 4.46 and 4.47, the called party subscribes to a 'buddy list' service. Calls from subscribers that are part of the buddy list are treated in a designated manner, e.g. by providing a special alerting pattern to the MS. The alerting pattern is used in the MS to select the ring tone. Table 4.6 lists the alerting pattern values that may be used.

Support of the various alerting levels and alerting categories depends on the MS. A subscriber of a particular service such as the above-mentioned *buddy list* service may indicate in her service profile that calls from buddy members should be offered with category 1. The user then selects appropriate alerting handling in the MS corresponding to alerting category 1. Further details of the use of the alerting pattern are specified in GSM TS 02.07.

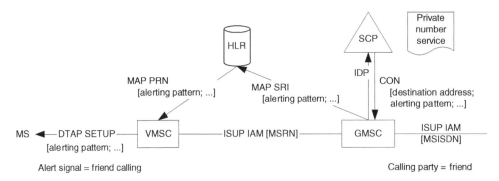

Figure 4.46 Architecture for setting the alerting pattern for MT call

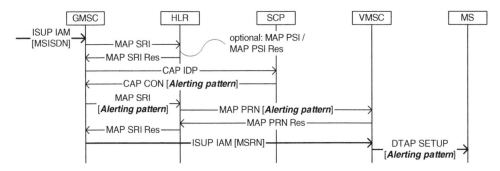

Figure 4.47 Signal sequence for setting the alerting pattern

Table 4.6 Alerting patterns

Value	Pattern	Designation	Description
0	1	Level 0	MS shall not apply audible ring tone
1	2	Level 1	Urgency level 1
2	3	Level 2	Urgency level 2 (more urgent than level 1)
3		Reserved	
4	5	Category 1	User-configurable in the MS
5	6	Category 2	User-configurable in the MS
6	7	Category 3	User-configurable in the MS
7	8	Category 4	User-configurable in the MS
8	9	Category 5	User-configurable in the MS
9–255		Reserved	

Value, value of the alerting pattern; pattern, pattern sequence number as specified in GSM TS 09.02 [54]; designation, type classification for the alerting pattern; Description, use of the alerting pattern.

4.3.6.5 Reset Timer

The reset timer (RT) operation is introduced in CAMEL phase 2. RT is used to refresh the Tssf timer. The Tssf timer is used in CAMEL phase 1 already; see Chapter 3. A CAMEL phase 1 service does not have the capability to reset this timer. A CAMEL phase 2 service may, however, require more time to respond to a CAP operation from the gsmSSF. An example is initial DP: a CAMEL phase 2 on-line charging service needs to contact the SDP to obtain charge-related information. In that case, the CAMEL service may use RT to set the Tssf to a service-defined value. The RT operation would in that case be sent as immediate response to the initial DP operation (Figure 4.48).

The gsmSSF has two default values for Tssf: default non-user interaction value in the range 1–20 s; and default user interaction value in the range 1–30 min. In the example in Figure 4.48, the default value of Tssf (for non-user interaction control) is 5 s. The service 'buys' itself another 15 s to complete its internal processing. In this example, the service queries a MNP database and a credit database, residing in the SDP. The RT operation may also be used during user interaction, in the case that an announcement with long duration is played. The syntax of the RT operation in CAP allows for setting the Tssf timer with a value in the range 0–2 147 483 647 s. However, a service logic designer should not use higher values than needed.

The rules for the gsmSSF or the assisting gsmSSF for starting the Tssf timer are as follows:

- When the gsmSSF FSM transits to waiting for instructions (WFI), the Tssf is started with the default non-user interaction value.
- When the gsmSSF FSM transits to monitoring (MON) or idle, the Tssf is stopped.
- When the gsmSSF FSM transits to one of the user interaction states, the Tssf is started with the default user interaction value.
- When the gsmSSF receives CAP RT when the FSM is in the state WFI or in one of the user interaction states, the Tssf is loaded with the value contained in CAP RT. The validity of the Tssf timer value supplied with CAP RT is restricted to one gsmSSF FSM state. When the gsmSSF FSM transits to WFI or to one of the user interaction states, the gsmSSF will use the default Tssf timer value, irrespective of previous CAP RT operations in the current CAMEL dialogue (Figure 4.49). The gsmSSF FSM transits from WFI to MON and then back to WFI. The second time the FSM transits to WFI, Tssf is again started with the default non-user interaction value.
- When the gsmSSF receives any other CAP operation when the FSM is in the state WFI or in one of the user interaction states and the processing of this CAP operation does not result in a state transition, then the Tssf is reloaded with the *current value* and restarted. This *current value* may be the default value that was used when the FSM transited to WFI or the user interaction state, or the value from the CAP RT that was received in the current FSM state.

CAMEL phase 3 specifies a condition for the CAP RT. This condition ensures that the gsmSSF will not accept successive CAP RT in the same FSM state. However, not all MSCs have implemented this check, but always accept the CAP RT when received in correct FSM state. For this reason, this condition is removed from the gsmSSF process in CAMEL phase 4.

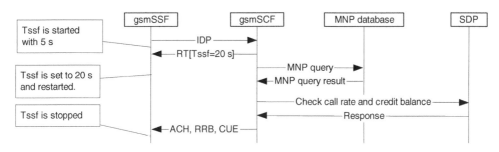

Figure 4.48 Setting the Tssf timer

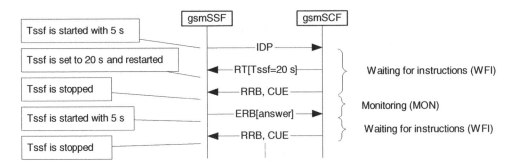

Figure 4.49   Setting Tssf in successive WFI states

4.3.6.6 Additional Detection Points

The following detection points (DPs) are introduced in CAMEL phase 2: for O-BCSM – O-Abandon, O_No_Answer, O_Busy, Route_Select_Failure and for T-BCSM – T_Abandon, T_No_Answer and T_Busy. These DPs may be used as event detection points (EDP). The introduction of these EDPs results in improved control over the call. For example, call establishment failure may be reported to the SCP.

4.3.7 Supplementary Service Invocation Notification

SSIN is a feature that relates to the Fraud Information Gathering System (FIGS). FIGS is specified in GSM TS 02.31 [6] and GSM TS 03.31 [32].[34] FIGS is a network architecture that enables an operator to monitor subscribers for fraudulent use of the network. FIGS involves a fraud detection system (FDS), which collects call-related information on suspected users. Various FIGS monitoring levels are specified.

- *FIGS level 1* – accelerated processing of transfer account procedure (TAP) files; see Chapter 7. TAP files are created from the CDRs that are generated in the MSC. For designated subscribers, the TAP files are processed with priority. This prioritized handling of TAP files enables the operator to receive early warnings when fraud is suspected. FIGS level 1 has no requirement on CAMEL.
- *FIGS level 2* – on-line call monitoring (start time and stop time). CAMEL is used for FIGS level 2. The CAMEL service monitors the start and end of the call and forwards the call-related information to the FDS. FIGS level 2 requirements may be fulfilled with CAMEL phase 1 (partially) or with CAMEL phase 2 (fully).
- *FIGS level 3* – on-line call monitoring and monitoring the use of supplementary services. The following supplementary services are monitored: multi party service (MPTY); GSM TS 03.84 [42]; explicit call transfer (ECT); GSM TS 03.91 [47]; call deflection (CD); GSM TS 03.72 [37]; call forwarding (CF); GSM TS 03.82 [40]; and call hold (CH); GSM TS 03.83 [41]. Refer to GSM TS 02.31 [6] and GSM TS 03.31 [32] for a detailed description of FIGS.

The SSIN feature in CAMEL phase 2 fulfils the FIGS requirement for the monitoring of MPTY, ECT and CD.[35] Refer to Figure 4.50 for an example call flow. The B-party deflects an incoming

---

[34] For GSM TS 03.31, there is no GSM R97 version available.

[35] Call deflection is introduced in GSM R98; the related SSIN feature for CD is already introduced in GSM R97.

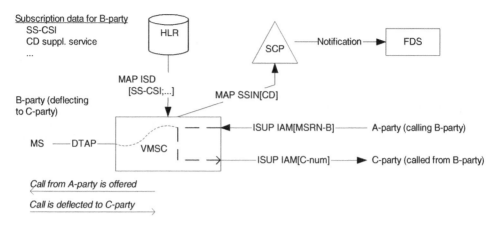

Figure 4.50 Supplementary service invocation notification for call deflection

call to a C-party. An operator may provision a subscriber with SS-CSI in her HLR subscription profile. The SS-CSI is a CAMEL subscription information element, comparable to O-CSI and T-CSI. SS-CSI consists of the following information: *gsmSCF address* – address of the entity that shall receive the notifications; and *SS event list* – list with indications of the supplementary services for which the feature applies. The list may contain any combination of MPTY, ECT and CD.

The gsmSCF address in SS-CSI may be the GT of an FDS. This has the effect that SSI notifications are sent directly to the FDS. Alternatively, notifications may be sent to the gsmSCF, which forwards the notifications to the FDS.

The HLR sends SS-CSI to the VLR during location update or when SS-CSI is provisioned for a subscriber. The HLR should not include a particular notification marking in SS-CSI if the subscriber does not have a subscription to the related supplementary service. When the VLR receives SS-CSI, it should not have to check whether the subscriber has a subscription to the GSM supplementary services that are included in SS-CSI.

A subscriber may invoke a supplementary service by entering designated command strings on the keypad. The required command strings for MPTY, ECT or CD are specified in GSM TS 02.30 [5].

When a subscriber invokes MPTY, ECT or CD, the MSC performs subscription checks and other checks as specified in the appropriate specification for the invoked service. When all conditions are fulfilled, the supplementary service is executed. If the supplementary is included in SS-CSI, then the MSC sends an SSI notification to the gsmSCF after the execution of the supplementary service is completed. The sending of an SSI notification to the gsmSCF does not affect the handling of the supplementary service.

4.3.7.1 Rationale

The rationale of SSIN is that MPTY, ECT and CD may be used fraudulently. These features have in common that one party may establish an (international) call connection between two other parties, on the account of this one party.

- *MPTY* – a user may set up a multi-party call with a B-party and a C-party; A, B and C are in speech connection with one another. The A-party pays for the two calls and indirectly for the speech connection between B and C.
- *ECT* – a user may place an incoming or outgoing call and establish a second call. By invoking ECT, these two calls are connected to one another and the ECT-invoking party disconnects

herself from the call. In this way, an incoming (local) call may be connected to an international destination.

- *CD* – a user may deflect an incoming call to a forwarding destination. The calling party is effectively connected to the deflected-to destination. In this manner, a local call may be deflected to an international destination.

FIGS level 3 also specifies monitoring for call forwarding and call hold. The invocation of call forwarding may be monitored by means of regular CAMEL call control; there is no explicit notification defined for call forwarding. For the monitoring of the invocation of call hold, there is no CAMEL mechanism defined.

4.3.7.2 Notification

The SSI notification to the gsmSCF is the *SS invocation notification* MAP message. SSIN contains the following elements:

- *Event indication* – this element indicates which GSM supplementary service was invoked.
- *Destination number(s)* – this element indicates the destination(s) of the <u>outgoing</u> call(s) involved in the ECT or CD service.
    - o For ECT, both involved call legs may be incoming calls or outgoing calls. When a call leg involved in ECT is an outgoing call, then the number that is are reported is the number that was received over the radio access network to set up the call. This number may be a VPN number that is translated by a VPN service invoked from the serving MSC.
    - o For CD, the number that the call is deflected to is reported to the gsmSCF; as is the case with ECT, this number may be a VPN number.
- *IMSI* – IMSI of the subscriber that invoked the supplementary service.
- *MSISDN* – MSISDN of the subscriber that invoked the supplementary service.

4.3.8 Short Forwarded-to Numbers

When a GSM subscriber wants to have her calls (conditionally) forwarded to an alternative destination, she needs to register a FTN in the HLR. FTN registration is performed with designated signalling messages between MS, MSC and HLR. An example of a forwarding setting is:

Condition = not reachable
Basic service = TS11 (speech)
FTN = +31 65 161 3900

Within the context of a VPN service, a subscriber may wish to forward her calls, when she is not reachable, to a VPN destination. In that case, the FTN would be:

FTN = 9012

9012 is a VPN number; when a forwarded call is established to that number, a VPN service will be invoked to translate this FTN into a public number, e.g. +31 161 24 9012. The invocation of a VPN service for this forwarded call follows the rules that apply to CAMEL invocation for mobile originated calls.

However, an HLR normally does not allow the registration of such FTN (9012), because it violates certain FTN registration rules, e.g. the HLR must be able to translate the FTN into a number in international format, in accordance with the number plan rules that are defined in ITU-T E.164 [128].

For the above reason, CAMEL phase 2 has introduced the Translation Information Flag CAMEL Subscription Information (TIF-CSI). The TIF-CSI is a flag (BOOLEAN); the presence of TIF-CSI in a subscriber's HLR profile defines that that subscriber is allowed to register 'short FTNs'.[36] 9012 is an example of a short FTN. The following checks in the HLR are skipped when a subscriber with TIF-CSI registers an FTN:

- *FTN validity* – the HLR will not check whether the FTN corresponds to a special number as defined in that operator's numbering plan. For example, 1233 may be the number that is used for voicemail box access and is therefore normally not allowed as FTN. For a VPN subscriber, 1233 may, however, be a VPN number.
- *Conversion to international format* – the HLR will not attempt the convert the FTN to international format.
- *Call barring checks* – the HLR will not check the FTN for possible violation of the setting of BOIC and barring of outgoing international calls except to the home country (BOIC-exHC). The BAOC check is, however, not skipped. For call barring functionality, refer to GSM TS 03.88 [45].

4.3.8.1 Call Forwarding with CAMEL Invocation

When call forwarding is performed for a subscriber who has registered a short FTN, a CAMEL service is required to translate the short FTN into the corresponding public destination. Hence, a subscriber who has TIF-CSI shall also have O-CSI.

The Short FTN may be used for the three kinds of call forwarding call cases: early call forwarding in GMSC; late call forwarding in VMSC; and optimal routing at late call forwarding (ORLCF).

Early Call Forwarding

When a GMSC receives an FTN in MAP SRI-Res, it normally verifies that the FTN has a valid format, i.e. complies with ITU-T E.164 [128] number format rules. A short FTN may be sent to a GMSC that supports CAMEL phase 2 or higher. A CAMEL phase 2 GMSC does not apply the regular FTN validity check. The GMSC should not have to verify that a short FTN is accompanied by an O-CSI; the HLR ensures that an O-CSI is sent in this case.

An operator that uses CAMEL phase 2 functionality such as short FTN, should ensure that the GMSC supports CAMEL phase 2. MT call handling and early MF call handling is normally performed by a GMSC in the HPLMN. Hence, the HLR should always be able to send the Short FTN to the GMSC (Figure 4.51). It is not recommended for an operator to have a mix of CAMEL phase 1 and CAMEL phase 2 GMSCs. When BOR is used, a mix of CAMEL phase 1 and CAMEL phase 2 GMSCs may occur.

Although a precondition for sending a short FTN to the GMSC is that the GMSC supports CAMEL phase 2, the O-CSI that accompanies a short FTN may actually be a CAMEL phase 1 O-CSI.

Late Call Forwarding

Analogous to the GMSC, when a VLR receives an FTN in MAP ISD, it normally verifies that the FTN has a valid format. The HLR may send a short FTN to a VLR that supports CAMEL phase 2 or higher. A CAMEL phase 2 VLR shall not apply the regular FTN validity check. Similar to the GMSC, the VLR should not have to verify that a short FTN is accompanied by an O-CSI; the HLR ensures that an O-CSI is sent to the VLR in this case (Figure 4.52).

---

[36] A 'short FTN' is generally regarded as an FTN that has no country code.

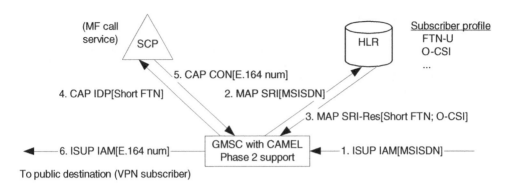

Figure 4.51 Early call forwarding with short FTN

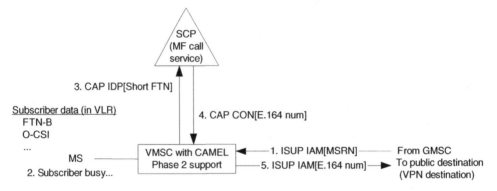

Figure 4.52 Late call forwarding with short FTN

Here we have the scenario where subscribers with short FTN, e.g. VPN subscribers, may roam in CAMEL phase 2 supporting networks, in CAMEL phase 1 supporting networks or in non-CAMEL networks. When the subscriber is not in a CAMEL phase 2 network, the HLR does not send the short FTN to the VLR. Hence, the VPN subscriber can use her short FTN only when registered in a CAMEL phase 2 network.

Optimal Routing at Late Call Forwarding
When ORLCF is initiated, the VMSC returns the FTN to the GMSC in MAP RCH. This FTN may be a short FTN. When the VMSC includes a short FTN in MAP RCH, it should verify that the GMSC supports CAMEL phase 2.[37] The short FTN in MAP RCH will be accompanied by the O-CSI for this subscriber. The GMSC, which will handle the optimally routed call forwarding, will invoke the CAMEL service for the translation of the Short FTN (Figure 4.53).

When the GMSC receives MAP RCH it normally applies an 'FTN validity check'. If the FTN in MAP RCH is a short FTN, the GMSC will not apply the FTN validity check. The reason is that the short FTN does not contain a country code. It is the responsibility of the CAMEL service to ensure that the forwarded call is not routed to a destination that would result in a cost penalty for the forwarding subscriber.

---

[37] This VLR behaviour, to check GMSC CAMEL phase 2 support before sending a short FTN in MAP RCH, is not formally specified in CAMEL. However, the VLR will in any case verify that the GMSC supports CAMEL phase 2 for the purpose of including CAMEL phase 2 O-CSI in MAP RCH.

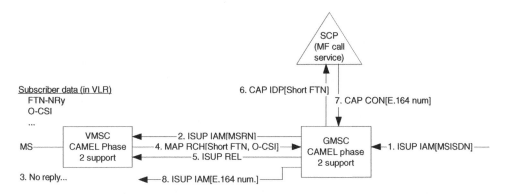

Figure 4.53 Optimal routing at late call forwarding with short FTN

Provisioning CAMEL Data

The operator should ensure that the subscription data in HLR remains consistent at all times. When, for a subscriber who has O-CSI and TIF-CSI, the O-CSI is removed at one point, the subscriber may end up with a short FTN but without a corresponding CAMEL service. Call forwarding to the short FTN would in that case fail. Hence, when O-CSI is removed for a subscriber, TIF-CSI will also be removed. If that subscriber has a short FTN registered at that moment, then the short FTN should also be removed.

4.3.9 Conditional Triggering

In CAMEL phase 1, triggering of CAMEL services for call control is *unconditional*. This implies that, for the various call cases, the availability of O-CSI or T-CSI will always result in triggering the related CAMEL service.

- *Mobile originated and forwarded calls* – a subscriber may receive O-CSI in the VLR, at the moment of registration (location update, data restoration procedures). For every MO call that is initiated by the subscriber and for every MF call that is initiated on behalf of the subscriber, an O-CSI based CAMEL service triggering takes place. Likewise, when O-CSI is sent to the GMSC during terminating call handling, and a forwarding call leg is created in the GMSC, then an O-CSI-based CAMEL service is triggered for the forwarding call leg. Finally, O-CSI may be sent from MSC to GMSC, during ORLCF handling. As a result, an O-CSI based CAMEL service is triggered for the forwarding call leg that is created in the GMSC.
- *Mobile terminated calls* – when a subscriber has T-CSI, the HLR will send the T-CSI to the GMSC for every terminating call for the subscriber. As a result, a T-CSI based CAMEL service is triggered for every terminating call.

For certain calls, however, CAMEL control is not necessary. Examples include:

- personal number translation – if the personal numbers have a length of three or four digits, then triggering may be restricted to numbers of three or four digits;
- on-line charging – if calls to 0800 numbers do not require on-line charging, then it may not be necessary to invoke a CAMEL service for calls to these numbers;
- speech services – services that are required for speech calls only do not need to be triggered when a data call or Fax call is established.

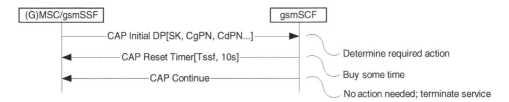

Figure 4.54   Immediate CAMEL service termination

For these calls that do not require CAMEL control, the CAMEL service may simply terminate immediately and give control of the call back to the (G)MSC/gsmSSF (Figure 4.54).

The CAP reset timer operation in the example in Figure 4.54 may be generated by the SCP platform, rather than by the CAMEL service. The service concludes that no CAMEL control is needed and terminates the service logic. The call continues without CAMEL control.

For these services, there should not be CAMEL service triggering if the call does not satisfy the trigger criteria. If the trigger criteria are not matched, then the call will continue without CAMEL control. Conditional triggering enables the operator to define trigger conditions that result in CAMEL service triggering for selected calls only. For calls that do not satisfy the trigger criteria, call handling will proceed without CAMEL control.[38] Conditional triggering reduces signalling load, (G)MSC load and load on the SCP.

When trigger criteria are checked and fulfilled, CAMEL triggering takes place; the checking of the trigger criteria has no effect on the further handling of the call or on the CAMEL service processing; the CAMEL service may not be aware that conditional triggering check took place. CAP IDP contains no indication about the conditional triggering.

In line with the CAMEL principles, each subscriber may have her own set of trigger criteria in the subscriber profile in the HLR, e.g. for pre-paid subscribers, the operator may define criteria that prevent CAMEL triggering for calls to 0800* and, for VPN subscribers, the operator may define trigger criteria that result in triggering for calls to numbers with three, four or five digits. Each subscriber may have two sets of trigger conditions: trigger conditions associated with O-CSI; and trigger conditions associated with T-CSI. The association of trigger conditions with O-CSI or T-CSI requires that the CAMEL capability handling parameter in O-CSI or T-CSI has a value of CAMEL2 or higher.

4.3.9.1 Mobile Originated and Mobile Forwarded Calls

Trigger criteria that are associated with O-CSI may consist of a combination of the following categories.

Category 1: Number-based Criterion
This criterion consists of either or both of the following sub-criteria:

(1) A set of (maximum 10) number strings; this sub-criterion is fulfilled if the leading digits of the dialled number match with the digits of one of the number strings. An example is the following:

Criterion contains the following list:
0800 123

---

[38] A common mistake is to associate the 'default call handling' (DCH) in O-CSI or T-CSI with the action an MSC takes when trigger criteria are not satisfied. However, DCH is not related to conditional triggering.

0800 125
0900
Dialled number:
0800 → does not lead to triggering (no match found; probably not a valid number)
0800 1234 → does lead to triggering (matches 0800 123)
0800 126 → does not lead to triggering (no match found)
0900 889 → does lead to triggering (matches 0900)

(2) A set of (maximum 3) number length indicators; this sub-criterion is fulfilled if the dialled number has a length that matches any of the number length indicators. An example is the following:

Criterion contains the following list:
4 digits
5 digits
Dialled Number:
5678 → shall lead to triggering (number has 4 digits)
0161 24 5678 → shall not lead to triggering (number has 10 digits)

For the number-based criterion to be fulfilled, both sub-criteria must be fulfilled. If one of the sub-criteria is not present, then that sub-criterion is considered to be fulfilled. The number based criterion may be defined as *enabling* or *inhibiting*:

- *enabling* when the criteria are met, so triggering may take place, unless any of the other criteria prohibits triggering; this method enables an operator to trigger *only* for numbers with a length of four or five digits, for example;
- *inhibiting* when the criteria are met, so triggering shall not take place, even when all of the other criteria permit triggering; this method enables an operator to prohibit triggering for calls to numbers starting with 0800, for example.

For MO calls, the conditional triggering check is performed on the number string that is received from the radio access network. For mobile forwarded calls, the conditional triggering check is performed on the FTN. In the case of early call forwarding, the FTN may have been provided by the mobile terminating CAMEL service, by means of CAP CON.

Category 2: Basic Service Code
This criterion consists of a set of (maximum 5) basic service code values. Each basic service code may represent a single basic service or a group of basic services. The basic services are specified in GSM TS 02.01 [1]. A basic service may be a bearer service (GSM TS 02.02 [2]) or a tele service (GSM TS 02.03 [3]). The basic service codes that are used for the basic service trigger list are specified in GSM TS 09.02 [54], parameter '*Ext-BasicServiceCode*'. An example of a basic service is: '*telephony*'. An example of a basic service group is: '*allFacsimileTransmissionServices*'.

This conditional triggering check entails that the basic service of the mobile originating or mobile forwarded call is compared against the list of basic service code values. Triggering takes place when the basic service for the call is included in the list; either explicitly or implicitly. In the latter case, the basic service for the call is part of a basic service group code[39] in the trigger list.

Not all Basic Service codes may be used for conditional triggering. Exceptions include, amongst others: tele service 12 (emergency calls) – CAMEL control is not applicable to emergency calls;

---

[39] GSM TS 09.02 specifies that compound basic service group codes may not be used in InsertSubscriberData. That statement does, however, not apply to the O-CSI trigger criteria.

and tele service 20 group (SMS) – CAMEL call control capability is not applicable to SMS; see Chapters 5 and 6 for CAMEL control of SMS.

The basic service code for the call, to be used for the conditional triggering check, is derived as follows:

- *MO call* – the basic service is derived from the bearer-related parameters that are received from the mobile station. These parameters may be a combination of BC, LLC and HLC. These parameters are specified in GSM TS 04.08 [49].
- *MF call* – the basic service is derived from the BC, LLC and HLC from the incoming ISUP message (ISUP IAM) or is received from the HLR in the MAP SRI Result message.

Rules for deriving the basic service from BC, LLC and HLC are specified in GSM TS 09.07 [55].

Category 3: Call Type Criterion

The third criterion for MO and MF calls is the call type, more specifically, whether the call is a forwarded call or an originating call. This criterion may be used to restrict CAMEL service triggering to originating calls or to restrict CAMEL service triggering to forwarded calls.

Executing the Triggering Check

When MO call establishment takes place or when call forwarding occurs, the (G)MSC verifies whether the MO or MF call satisfies the trigger criteria. For triggering to occur, all enabling criteria must be met and none of the inhibiting criteria shall be met. Even though the trigger criteria for O-CSI may consist of 10 numbers, there is still only one CAMEL service associated with these numbers. As a result, the order of the numbers in this list is not relevant.

When early call forwarding takes place, the HLR will send the FTN plus O-CSI, including trigger criteria, to the GMSC, so GMSC can initiate the forwarding leg. If the HLR can deduce from the FTN and the basic service for the call that the trigger criteria are met, then the HLR may decide to send O-CSI to the GMSC, but not the trigger criteria. In that case, a CAMEL service will be invoked in the GMSC for the forwarded call. Likewise, if the HLR can deduce from the FTN and the basic service for the call that the trigger criteria are not met, then the HLR may decide not to send O-CSI to GMSC. In that case, no CAMEL Service will be invoked in GMSC for the forwarded call.

Note: if HLR sends/has sent T-CSI to the GMSC, then care should be taken regarding performing the O-CSI trigger check in HLR instead of GMSC. The MT CAMEL Service, triggered with T-CSI, may provide alternative routing information. In that case, the O-CSI trigger check in HLR may give an incorrect result. In this case, HLR may rather leave the O-CSI trigger check over to GMSC. Figure 4.55 represents an example of conditional triggering for an MO call.

Triggering Check for ORLCF

When ORLCF takes place in the VMSC and the forwarding subscriber has O-CSI with trigger criteria, then the triggering check shall be performed in the VMSC (prior to sending MAP RCH), rather than in the GMSC (during forwarding call leg establishment). The rationale is twofold:

(1) *Signalling efficiency* – when the forwarding condition is met in the VMSC, the destination number and basic service for the forwarded call are known. Hence, the VMSC may perform the triggering check. If the triggering check indicates that triggering shall take place, then VMSC includes O-CSI in MAP RCH. The GMSC will apply O-CSI to the forwarded leg, without checking criteria. If the triggering check indicates that no triggering takes place, then VMSC omits O-CSI from MAP RCH. As a result, the GMSC will not apply O-CSI to the forwarded leg.
(2) *SCCP message length* – the MAP RCH message may exceed a size limit when it contains O-CSI trigger criteria. MAP RCH in GSM R97 uses application context (AC) v3. For this

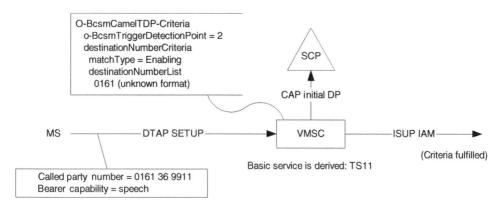

Figure 4.55   Conditional triggering for an MO call

version of RCH, no SCCP segmentation is specified. As a result, the VMSC does not have the possibility to transfer the RCH argument in a segmented MAP message. The size of the RCH argument is therefore limited and so the O-CSI criteria cannot be included. GSM R98 and onwards specify a segmentation mechanism for MAP RCH, using AC v4. However, for consistency reasons, when MAP RCH with AC v4 is used, the VMSC still performs the O-CSI triggering check.

In the example in Figure 4.56, MAP RCH contains no O-CSI, because the triggering check in the VMSC indicated that no triggering takes place for a forwarded call.

4.3.9.2 Mobile Terminated Calls

The trigger criteria for T-CSI consist of a set of (maximum five) basic service codes. The basic service codes in the trigger list may relate to a specific basic service or a basic service group. This conditional triggering mechanism allows an operator to trigger a CAMEL service for voice calls only or for data calls only, for example (e.g. calls with Bearer Service 30, for video telephony).

The trigger criteria associated with T-CSI are checked internally in HLR. They are not transported to the GMSC during terminating call handling. Therefore, MAP SRI Res does not contain T-CSI

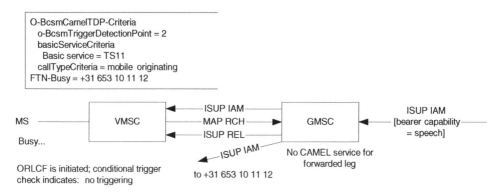

Figure 4.56   Conditional triggering check for ORLCF

trigger criteria (GSM TS 09.02 [54] R97).[40] The reason is that the basic service that is requested for a call is known in the HLR, at the time of terminating call handling.

The basic service that is requested for the call is derived by the HLR from the BC, LLC and HLC. BC, LLC and HLC are included in MAP SRI, sent from GMSC to HLR.

Figure 4.57 shows an example of conditional triggering check for a mobile terminating call. If the result of the trigger check indicates that triggering shall take place, then HLR will apply the normal CAMEL terminating call handling, which entails sending T-CSI to the GMSC. If the result of the trigger check indicates that no triggering will take place, then HLR will continue call processing without any CAMEL handling for the terminating part of the call.

Should a call forwarding condition apply for this call, then HLR may still send O-CSI to the GMSC, including trigger conditions, for the invocation of a CAMEL service for the forwarded leg.

4.3.10 USSD control

4.3.10.1 Introduction

Unstructured supplementary service data (USSD) is a generic control mechanism in the GSM network. USSD facilitates operator-specific message exchange between an MS and a service in the network. See Figure 4.58 for a node overview.

GSM specifications make a distinction between USSD functionality performed by the VLR and USSD functionality performed by the MSC. In practice, however, MSC and VLR are integrated in one node and the distinction may not be reflected in implementations of USSD.

USSD uses non-call-related messages over DTAP and MAP. USSD is specified by the following specifications:

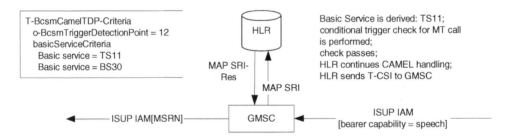

Figure 4.57 Conditional triggering check for mobile terminating call

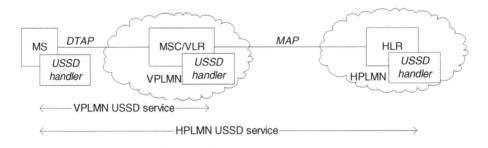

Figure 4.58 USSD network overview

---

[40] In CAMEL phase 3, T-CSI trigger criteria for MT call handling may be sent from HLR to GMSC; these trigger criteria are used for triggering at call set up failure events (3GPP TS 29.002 R99).

GSM TS 02.30 [5] general MMI requirements, including USSD aspects
GSM TS 02.90 [21] USSD service requirements and USSD MMI
GSM TS 03.38 [34] alphabets used for USSD
GSM TS 03.90 [46] technical implementation
GSM TS 04.08 [49] DTAP
GSM TS 04.80 [51] supplementary services-related signalling
GSM TS 04.90 [53] USSD signalling
GSM TS 09.02 [54] USSD MAP signalling

USSD is used to initiate the execution of a USSD handler instance. A USSD handler may be located in the MS, in the MSC/VLR or in the HLR (or external to the HLR). The USSD handler that receives a USSD service request executes the instruction as indicated in the service request and responds to the sender of the USSD service request.

Two USSD service categories exist: (1) MS-initiated USSD services; and (2) network-initiated USSD services.

4.3.10.2 MS-initiated USSD Services

This category of USSD services is initiated by entering a USSD command on the MS. A USSD command is a set of characters in accordance with a standardized man–machine Interface (MMI) format. The format of a USSD command is specified in GSM TS 02.90 [21]. When a USSD command is entered on the MS keypad, the MS recognizes the character string as a USSD command and starts a USSD process. Examples of USSD commands include:

*101*00 31 161 36 22 11# <SEND>

In this example, '101' is the USSD service code, e.g. a USSD callback request.[41] The characters '00 31 161 36 22 11' constitute the USSD service argument, e.g. a destination number. The MSC forwards this string to the HLR of the subscriber. The HLR initiates the USSD handler that is associated with USSD code 101.

*151*25# <SEND>

In this example, '151' is the USSD service code and '25' the USSD service argument. For this example, the MSC starts the USSD handler that is associated with USSD service code 151. This USSD service could be used to request the current cell Id.

The USSD service code indicates whether the USSD service request will be handled by the VPLMN or by the HPLMN. Examples include:

- 100–149 – HPLMN USSD service; the VMSC passes the USSD request on to the HLR of the served subscriber. If an operator wants to provide a USSD service to his subscribers when they are registered to other networks, then a service code in the range 100–149 should be used. Examples of MS-initiated USSD services include: pre-paid balance inquiry; pre-paid voucher update; and IMSI query.
- 150–199 – VPLMN USSD service; the VMSC may handle the USSD request locally. An operator may use this USSD service code range to offer a service to any registered subscriber.

A USSD-like service which is executed in the MS uses service code 06.

---

[41] USSD callback is a method of establishing calls by roaming pre-paid subscribers.

*#06#

This service code is used to instruct the MS to display the IMEI.

4.3.10.3 Network-initiated USSD

The HLR or a VLR may at any time initiate a USSD service session with the MS. The network includes a USSD feature code in the USSD message that is sent to the MS. The MS executes the requested USSD service. The USSD request from the network may entail the placing of a text string on the user's MS display.

Network-initiated USSD may originate from the HLR or from the VLR. With network-initiated USSD, the intended recipient of the USSD information flow is always the MS. For that reason, there are, unlike for MS-initiated USSD, no defined USSD feature code ranges to classify the recipient of network-initiated USSD.

Network-initiated USSD may have the function of communicating with the user or may be used to communicate with an application. Communicating with the user is standard functionality for USSD-supporting GSM phones. Other USSD applications must be pre-installed in the MS.

4.3.10.4 CAMEL Interworking with USSD

CAMEL phase 2 introduces the possibility for an operator to extend USSD signalling with a USSD interface between HLR and an external application. Effectively, the USSD handler is placed in an external node. The architecture for CAMEL interaction is represented by Figure 4.59.

Conceptually, the entity to which the USSD signalling may be extended is the gsmSCF. Practically, this external entity may also be another entity than a gsmSCF, such as pre-paid system, information system, service provisioning system, etc.

When the USSD flow between MS and HLR is extended to run between MS and gsmSCF, the HLR functions as a USSD relay, in the same manner that the MSC/VLR may function as USSD relay. The USSD flow between MS and gsmSCF applies to both MS-initiated USSD and network-initiated USSD. The USSD signalling between HLR and gsmSCF uses the same MAP messages as the USSD signalling between MSC/VLR and HLR.

4.3.10.5 MS to gsmSCF USSD Signalling

The MS to gsmSCF USSD signalling is a subscription-based service. An operator may offer a USSD service in the gsmSCF to selected subscribers or to the entire subscriber base. The CAMEL subscription information elements used for this purpose are the *U-CSI* and the *UG-CSI*.

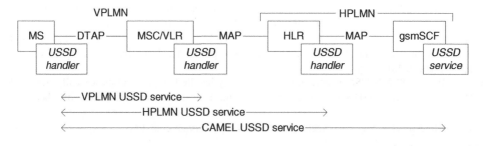

Figure 4.59   CAMEL interaction with USSD

U-CSI

This is a list consisting of a set of USSD service identifiers. Each service identifier is a combination of USSD service code and gsmSCF address (Table 4.7). U-CSI may be provisioned for individual subscribers. The USSD code in U-CSI corresponds to the USSD service that may be requested by a subscriber. The gsmSCF address represents the address of the gsmSCF. CAMEL does not specify the format of this address; it may be GT or signalling point code. CAMEL does not specify the length of U-CSI.

UG-CSI

UG-CSI has the same structure as U-CSI. Whereas U-CSI may be provisioned per subscriber, UG-CSI is generic for the HLR.

When a subscriber requests a HPLMN USSD service, then the HLR applies the following checks:

If *the requested USSD code indicates an HLR-internal USSD service,* **then** *execute this USSD service;*

Otherwise, if *the subscriber has U-CSI* **and** *the requested USSD code is contained in U-CSI,* **then** *forward the USSD service request to the gsmSCF assigned to the USSD code in U-CSI;*

Otherwise, if *UG-CSI is present and the requested USSD code is contained in UG-CSI,* **then** *forward the USSD service request to the gsmSCF assigned to the USSD code in UG-CSI;*

Otherwise *return an error to the MS.*

The above sequence of checks is recommended behaviour. Operators may choose to apply different USSD-related procedures.

USSD communication from MS to gsmSCF is started with the MAP message process unstructured SS request. The gsmSCF responds with process unstructured SS request ack (see Figure 4.60). When the MS receives the process unstructured SS request ack, the USSD session is closed.

Instead of responding directly to the process unstructured SS request, the gsmSCF may send notification(s) to the MS or may request further input. Here, the gsmSCF initiates a MAP unstructured SS request message or a MAP unstructured SS notify message to the MS. The MS responds with unstructured SS request ack or unstructured SS notify ack. The gsmSCF may subsequently conclude the USSD session by sending the process unstructured SS request ack to the MS.

Table 4.7   U-CSI structure

U-CSI	
USSD code	gsmSCF address
< USSD code1 >	< address1 >
< USSD code2 >	< address2 >
. . .	. . .

Figure 4.60   Elementary message sequence for MS-initiated USSD session

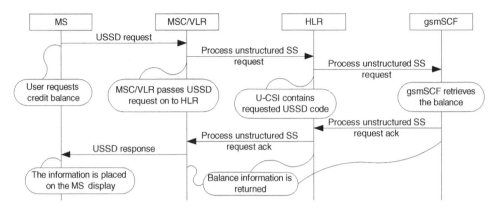

Figure 4.61   MS-initiated USSD service; single request

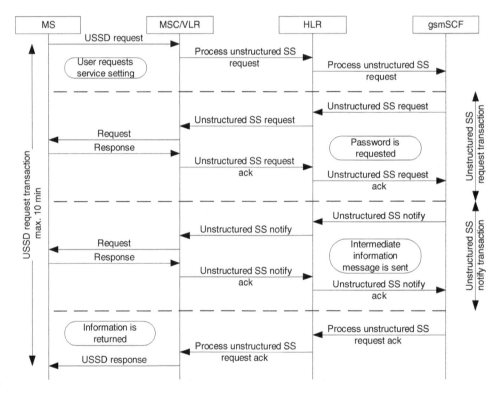

Figure 4.62   MS-initiated USSD service; further information requested

Figures 4.61 and 4.62 represent two examples of USSD signalling flows between MS and gsm-SCF: Figure 4.61, MS requests information from gsmSCF; Figure 4.62, MS requests information from gsmSCF, gsmSCF requests further input from MS.

The gsmSCF may apply the unstructured SS request transaction and the unstructured SS notify transaction zero or more times. However, whilst the gsmSCF is requesting further information from the MS or sending notifications, the initial process unstructured SS request MAP message from the MS is still pending. That is to say, the gsmSCF has not yet responded to that request. The MS-initiated USSD session as a whole is therefore limited to the operation time associated with process unstructured SS request. This operation timer is defined as 10 min (GSM TS 09.02 [54]).[42]

4.3.10.6 gsmSCF to MS USSD Signalling

For USSD signalling from gsmSCF to MS, no subscription data in HLR is required. The gsmSCF may at any time start a USSD session with an MS. The gsmSCF sends a USSD service request to the HLR, which passes the USSD service request on to the MSC/VLR (using the VLR address stored in the HLR). The MSC/VLR passes the USSD service request on to the MS.

The elementary message sequence for gsmSCF-initiated USSD signalling consists of the MAP unstructured SS request message and the MAP unstructured SS notify message (Figure 4.63).

The gsmSCF is responsible for releasing the USSD session when it has received the acknowledgement to the request or the notification. The gsmSCF may use one or more USSD transactions within a single USSD session. Figure 4.64 presents an example sequence flow for the gsmSCF-initiated USSD session.

When the gsmSCF initiates a USSD session with the MS, it may include an alerting pattern in the initial USSD message. In this manner, an operator may associate distinctive audible warnings with its network-initiated USSD services.

4.3.10.7 USSD Phases

Two USSD phases exist:

- *USSD phase 1* – this USSD phase is specified in GSM phase 1. USSD phase 1 uses MAP v1 messages. USSD phase 1 is the only feature in CAMEL that uses MAP v1 signalling with the SCP.
- *USSD phase 2* – this USSD phase is specified in GSM phase 2 onwards. USSD phase 2 uses MAP v2 messages. USSD phase 2 is the only feature in CAMEL that uses MAP v2 signalling with the SCP.

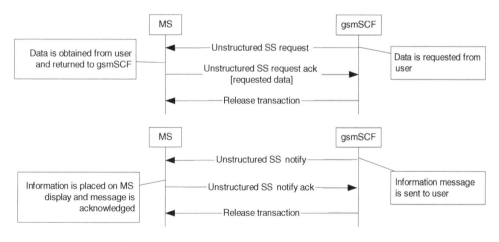

Figure 4.63 Elementary message sequence for gsmSCF-initiated USSD session

---

[42] For most MAP Operations, a value *range* is specified; 'process unstructured SS request' is an exception, in the sense that a *fixed* Operation timer value is specified.

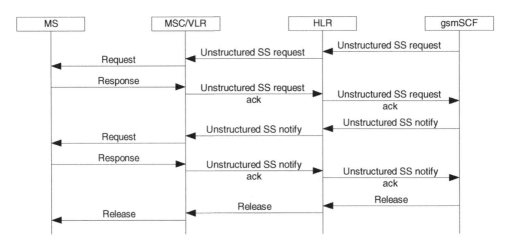

Figure 4.64   gsmSCF-initiated USSD session

The MAP messages that are mentioned in the present section relate to USSD phase 2. For USSD phase 1 signalling, refer to GSM TS 03.90 [46] and GSM TS 09.02 [54]. gsmSCF-initiated USSD may be used with USSD phase 2 only.

4.3.10.8 Subscriber Identification for USSD

Subscriber identification for USSD may be done with MSISDN or with IMSI, depending on the initiating entity.

- *gsmSCF-initiated USSD* – the USSD messages unstructured SS request and unstructured SS notify, both sent from gsmSCF to HLR, contain the subscriber's IMSI.
- *MS-initiated USSD* – the USSD message process unstructured SS request contains IMSI and optionally MSISDN. Since USSD signalling between MSC and HLR is based on IMSI, the IMSI can always be included in USSD message relay from HLR to gsmSCF. As an implementation option, U-CSI and UG-CSI could contain an additional field per service entry (i.e. the service entry containing USSD code and gsmSCF address). This additional field could contain an indication whether the subscriber's MSISDN should be included in process unstructured SS request.

4.4 Subscription Data

CAMEL Phase 2, analogous to CAMEL phase 1, makes use of CAMEL subscription information (CSI). Each CAMEL phase 2 feature has an associated CSI. See Table 4.8 for an overview of the CAMEL Phase 2 CSIs.

The exact structure of each CSI is specified in GSM TS 09.02 [54], by means of ASN.1 definition. Exceptions are U-CSI and UG-CSI; these elements are not specified in GSM TS 09.02 [54]. The reason is that U-CSI and UG-CSI are defined for use in the HLR only; they are not included in any MAP message flow. The trigger criteria for DP terminating attempt authorized in T-CSI are not specified in ASN.1 either, since these criteria are used in the HLR only. T-CSI criteria for Terminating Attempt Authorized are introduced in ASN.1 in 3GPP R99 for Any Time Subscription Interrogation; see Section 5.8. TIF-CSI is specified in GSM TS 09.02 [54] in GSM R98. The use of the various CSIs is specified in GSM TS 03.78 [38].

Table 4.8 CAMEL phase 2 subscription information elements

Element	Description	Entities in which it is used	CAMEL phase[a]	CAMEL capability[b]
O-CSI	Originating CSI	MSC, GMSC	Phase 1	Defined by CCH
T-CSI	Terminating CSI	GMSC	Phase 1	Defined by CCH
SS-CSI	Supplementary service CSI	MSC	Phase 2	Phase 2
TIF-CSI	Translation information flag CSI	HLR, MSC (GSM R98)	Phase 2	Phase 2
U-CSI	USSD CSI	HLR	Phase 2	Phase 2
UG-CSI	USSD generic CSI	HLR	Phase 2	Phase 2

[a] This column indicates the CAMEL phase in which this CSI is introduced.

[b] This column indicates which capability set is used for this CSI. For O-CSI and T-CSI, the value of CCH in the CSI indicates which CAMEL protocol should be used. For the other CSIs, a generic feature handling is specified that applies to each CAMEL phase in which this CSI may be used, e.g. the handling of SS-CSI in the MSC for a CAMEL phase 3 subscriber is identical to the handling of SS-CSI in the MSC for a CAMEL phase 2 subscriber.

4.4.1 Originating CSI

Originating CSI (O-CSI) may be used in the MSC/VLR for the invocation of a CAMEL service for MO and MF calls; O-CSI may also be used in the GMSC for the invocation of a CAMEL service for MF calls. The following two elements are introduced in O-CSI in CAMEL phase 2.

(1) *CAMEL capability handling* – from CAMEL phase 2 onwards, the O-CSI has a CAMEL phase associated with it, the *CAMEL capability handling* (CCH). When the MSC or GMSC uses O-CSI to invoke a CAMEL service, the value of CCH in O-CSI determines which CAMEL protocol will be used for that CAMEL service: CAMEL phase 1 or phase 2. An HLR may have internal data storage mechanisms that allow an operator to commission two CAMEL services, for MO and MF calls, for a subscriber: a CAMEL phase 1 service and a CAMEL phase 2 service. As an example, a subscriber may have a CAMEL phase 1 O-CSI (CCH = CAMEL1) and a CAMEL phase 2 O-CSI (CCH = CAMEL2). This dual O-CSI mechanism is used for fallback to a lower CAMEL phase where a CAMEL phase 2 subscriber registers in a CAMEL phase 1 network. When HLR sends CAMEL subscription data to VLR or GMSC, at the most one O-CSI may be sent to that node.

(2) *Trigger criteria* – an O-CSI may have trigger criteria associated with it.

4.4.2 Terminating CSI

Terminating CSI (T-CSI) is used for invocation of CAMEL service for MT calls by GMSC. From CAMEL phase 2 onwards, T-CSI also has a CCH parameter associated with it. When the GMSC uses T-CSI to invoke a CAMEL service, the CCH of T-CSI determines whether a CAMEL phase 1 or CAMEL phase 2 service should be invoked.

Similar to O-CSI, a subscriber may have multiple T-CSIs in the HLR. For T-CSI it is less likely that fallback to CAMEL phase 1 is required, since terminating call handling is normally performed in a GMSC in the HPLMN. The HLR may send no more than one T-CSI to the GMSC for that subscriber. From CAMEL phase 2 onwards, T-CSI may have trigger criteria associated with it.

4.4.3 Supplementary Service CSI

SS-CSI is part of the CAMEL subscription information in HLR and may be sent to the VLR. SS-CSI is used for supplementary services invocation notification (SSIN). Invocation of supplementary services is done by means of signalling between MS and the serving MSC. For that reason, SS-CSI may be sent to MSC only and not to GMSC.

4.4.4 Translation Information Flag CSI

Translation information flag CSI (TIF-CSI) may be used in the HLR to allow CAMEL subscribers to register a short FTNs whereby certain FTN number checks are not executed by the HLR. TIF-CSI has no data contents; it is a *flag* only.

MAP in GSM R97 does not specify the capability to transport TIF-CSI from HLR to VLR; TIF-CSI is used internal in the HLR only. However, due to the introduction of call deflection in GSM R98, TIF-CSI may be transported to VLR as from GSM R98 onwards. Refer to Section 4.7.11 for the interaction with TIF-CSI.

4.4.5 Unstructured Supplementary Service Data CSI

Unstructured supplementary service data CSI (U-CSI) may be used in the HLR for the purpose of USSD communication between HLR and gsmSCF. U-CSI consists of a set of data pairs; each data pair contains a USSD service code and a gsmSCF address. The number of data pairs contained in U-CSI is not specified by CAMEL; implementers may decide on a suitable length of U-CSI.

4.4.6 USSD Generic CSI

USSD generic CSI (UG-CSI) may be used for the same purpose as U-CSI. Unlike all other CAMEL subscription elements (O-CSI, T-CSI, SS-CSI, TIF-CSI, U-CSI), UG-CSI is not subscriber-specific. Rather, it is generic in the HLR; UG-CSI may be used for all subscribers.

UG-CSI has the same structure as U-CSI, i.e. a set of data pairs, each data pair containing a USSD Service Code and a gsmSCF Address.

4.5 Basic Call State Model

The basic call state models (BCSM) of CAMEL phase 2 have more capability than the BCSMs of CAMEL phase 1. The principles of the BCSM and the use of BCSMs in CAMEL are described in Chapter 2. The process of establishing a circuit-switched call with CAMEL phase 2 control is no different from the process of establishing a circuit-switched call with CAMEL phase 1 control. The BCSMs in CAMEL phase 2, however, specify additional DPs, where interaction may take place between the gsmSSF and the gsmSCF. These additional DPs allow for tighter control over the call and additional capability, like follow-on calls.

4.5.1 Originating Basic Call State Model

Figure 4.65 depicts the O-BCSM. For each MO call and for each MF call, an instance of the O-BCSM is invoked in the serving MSC or GMSC.

In CAMEL phase 1, the gsmSCF has limited capability to alter the basic call handling as defined by the BCSM. The only way in which a CAMEL phase 1 gsmSCF may alter the basic call handling is by force-releasing the call during establishment or during the active phase of the call. When a CAMEL service provides alternative call information during call establishment, this does not constitute alteration of the basic call handling. The BCSM will in that case still follow the state transitions that are defined for basic call handling.

A CAMEL phase 2 gsmSCF may manipulate the call handling. The gsmSCF may do this by inducing follow-on calls. Follow-on calls may be induced when call establishment fails or when an active call is disconnected by the called party.

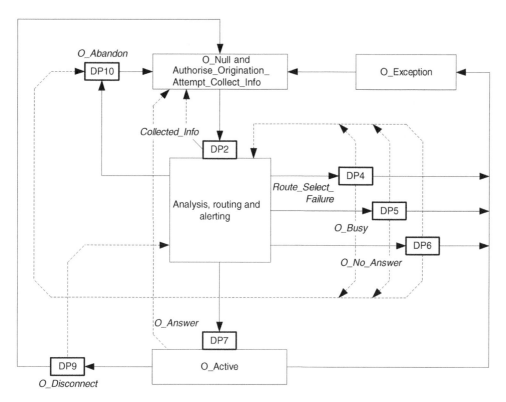

Figure 4.65 CAMEL phase 2 O-BCSM. Reproduced from GSM TS 03.78, v6.11.1, figure 3, by permission of ETSI

In addition, the gsmSCF may suspend call processing when a certain point in call is reached. An example of this is the case whereby call establishment is suspended at the beginning of the call, to play an announcement to the calling subscriber. The BCSM makes a distinction in *basic call transition* and *transition beyond basic call.*

- *Basic call transitions* – these transitions are in accordance with the BCSM. If gsmSCF does not manipulate the call handling process, then the BCSM follows the basic call transitions. These transitions are indicated with a solid line in the BCSM. An example is the state transition from DP O_Disconnect to PIC O_Null. This state transition represents the normal call flow when the called party disconnects from the call.
- *Transition beyond basic call* – these transitions are followed by the BCSM under instruction from the gsmSCF. The gsmSCF forces the BCSM to continue in a particular PIC. These transitions are indicated with a dotted line in the BCSM. An example is the state transition from DP O_Busy to PIC analysis, routing and alerting.

The DP numbers in the BCSM are used for identification purposes only.

4.5.1.1 Detection Points

The call establishment failure DPs are added in CAMEL phase 2. In CAMEL phase 1, call establishment failure is signalled to the gsmSCF by means of CAP dialogue abort, since the CAMEL phase 1 O-BCSM has no DPs associated with these events and hence the gsmSSF has no means of informing the gsmSCF about the call establishment failure. In CAMEL phase 2, these failure events can be reported to the gsmSCF in a proper manner. The principles of arming and reporting of detection points are described in Chapter 2.

The following enhancements are made to the DP definitions in the O-BCSM.

Collected_Info (DP2)

In CAMEL phase 2, the time that the O-BCSM may be in DP Collected_Info is much longer than in CAMEL phase 1. This is caused by the user interaction that may take place at this point. Although the user interaction may entail the exchange of various CAP operations between the gsmSSF and the gsmSCF and various state transitions in the gsmSSF FSM, the O-BCSM remains in this initial DP Collected_Info state.

The MSC may in fact not be aware of this CAMEL interaction. User interaction at the beginning of the call introduces call establishment delay. When the MSC establishes the call and hands the control of the call over to the gsmSSF, it starts the internal T7 timer (T7 timer is used when waiting for ISUP ACM). The increased call establishment time could have the effect that T7 expires, since it will take longer before the MSC sends ISUP IAM and receives ISUP ACM. To prevent T7 expiry as a result of user interaction, the gsmSSF shall generate an 'early ACM' when user interaction is applied. This early ACM has the effect that: the T7 timer is stopped; and T9 is started – the MSC now starts waiting for an answer. Service designers should therefore take care not to use long announcements at the beginning of the call. The use of the T7 and T9 timers is defined in ITU-T Q.764 [138].

O_Busy (DP5)

When the called subscriber is busy, the BCSM will transit to the O_Busy DP. The busy condition may be reported to the gsmSCF. Depending on the way in which the O_Busy DP is reported (EDP-N, EDP-R) the gsmSCF may take action, such as creating a follow-on call or releasing the call. The busy event that is reported may be caused by NDUB or UDUB.

The O_Busy DP is also used to signal the not reachable condition to the gsmSCF. The distinction between the busy condition and the not reachable condition is made through the *busyCause* parameter in the *EventSpecificInformationBCSM* in the event report BCSM operation.

If the busy event is the result of the reception of an ISUP REL, then the *busyCause* is in principle a copy of the cause code received in the ISUP release message. The *busyCause* reported to the gsmSCF does not include the diagnostics information that may be received in ISUP REL. Refer to ITU-T Q.850 [142] for an overview of the *busyCause* values that may be reported to the gsmSCF.

The reported *busyCause* may be used by the gsmSCF for internal processing, e.g. to decide whether a follow-on call shall be created. If the gsmSCF creates a follow-on call, then the ISUP REL is not propagated towards the calling subscriber. The MS will therefore not get a release indication. Instead, the gsmSCF may apply an audible forwarding indication to the calling subscriber.

The arming mode of DP O_Busy determines the action that may be taken by the service logic.

- *Transparent* – the busy event is not reported to the gsmSCF. The gsmSSF sends any pending charging reports and terminates the CAMEL dialogue.
- *EDP-N* – the busy event is reported to the gsmSCF, followed by any pending charging reports. The gsmSSF then terminates the CAMEL dialogue. The gsmSCF cannot generate a follow-on call in this scenario.
- *EDP-R* – any pending charging reports are sent to the gsmSCF, followed by the busy notification. The gsmSSF FSM transits to the waiting for instructions state. The gsmSCF may respond as follows:
  - ○ *CAP CUE* – by sending CAP CUE, the gsmSCF instructs the gsmSSF to continue clearing the call. The busy event is propagated to the calling party. The CAMEL dialogue is terminated.

- *CAP CON* – the sending of CAP CON results in the creation of a follow-on call. The busy event is not propagated to the calling party; the CAMEL dialogue may be retained, depending on the arming state of the DPs.
- *CAP RC* – by sending CAP RC, the gsmSCF instructs the gsmSSF to clear the call. This has similar effect to sending CAP CUE. With CAP RC, the gsmSCF may include a cause code. This cause code is used to replace the ISUP cause code in the backwards direction.

O_No_Answer (DP6)

If the called subscriber does not reply, the MSC where the O-BCSM is instantiated (VMSC or GMSC) will receive the ISUP REL indicating the no reply condition (ISUP cause = 19). As a result, the O-BCSM will transit to the O_No_Answer DP. The gsmSCF may decide to generate a follow-on call. The further handling of the O_No_Answer DP is similar to the handling of he O_Busy DP.

The gsmSSF does not include the ISUP cause in the No_Answer notification to the gsmSCF. The rationale is that the O_No_Answer DP may be caused only by ISUP cause 19 ['no answer from user (user alerted)']. A minor shortcoming of the absence of cause in the O_No_Answer notification is the following. The ISUP cause contains two main information fields: (1) *cause value* – this field contains the ISUP release cause; (2) *location* – this field contains an indication of the network entity in which the release was initiated. Hence, when the no answer event is reported, the gsmSCF does not get the location value. In addition, the release cause in CAP v3 may also contain the ISUP diagnostics information. The diagnostics information for the no answer event cannot, however, be reported to the gsmSCF.

Route_Select_Failure (DP4)

When the MSC fails to select an outgoing route (trunk) for a call, the Route_Select_Failure event occurs. The event may occur internally in the MSC or occur in a next exchange; in that case, the event is signalled over ISUP. The BCSM transits to the Route_Select_Failure DP. The Route_Select_Failure condition may be reported to the gsmSCF as EDP-N or EDP-R, or may not be reported, depending on the arming condition for that event. The gsmSCF may generate a follow-on call or may release the call. The cause that is received in ISUP REL is included in the event report to the gsmSCF. The gsmSCF may use the cause code for internal processing. The further handling of the Route_Select_Failure DP is similar to the handling of he O_Busy DP.

O_Abandon (DP10)

The O_Abandon event indicates that the calling party releases the call before the call reaches the active phase. This event may occur whilst call establishment is ongoing (BCSM is in PIC O_Null) or during user interaction at the beginning of the call (BCSM is in DP Collected_Info). The gsmSSF does not include a cause code in the abandon notification.

The abandon event cannot be armed as EDP-R.[42] When the abandon event is reported to the gsmSCF, this will result in immediate termination of the CAMEL dialogue. The gsmSCF does not have the possibility at this point of including free format data in the CDR of the call.

O_Answer (DP7)

In CAMEL phase 1, the answer event may be armed as EDP-N only; in CAMEL phase 2, it may also be armed as EDP-R. Arming the answer event as EDP-R may be required to provide the gsmSSF with charging instructions before the answer event is propagated to the calling party.

When the answer event is reported as EDP-R, the gsmSCF needs to reply with CAP CUE. When the event is reported as EDP-N, the gsmSCF is notified but does not need to respond. The gsmSSF propagates the answer event immediately to the calling party.

---

[42] The capability to arm the abandon event as EDP-R is introduced in CAMEL phase 3.

O_Disconnect (DP9)

The O_Disconnect DP may be armed on leg 1 and leg 2 individually. The capabilities for the SCP, when the O_Disconnect is reported, are as follows:

O_Disconnect on Leg 1

- EDP-N – the event is reported, followed by any pending charging reports. The disconnect event is propagated in the forwards direction (i.e. to the called party); the gsmSSF closes the CAMEL dialogue.
- EDP-R – the gsmSSF sends CAP ACR (if pending), followed by CAP CIRp [leg1] (if pending) and then followed by the event notification. The gsmSCF may respond with:
  ○ CAP CUE – this has the effect that the disconnect event is propagated to the called party.
  ○ CAP RC – this also has the effect that the disconnect event is propagated to the called party, but the gsmSCF may provide a different cause code that will be used in the ISUP REL to the called party.

CAP CIRp [leg2] is sent after the reception of CAP CUE or CAP RC. User interaction is not possible at that this point, since the calling party has already disconnected from the call.

O_Disconnect on Leg 2

- EDP-N – the event is reported, followed by any pending charging reports. The disconnect event is propagated in the backwards direction (i.e. to calling party); the gsmSSF closes the CAMEL dialogue.
- EDP-R – the gsmSSF sends CAP ACR (if pending), followed by CAP CIRp [leg2] (if pending) and then followed by the event notification. The gsmSCF may respond with:
  ○ CAP CUE – this has the effect that the disconnect event is propagated to the calling party. The gsmSSF sends CAP CIRp for leg 1, if pending.
  ○ CAP RC – this also has the effect that the disconnect event is propagated to the calling party, but the gsmSCF may provide a different cause code that will be used in the ISUP or DTAP signalling to the calling party. The gsmSSF sends CAP CIRp for leg 1, if pending.
  ○ CAP CON – using the CON operation results in the creation of a follow-on call.
  ○ User interaction – before the gsmSCF creates a follow-on call, an announcement may be played or information may be collected from the user. The charging reports related to the disconnected leg have already been sent to the gsmSCF. The gsmSCF may use dedicated charging operations for the user interaction. For the follow-on call itself, new charging operations may be sent.

No-reply Timers

The No_Answer event is the only event that may be armed with an event condition. When the gsmSCF arms the No_Answer event, it may provide a no reply timer value to the gsmSSF. The gsmSSF uses this timer value to monitor the answer of the called party. This gsmSCF-supplied timer functions as an gsmSSF-internal timer and is complementary to the MSC-internal no reply timer, the T9 timer, which is described in ITU-T Q.764 [138]. Figure 4.66 depicts which processes may generate a no reply event, i.e. no answer DP in the O-BCSM.

The following no reply events may occur for an MO call.

- *ISUP REL* – ISUP REL with cause code 19 (1) is received from the remote exchange or from an intermediate exchange. The ISUP REL results in the occurrence of the No_Answer event in the gsmSSF.
- *T(No Reply) expiry* – the gsmSCF has supplied a no reply timer to the gsmSSF. When this no reply timer expires, an ISUP REL is sent towards the destination exchange (2b) and the No_Answer event is generated in the BCSM (2a).
- *T9 expiry* – the MSC starts the T9 timer when it receives ISUP ACM from the remote exchange or from another process internal in the MSC. When T9 expires, the MSC generates a DTAP Release towards the MS (3a) and generates an internal ISUP Release in the forwards direction

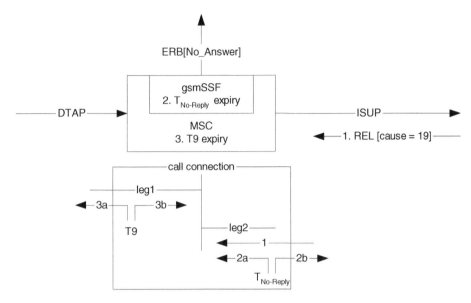

Figure 4.66   No reply causes

(3b). This ISUP REL is associated with leg1. As a result, the T9 expiry does not lead to the occurrence of the No_Answer event but to the occurrence of the O_Abandon event.

The No-Reply timer that may be provided by the gsmSCF has a value range of 10–40 s. The T9 timer in the MSC has a recommended value range of 90–180 s, as defined in ITU-T Q.118 [132]. When the gsmSCF does not provide a No Reply timer, the gsmSSF may, as a vendor's option, use a default value for the no reply timer. This default value has a recommended range of 10–180 s.

Source of Call Establishment Failure
The call establishment failure for an MO call may be caused by various events inside or outside the MSC. The respective call establishment failure events are mapped on a corresponding DP in the O-BCSM. The ISUP cause code is used to select the corresponding DP. Figure 4.67 depicts the various causes of call establishment failure.

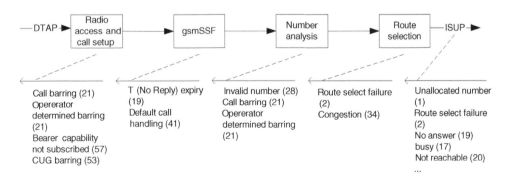

Figure 4.67   Causes of call establishment failure

The figures in brackets in Figure 4.67 refer to the ISUP cause code as defined in ITU Q.850 [142]. The exact usage of ISUP cause codes may differ per network and per vendor.

The following classification of call establishment failure can be made.

MSC-internal Event Before CAMEL Invocation
When the subscriber sets up an outgoing call, the MSC performs regular checks on the requested service. The MSC may determine that the call is not allowed due to call barring,[43] requested basic service not subscribed or CUG, for example.

MSC-internal Events after CAMEL Invocation
The number analysis and route selection processes that take place in the MSC may fail to establish the outgoing call. Figure 4.67 gives various possible failure causes. For these call cases, the MSC shall set the cause code (to be included in the backwards release message) to an appropriate value.

A sub-category of this class of events consists of the events that occur in the call handling process prior to the gsmSSF; these events are not reflected in Figure 4.67. These events are propagated in the forwards direction and are reported on the incoming leg in the gsmSSF, i.e. on leg 1. The resulting DP is O_Abandon.

These events may be:

- calling party abandon;
- T7 timer expiry; T7 (*'awaiting address complete timer'*) is the timer that is used when the MSC is waiting for ISUP ACM; this timer expires when ISUP ACM is not received within a predefined period (see ITU-T Q.764 [138] for more details of T7);
- T9 timer expiry; T9 ('awaiting answer timer') is the timer that is used when the MSC is waiting for ISUP ANM; this timer expires when ISUP ANM is not received within a predefined period (see ITU-T Q.764 [138] for more details of T9).

ISUP events
The MSC may receive ISUP REL before call answer. The cause code may indicate that the called subscriber is busy or not reachable, for example.

Any call establishment failure event that occurs after the gsmSSF and that is propagated in the backwards direction may have the effect that the O-BCSM transits to the corresponding DP. The relation between ISUP cause code and DP in the O-BCSM is presented in Table 4.9. When the gsmSCF receives an event report BCSM for DP O_Busy or Route_Select_Failure, it may evaluate the cause code to ascertain the real cause of call establishment failure.

Location Information in Cause Code
When an entity in the call establishment link releases an ISUP circuit, it generates the corresponding cause and includes a *location* in the cause. This location indicates to any receiver of the cause what kind of entity initiated the release of the circuit. A receiver of the cause may use the location value to decide on further action or may use it for statistics and diagnostics. The location is included in the call establishment failure indication to the SCP.

Table 4.9 Mapping between ISUP cause code and O-BCSM DP

ISUP cause	Description	DP in O-BCSM	Cause code
19	No answer from user (user alerted)	O_No_Answer	–
17, 20	User Busy, Subscriber Absent	O_Busy	Included in ERB
Other	refer to ITU-T Q.850 [142]	Route_Select_Failure	Included in ERB

---

[43] Refer to Section 4.7.9.

Figure 4.68 and Table 4.10 depict which location codes may be generated in various kinds of network elements. The dotted line for *private network* box in Figure 4.68 denotes that, for MO and MT calls in the GSM network, the MS is connected to the MSC, which is part of the public network. Hence, there is no private network involved for GSM access. 3GPP Rel-7 specifies CAMEL control for GSM access through PABX. In that case, the PABX is the private network.

The thick line for the *public network* box in Figure 4.68 denotes that the public network may be a PLMN. When an MSC generates an ISUP cause code, the location should therefore be set to *public network serving the local user* or *public network serving the remote user*.

The coding of the various location values is defined in ITU-T Q.850 [142].

4.5.2 Terminating Basic Call State Model

Figure 4.69 depicts the T-BCSM. For each MT call, an instance of the T-BCSM is invoked in the GMSC. In CAMEL phase 2, there are two ways in which the gsmSCF may manipulate the call flow as defined by the T-BCSM: (1) inducing call forwarding; and (2) generating follow-on call.

4.5.2.1 Call Forwarding

The gsmSCF has the capability to induce early call forwarding in the GMSC. The forwarding types that may be induced by the gsmSCF include:

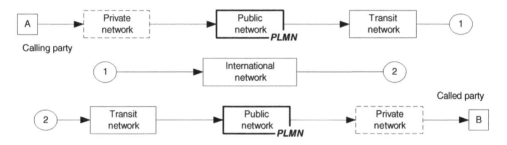

Figure 4.68   Generation of location codes for ISUP cause

Table 4.10   ISUP location codes for network elements

Network element	Permissible location code(s)	Comment
Private network	*Private network serving the local user, Private network serving the remote user*	
Local network	*Public network serving the local user, Public network serving the remote user*	The value *Public network serving the local user* may normally be generated in an originating MSC; the value *Public network serving the remote user* may normally be generated in a GMSC or in a terminating MSC
Transit network	*Transit network*	
International network	*International network*	
Called party	*User*	

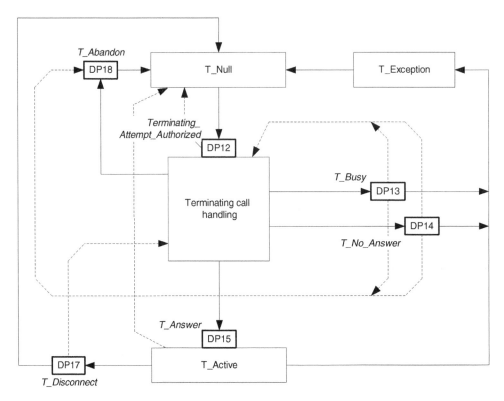

Figure 4.69   CAMEL phase 2 T-BCSM. Reproduced from GSM TS 03.78 v6.11.1, figure 4, by permission of ETSI

- *Unconditional* – when the gsmSCF receives CAP IDP operation, it may send CAP CON, containing a destination routing address that is different from the called party number that the gsmSCF received in CAP IDP. The call is now forwarded to the destination indicated by the destination routing address. If GSM call forwarding – unconditional is active for the called subscriber, then the gsmSCF-induced call forwarding overrides the GSM call forwarding. When the gsmSCF has induced call forwarding at this point, the T-BCSM follows its normal state transitions.
- *Not reachable (MS detached)* – if the subscriber is (temporarily) detached from the VMSC, then the HLR will fail to obtain an MSRN for this call. The failure to obtain an MSRN from the VLR is reported to the gsmSCF through the T_Busy event, with busyCause indicating *not reachable*. In addition, if the HLR provides an FTN to the GMSC, then the T_Busy event notification contains also the *callForwarded* parameter. The gsmSCF may now induce call forwarding by sending CAP CON containing a destination routing address that is different from the called party number that the gsmSCF received in CAP IDP. Here as well, the gsmSCF-induced call forwarding overrides GSM based call forwarding.

The unconditional call forwarding and the not reachable (MS detached) call forwarding are early call forwarding categories; these kinds of call forwarding occur in the GMSC. These call forwarding types may therefore be induced by the gsmSCF through the terminating CAMEL dialogue. The late call forwarding categories [no reply, busy and not reachable (no paging response)] occur in the VMSC. If any of these call forwardings occur, then the CAMEL service that is controlling the call through the GMSC is not informed about this.

One method for the gsmSCF to induce late call forwarding is to suppress the GSM call forwarding. A means to suppress the late call forwarding is setting the *forwarding counter* to the maximum value. The forwarding counter is an information element that may be carried in the ISUP information flow during call establishment. This counter is incremented every time a call is forwarded. If a gsmSCF sets the forwarding counter to the maximum value, then no further call forwarding will take place for this call.[44] Hence, if the called party is busy, for example, then the VMSC will not initiate call forwarding, even if the subscriber has call forwarding busy registered in the VMSC.

Instead, the failed call establishment results in an ISUP REL to be sent back to the GMSC. The GMSC reports the call release to the gsmSCF with the T_Busy event. The *callForwarded* parameter will not be present in the T_Busy event notification. The gsmSCF may now induce call forwarding.

4.5.2.2 Follow-on Call

The generation of follow-on calls from the T-BCSM follows the rules that apply to the follow-on calls from the O-BCSM. When the called party has disconnected from the call, the gsmSCF may create a follow-on call.

4.5.2.3 Detection Points

The main addition to the T-BCSM in CAMEL phase 2 consists of the call establishment failure DPs.

Terminating_Attempt_Authorized (DP12)
The handling of DP Terminating_Attempt_Authorized in CAMEL phase 2 may take longer than in CAMEL phase 1 due to user interaction. When the gsmSCF applies user interaction for MT call handling, the gsmSSF will generate an ISUP ACM in the backwards direction. The rationale of this ISUP ACM is twofold:

(1) Opening up the speech path in the backwards direction; in order for any announcement device to transmit speech to the calling party, an ISUP ACM will traverse through the entire ISUP link towards the calling party. The ISUP ACM indicates the availability of in-band information.
(2) Stopping the T7 timer in the originating MSC.

T_Busy (DP13)
T_Busy DP in T-BCSM is used for reporting the busy event and the not reachable event. The busyCause in the event notification indicates which event has occurred.

Any other call establishment failure event, except no reply, is also reported through the T_Busy DP. The T-BCSM does not have a dedicated Route_Select_Failure DP. The rationale is that the call leg between GMSC and VMSC is based on the MSRN that is allocated by the VMSC. That leg should therefore never lead to route select failure. However, GSM call forwarding or SCP-induced call forwarding may occur for the terminating call. The forwarding leg may lead to route select failure. As a result, the T_Busy DP may also report the route select failure event.

If T_Busy event notification contains the CallForwarded parameter, then the call establishment has not failed *yet*, but the call will be routed to another destination.

---

[44] The maximum value for the forwarding counter in ETSI ISUP differs from the maximum value for the forwarding counter in ANSI ISUP. ETSI ISUP has a 3-bit redirection counter (in redirection information), whereas ANSI ISUP has a 4-bit redirection counter. When a call is routed from an ETSI ISUP region (e.g. Europe) to an ANSI ISUP region (e.g. USA), then the forwarding counter may have the maximum value for ETSI ISUP, but since this is below the limit for ANSI ISUP, forwarding may still occur in the VMSC in USA. CAMEL phase 3 has a transparent solution for this.

T_No_Answer (DP14)

The T_No_Answer DP is used only for reporting the no reply event. For that reason, there is no cause code associated with this event.

As is the case for the T_Busy event, if the T_No_Answer event notification contains the Call-Forwarded parameter, then the call establishment has not failed *yet*, but the call will be routed to another destination.

T_Answer (DP15)

The T_Answer event has the same function as the O_Answer event in the O-BCSM. If a charging instruction was sent to the gsmSSF prior to the answer event, then charging will now start. The answer event may be generated by the forwarded call. If the call forwarding took place in the GMSC, then the CAMEL service was already informed and would probably have stopped the charging already.

T_Disconnect (DP17)

The handling of the T_Disconnect event is similar to the handling of the O_Disconnect event. If disconnect is reported for leg2, then gsmSCF may create a follow-on call.

T_Abandon (DP18)

The handling of the T_Abandon event is similar to the handling of the O_Abandon event. The occurrence of T_Abandon may e.g. be caused by the calling party abandoning the call set up or by expiry of the No Reply timer or the T9 timer in originating MSC.

4.5.2.4 Failure Reasons

For the terminating call handling, there are additional sources of the call failure event. The reasons are: (1) terminating call handling is logically divided over (at least) two functional entities; (2) terminating call handling includes interaction with HLR; (3) ORLCF; refer to Figure 4.70.

4.5.2.5 MAP SRI Error or FTN

When the CAMEL service for terminating call is activated and has instructed the GMSC to continue call establishment, the GMSC re-interrogates the HLR for a MSRN. If the HLR fails to obtain a MSRN from the VLR, it returns a MAP SRI-Error or a FTN to the GMSC. The MAP error or the FTN is translated into the corresponding detection point for the T-BCSM and corresponding cause code, e.g.

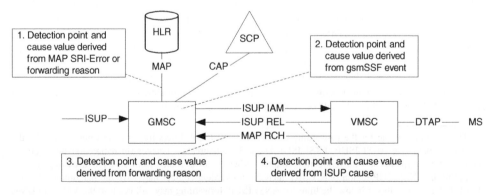

Figure 4.70 Failure reasons for terminating call

MAP Error "AbsentSubscriber" → DP T_Busy[cause = 20]

gsmSSF-internal Event
The gsmSSF internal event may be the expiry of the no reply timer. This results in DP T_No_Answer (no cause code included). A call diversion service that is monitoring the call establishment will take care of the following. When the terminating call is routed to the called subscriber in the VMSC, then the following no reply timers may be running in parallel:

(1) SSF-internal no-reply timer, loaded with a value received in the No_Answer event arming instruction;
(2) T9 timer in the VMSC, to monitor the answer from the called party, loaded with an MSC default value;
(3) NoReplyConditionTime in VMSC; the NoReplyConditionTime in VMSC may be part of the called subscriber's call forwarding subscription.

The service logic designer may want to ensure that the SSF-internal no-reply timer has the shortest setting, to force a No_Answer event in the GMSC if the called subscriber does not respond within a predefined time.

ORLCF
When the GMSC receives MAP resume call handling (RCH), then the *forwarding reason* carried in RCH is used to select the DP and cause code, e.g.

Forwarding reason = No Reply → DP T_No_Answer
Forwarding reason = Busy → DP T_Busy[cause = 17]

ISUP Release
When the GMSC receives ISUP REL, then the ISUP cause value is used to derive the DP in the T-BCSM. ISUP cause 19 is mapped to DP T_No_Answer. All other ISUP cause values are mapped to T_Busy.

4.6 CAMEL Phase 2 Relationship

Figure 4.71 reflects which CAMEL phase 2 operations may be used in the different CAMEL phase 2 relationships. The CAP operations are listed in Table 4.11. As follows from Figure 4.71, the different CAMEL relationships have different sets of CAMEL operations. Each relationship is established with a different application context (AC). The ACs that are used in CAMEL phase 2 are:

- *CAP-v2-gsmSSF-to-gsmSCF-AC* – this AC is used by the gsmSSF when it establishes a CAMEL relationship with the SCP, as a result of receiving an O-CSI or T-CSI from VLR or HLR respectively. This relationship is started with the initial DP operation.
- *CAP-v2-assist-gsmSSF-to-gsmSCF-AC* – this AC is used by the assisting gsmSSF when it establishes an assisting dialogue. This relationship is started with CAP ARI.
- *CAP-v2-gsmSRF-to-gsmSCF-AC* – this AC is used by an intelligent peripheral when it establishes an assisting dialogue. This relationship is started with CAP ARI.

4.6.1 CAP v2 operations

Table 4.11 contains an overview of the CAP v2 operations and a brief description per operation. The commonly used abbreviation is indicated in parentheses. The CAP operation flows marked [1] are introduced in CAMEL phase 2.

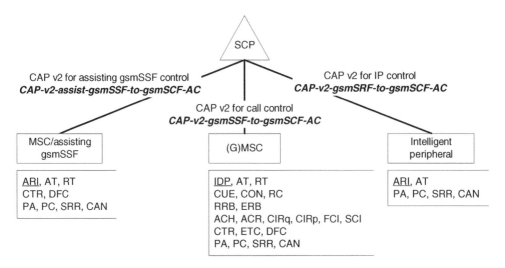

Figure 4.71   Overview of CAP operations per CAMEL relationship

4.7 Interaction with GSM Supplementary Services

Chapter 1 lists the various GSM supplementary services. The present section describes the interaction between CAMEL phase 2 and those supplementary services. For some supplementary services, the interaction has already been described in previous sections of this chapter.

4.7.1 Line Identification

Line Identification consists of the services CLIP, CLIR, COLP and COLR. Refer to GSM TS 02.81 [13] for a description of these services. The only service that CAMEL phase 2 may influence is CLIP. A CAMEL service may provide or modify the additional calling party number (A-CgPN) during call establishment. The A-CgPN is transported over the ISUP signalling flow and is used in the destination exchange of the called party to identify the calling party. When the destination exchange receives the ISUP IAM and the ISUP IAM contains an A-CgPN, then the destination exchange uses the A-CgPN in the DTAP (or DSS1) Setup message instead of the calling party number (CgPN).

A CAMEL service may use this method to replace the public E.164 number of the calling party by a VPN number. The called party receives the VPN number of the calling party on her display. Figures 4.72 and 4.73 present use cases of the A-CgPN.

The VPN service in Figure 4.72 has determined that the called party belongs to the same VPN group. Therefore, the calling party may be identified to the called party by means of her VPN number instead of her public number. The CAMEL service in Figure 4.73 is an MT call service controlling the call to the called party. In this example scenario, the service uses the location information that is present in CAP IDP to determine whether the called party will receive the A-CgPN. If the called party is currently located in a non-CAMEL network, then the service may replace the VPN A-CgPN with the calling party's public number. The rationale is that the called party would otherwise receive a CLI representing a VPN number. Should the called party return a call to that CLI, then the call establishment fails, since the CAMEL VPN service, which is required to translate the VPN number into the public number, cannot be called from that VPLMN.[45]

---

[45] The VPLMN in question may in fact support CAMEL, but no CAMEL agreement is in place between HPLMN and VPLMN.

Table 4.11 Overview of CAP v2 operations

Operation	Description
Activity test (AT)	This operation is used between gsmSCF and gsmSSF, assisting gsmSSF or intelligent peripheral. It is used to check for the existence of the CAMEL relationship
Apply charging (ACH)[1]	This operation is used to instruct the gsmSSF to apply on-line call duration control. It may be used for an outgoing call or during user interaction
Apply charging report (ACR)[1]	This operation contains the result of the on-line call duration instruction
Assist request instruction (ARI)[1]	This operation is used by the assisting gsmSSF or the intelligent peripheral to establish an assisting dialogue for user interaction
Call information report (CIRp)[1]	This operation contains the call-related information that was previously requested for this call
Call information request (CIRq)[1]	This operation is used to request call-related information
Cancel (CAN)[1]	This operation has a dual purpose: (1) It may be used to disarm armed detection points and to cancel requests for reports. It is normally used when a CAMEL service wants to terminate the relationship (2) It may be used to prevent or stop the execution of a user interaction operation, which was previously sent to the gsmSRF or to the intelligent peripheral
Connect (CON)	This operation is used to instruct the gsmSSF to continue call establishment with modified information. This operation may also be used to generate a follow-on call
Connect to resource (CTR)[1]	This operation is used to instruct the gsmSSF or assisting gsmSSF to connect the call to a specialized resource, for user interaction
Continue (CUE)	This operation is used to instruct the gsmSSF to continue call processing at the DP where call processing was suspended
Disconnect forward connection (DFC)[1]	This operation is used to terminate the connection to a specialized resource or to terminate a temporary connection to an assisting gsmSSF or intelligent peripheral
Establish temporary connection (ETC)[1]	This operation is used to establish a temporary connection between the serving (G)MSC and an MSC with assisting gsmSSF
Event report BCSM (ERB)	This operation is used by the gsmSSF to inform the gsmSCF about the occurrence of an event
Furnish charging information (FCI)[1]	The gsmSCF may use this operation to place service-specific data in the CDR for the call
Initial DP (IDP)	This operation is used by the gsmSSF to start a CAMEL service
Play announcement (PA)[1]	A CAMEL service may use this operation to instruct the gsmSRF or intelligent peripheral to play an announcement
Prompt and collect user information (PC)[1]	A CAMEL service may use this operation to instruct the gsmSRF or intelligent peripheral to play an announcement and to collect digits from the user
Release call (RC)	This operation is used by the gsmSCF to release a call
Request report BCSM (RRB)	This operation may be used by the gsmSCF to arm or disarm detection points in the BCSM
Reset timer (RT)[1]	This operation may be used by the gsmSCF to reload and restart the Tssf timer
Send charging information (SCI)[1]	This operation may be used by the gsmSCF to send advice of charge information to the served subscriber
Specialized resource report (SRR)[1]	This operation is used by the gsmSRF to inform the gsmSCF that the playing of an announcement is complete

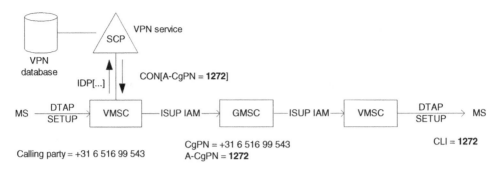

Figure 4.72 Setting the additional calling party number for VPN subscriber

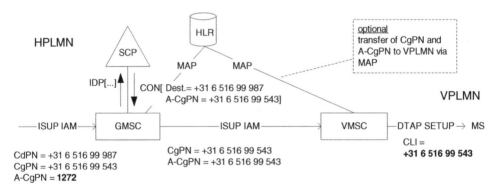

Figure 4.73 Restoring the additional calling party number

The CAMEL service cannot *remove* the A-CgPN from the ISUP message. Therefore, the A-CgPN may be set equal to the CgPN. The VMSC will still use the A-CgPN as CLI, but the A-CgPN now has a public number format.

CAMEL does not have the capability to modify the CgPN. The reason is that the CgPN will always be available in the ISUP information flow in an unmodified form. The CgPN is used, amongst other uses, for lawful intercept purposes. Refer to GSM TS 02.33 [8] for lawful intercept specifications.

An additional CLIP feature that is reflected in Figure 4.73 is the following. For the terminating part of a GSM call, the CgPN and A-CgPN are carried in the ISUP signalling between GMSC and VMSC. The international carrier may, however, not have the capability to convey these elements. In that case, the service level for the called party is negatively affected. To overcome this shortcoming, the CgPN and A-CgPN may be included in the MAP signalling between GMSC and HLR (MAP SRI) and in the MAP signalling between HLR and VMSC (MAP PRN), when requesting the MSRN. The VLR will receive the CgPN and A-CgPN. Hence, the VMSC can provide a CLI to the called party, even if the international carrier did not transport the CLI. This functionality is introduced in GSM R98.

4.7.2 Call Forwarding

The interaction between CAMEL and the various call forwarding scenarios and call deflection is described throughout the present chapter.

4.7.3 Explicit Call Transfer

An example of explicit call transfer (ECT) is depicted in Figure 4.74. ECT is specified in GSM TS 02.91 [22] and GSM TS 03.91 [47]. A subscriber may invoke ECT when she has two calls established: one call active and one call suspended. Invocation of ECT has the effect that (1) the active party and the suspended party are connected in speech and (2) the subscriber who had invoked the ECT is disconnected from the call. MMI commands for invoking ECT are described in GSM TS 02.30 [5]. When the execution of ECT is complete, the call party that had invoked ECT may establish or receive another call.

The individual calls in the ECT connection may be incoming or outgoing. In the example case in Figure 4.74, both calls are outgoing subject to CAMEL control. The CAMEL services are independent service instances. Although the two CAMEL services are invoked with the same O-CSI for this subscriber, the services may reside in different SCPs, as a result of dynamic load sharing (see Chapter 2). The CAMEL services that are involved in this call do not receive an indication about the ECT invocation. The MAP notification that results from SS-CSI in the subscriber's VLR profile is not meant for call control purposes and may be sent to a different node than the call control SCP.

When the A-party has a subscription to ECT supplementary service, the service logic shall take care with applying user interaction at the end of the call or with applying a pre-paid warning tone. When ECT was invoked for the call, the warning tone or announcement will be heard by the other connected call party, not by the served CAMEL subscriber. When one of the two parties disconnects from the call, this is reported as Disconnect[leg2] for the CAMEL service for that call and subsequently as Disconnect[leg1] for the other CAMEL service.

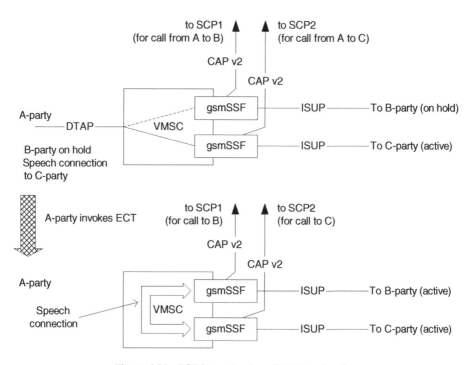

Figure 4.74 ECT invocation by a CAMEL subscriber

Ideally, CAP would contain a service notification, to inform the respective CAMEL services that ECT is invoked for this call. Such functionality, in the form of a designated detection point in the BCSM, was proposed for CAMEL phase 4, but was eventually not included in the standard.

4.7.4 Call Waiting

Call Waiting (CW) and Call Hold (CH) are specified in GSM TS 02.83 [15] and GSM TS 03.83 [41]. There is no direct interaction with the CW supplementary service. When a called subscriber is busy, then the VLR may still allocate an MSRN (for the second call to the called subscriber) when MAP PRN is received for that subscriber. A CAMEL service that is controlling the second incoming call at the GMSC, does not know that the call is offered as a waiting call. (Figure 4.75).

The second call may be accepted or rejected by the called subscriber or the no reply condition may occur. If the called subscriber has call forwarding active, then the rejected or timed-out waiting-call may lead to call forwarding, possibly in combination with optimal routing.

4.7.5 Call Hold

A subscriber who is engaged in a speech call may place the active call party on hold and establish a second call or accept an incoming call. In the latter case, CH is used in combination with CW (Figure 4.76). The active call in which the A-party is involved may be an incoming call or an outgoing call. In either case, a CAMEL service may be controlling the active call. In the example in Figure 4.76, A-party has an active call to B-party and receives a call from C-party. The invocation of Call Hold for the active call is not indicated to the CAMEL service for that call. Hence, the period during which the call is suspended is not taken cognizance of with respect to on-line charging of the call. This applies to both the call duration reported in the CAP ACR and the call duration reported in CAP CIRp.

Although the invocation of the call hold service may be recorded in an Event Record, the suspended duration is not reflected in the CDR for the call. Refer to GSM TS 12.05 [57]:

> The use of the call hold service shall be recorded either in-line in the appropriate call record or in a separate supplementary service "invocation" record as described above. For the avoidance of doubt, the duration for which the call is held, i.e. is inactive, is not recorded.

If the CAMEL pre-paid warning tone is played during suspended state, then the served subscriber will not hear the warning tone. As was suggested for ECT, CAMEL would have benefited from a notification at invocation of call hold.

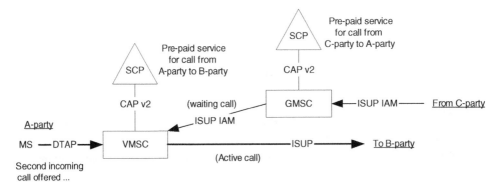

Figure 4.75 Call waiting in combination with CAMEL

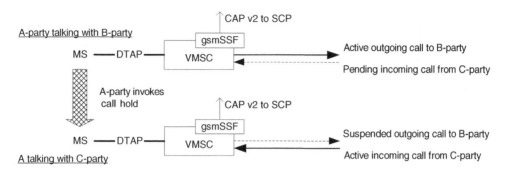

Figure 4.76   Call hold in combination with CAMEL

4.7.6 Completion of Calls to Busy Subscribers

Completion of calls to busy subscribers is specified in GSM TS 02.93 and GSM TS 03.93. When call establishment to a called party fails due to the called party being busy, the calling party may request CCBS, provided that CCBS is supported by the network. CCBS is requested by entering a designated digit string on the keypad, when receiving the busy tone. The MSC starts a process to monitor the state of the busy subscriber. This process involves also the HLR of the calling party, the HLR of the called party and the serving MSC of the called party. When the called party becomes available, the calling party receives an audible notification. The calling party may accept the notification, which results in the establishment of a call (a 'CCBS call') to that called party. If the calling party has O-CSI in the VLR, then the CCBS call may be subject to regular CAMEL handling. The CAMEL service does not know that this is a CCBS call.

4.7.7 Multi-party

Multi Party (MPTY) is specified in GSM TS 02.84 [16] and GSM TS 03.84 [42]. MPTY allows a subscriber to build a conference call with up to five remote parties. Figure 4.77 presents an example where a calling party adds two additional outgoing call connections to the call. The four call parties are in speech connection with one another.

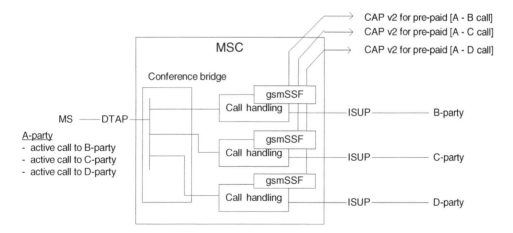

Figure 4.77   Multi-party call with CAMEL control

Each outgoing call may be subject to CAMEL control (based on O-CSI) for the calling party. The individual CAMEL services do not know that the call they are controlling is part of a multi-party call. If one outgoing call is released as a result of Tcp expiry, for example, then the multi-party may continue with the remaining call parties connected. If the CAMEL warning tone is played by the gsmSSF for the call to the D-party, then all other parties connected to the call will hear this tone. The reason is that gsmSSF may isolate the D-party from the incoming side, when playing the warning tone towards A-party, but the tone will then also be distributed to B-party and C-party. The SSI notification could be used by gsmSCF to be notified about the MPTY invocation.

4.7.8 Closed User Group

CUG is specified in GSM TS 02.85 [17] and GSM TS 03.85 [43]. CUG is a service that enables an operator to define user groups. Subscribers belonging to a user group have restrictions for outgoing calls and incoming calls.

When a CAMEL subscriber with CUG subscription establishes an outgoing call, then the CUG check is performed before CAMEL service invocation. If the check fails, then the call is released. For a terminating call in the GMSC, the SCP is not allowed to modify the destination number if the served CAMEL subscriber is a member of a CUG and the incoming call was marked as a call from a CUG subscriber. If the SCP attempts to modify the destination number in that case, then the GMSC will release the call and the CAMEL service will be terminated.

4.7.9 Call Barring

CB is specified in GSM TS 02.88 [20] and GSM TS 03.88 [45]. The CB interaction is split up in two categories: barring of outgoing calls and barring of incoming calls.

4.7.9.1 Barring of Outgoing Calls

Three outgoing call barring categories exist: (1) barring of all outgoing calls; (2) barring of outgoing international calls; and (3) barring of outgoing international calls, except to the home country. Figure 4.78 depicts how these CB categories are checked in combination with CAMEL.

The *unconditional call barring category* (BAOC) is checked before CAMEL invocation. A subscriber that has BAOC is barred from outgoing call establishment and, as a result, no CAMEL service is invoked. The *conditional outgoing call barring categories* (BOIC and BOIC-exHc) are checked after CAMEL invocation. When the gsmSCF modifies the destination address of the call, the modified destination of the call is still subject to CB check. Hence, a CAMEL service cannot violate the CB for an outgoing call.

For forwarded calls, the CB checks are performed during registration of the FTN in the HLR. Therefore, there is no CB check in MSC or GMSC at the time of call forwarding initiation. A

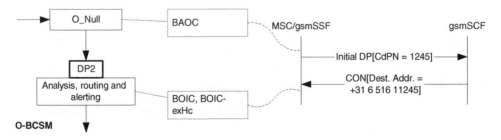

Figure 4.78   Barring of outgoing calls in combination with CAMEL

CAMEL service should therefore take care not to provide a destination number to the MSC, when controlling an MF call, that may violate the subscriber's CB settings.

CB may be applied to individual basic services or to a basic service group, e.g. BAOC may be applied to speech, data and Fax calls, but not to MO SMS. CB may be set by a subscriber, using MMI commands as specified in GSM TS 02.30 [5]. Alternatively, an operator may set certain CB categories for a subscriber. Common examples of operator-enforced BAOC settings include:

(1) Pre-paid international roaming with USSD callback – the subscriber is barred from establishing outgoing calls, but can still use USSD. USSD may be used to request the establishment of a network-initiated call.
(2) Pre-paid credit depletion – when the pre-paid credit is depleted, the operator may activate BAOC for the subscriber. The subscriber may use USSD to replenish her account.
(3) CAMEL fallback – when a CAMEL pre-paid subscriber roams in non-CAMEL network, the HLR induces BAOC for speech, data and fax calls.

4.7.9.2 Barring of Incoming Calls

Two incoming call barring categories exist: (1) barring of all incoming calls; and (2) barring of incoming calls when roaming (BICROAM). These two categories are checked in the HLR at the time of terminating call handling, before CAMEL service invocation. Hence, there is no interaction with CAMEL.

4.7.10 User-to-user Signalling

User-to-user signalling (UUS) is specified in GSM TS 02.87 [19], GSM TS 03.87 [44] and GSM TS 04.87 [52]. UUS enables calling party and called party to exchange information during call establishment and during the active phase of the call. The information that is transferred with UUS is referred to as user-to-user information (UUI) elements. The transfer of UUI elements through the ISUP signalling between calling and called party is transparent for the network. UUS specifies three service levels:

(1) UUS service 1 – UUI elements may be transferred in the ISUP call establishment messages (IAM, CPG, ACM, ANM, CON) and in the ISUP call release message (REL).
(2) UUS service 2 – in addition to UUS service 1 capability, additional UUI elements may be transferred in designated ISUP messages (ISUP UUI) during call establishment.
(3) UUS service 3 – in addition to UUS service 2 capability, additional UUI elements may be transferred in designated ISUP messages (ISUP UUI) during the call.

When the calling or called party disconnects from an active call and the disconnect is reported to the SCP in interrupt mode (EDP-R), the SCP may apply CAP RC to release the call. Using CAP RC at this point has the effect that the ISUP REL originating from the party that disconnected from the call is replaced by an ISUP REL that is generated by the MSC/gsmSSF. Any UUI elements that were present in the ISUP REL will in that case not be transferred to the other party in the call.

4.7.11 Call Deflection

CD is specified in GSM TS 02.72 [11], GSM TS 03.72 [37] and GSM TS 04.72 [50]. CD allows a subscriber to *deflect* an incoming call to an alternative destination. CD may be compared with call forwarding as follows:

- *Call forwarding* – the subscriber registers an FTN in the HLR. When an incoming call cannot be delivered to this subscriber, the MSC or GMSC forwards the call to the registered FTN. For example, when a subscriber does not want to answer an incoming and alerting call, the subscriber may press the 'no' key on the MS keypad. The call is forwarded to the FTN that is registered for the busy condition.
- *Call deflection* – when the subscriber receives a call, but does not wish to answer the call, the subscriber may *ad hoc* define the desired destination to which this call shall be forwarded ('deflected'). The subscriber may choose the destination, based on, for example, the CLI of the incoming call.

CD is invoked by entering (see GSM TS 02.30 [5]):

4 * <destination number> SEND.

When CD is invoked, the deflected call may be subject to CAMEL control in the same manner as a forwarded call may be subject to CAMEL control. O-CSI, including trigger conditions, are applied to the call.

If the HLR has sent TIF-CSI to the VLR during registration, then the subscriber is entitled to use 'short deflected-to-numbers'. For example:

4 * 8756 SEND.

A forwarded call to '8756' is now established. Presuming that the subscriber has O-CSI in the VLR, a CAMEL service is now invoked for this call. The CAMEL service translates the short code into the public directory number. The call is now routed to the translated destination number.

A deflected call may be subject to ORLCF just like other late forwarded calls. When the VMSC sends MAP RCH to the GMSC, to request the GMSC to resume the handling of this call, the forwarding reason in MAP RCH is set to unconditional. The rationale is that MAP RCH has the ability to distinguish between four different forwarding reasons: busy, no reply, not reachable and unconditional (GSM TS 09.02 [54], parameter 'ForwardingOptions'). Unconditional conditional call forwarding cannot occur in the VMSC. Therefore, this value is used to denote CD in MAP RCH.

The GMSC maps the unconditional forwarding reason to DP T_Busy in the forwarding notification feature. The cause value used in the forwarding notification in this scenario may be vendor-specific.

4.8 Interaction with Network Services

Network services do not apply to individual subscribers but are offered generically by an operator. Some network services have an impact on CAMEL; other services depend on the availability of CAMEL in the network.

4.8.1 Basic Optimal Routing

BOR is specified in GSM TS 03.79 [39]. BOR is one of the two categories of OR; the other OR category is ORLCF, which is described in other sections. BOR entails that terminating call handling may be initiated from a GMSC residing in a PLMN other than the called subscriber's HPLMN. The network where the GMSC is located for terminating call handling is the IPLMN. In the case of BOR, the IPLMN and the HPLMN are different networks. When BOR is applied for a call, the terminating leg for that call is routed more efficiently. Terminating call handling always involves a GMSC; the GMSC performs HLR interrogation and routes the call to the VMSC of the called subscriber, based on the MSRN received from HLR. The effect of applying BOR to a mobile terminating call is illustrated by the following example.

Calling party (A-party), who belongs to a UK-based operator, is roaming in France. This A-party establishes a call to a B-party, who belongs to the same UK-based operator and who is also roaming in France at this moment.

Non-BOR Case

The call from the A-party is destined for a UK subscriber (e.g. +44 20 xxx), so the VMSC in France routes this call to the UK. Within the UK, the call is routed to the HPLMN of the destination subscriber. A GMSC in that PLMN performs HLR interrogation and receives the MSRN for the call to the VMSC of the B-party. B-party is currently in France, so the call is routed to the VMSC in France.

As a result, there are *two international call legs*: (1) the A-party has established an MO call from France to the UK; (2) the B-party has a roaming terminating call from the UK to France (Figure 4.79).

BOR Case

When the A-party establishes the call, the VMSC determines that the call is destined for a subscriber that belongs to a PLMN with which the VPLMN operator has a 'BOR agreement'. The BOR agreement entails that this PLMN is entitled to perform HLR interrogation in the other PLMN. As a result, the VMSC of the A-party forwards the call to a GMSC in the same network. This GMSC may be a designated node or may be co-located with the VMSC. The GMSC derives the GT of the HLR from the B-party's MSISDN and then performs the HLR interrogation. The HLR returns the MSRN for this call. The GMSC may now route the call to the VMSC, using the MSRN.

As a result, there are two *local call legs*: (1) the A-party has established an MO call from a VMSC in the French network to a destination within that same network; (2) the B-party has a roaming terminating call within the French network (Figure 4.80).

The GMSC in the IPLMN shall be configured to know for which HPLMNs it is permitted to interrogate the HLR; the HLR in turn will be configured to allow BOR for selected IPLMNs, besides its own PLMN.

The calling party in Figure 4.80 has effectively established a local call. If the calling party is subject to on-line charging, then the CAMEL service will take care when BOR is used. CAMEL phase 2 has no mechanism to provide the SCP with an explicit indication that BOR was successfully

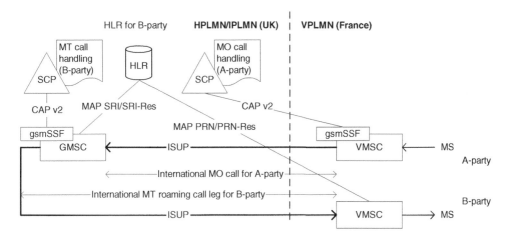

Figure 4.79   Roaming call without basic optimal routing

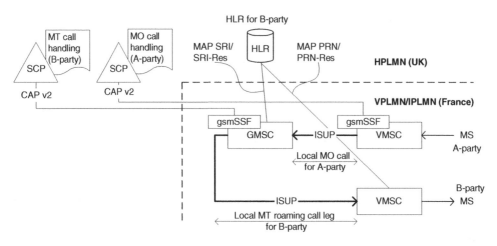

Figure 4.80   Roaming call with basic optimal routing

applied for the call. The location of the called subscriber determines, amongst others, whether BOR is permissible for the call. When the calling party in Figure 4.80 sets up the call to the UK-based destination, the SCP cannot know whether BOR will be applied for this call. As a result, the service cannot know how to rate this call.

CAMEL phase 4 has improved interaction with BOR. An SCP may determine whether BOR should be applied to the outgoing call. CAMEL phase 4 enables the SCP to instruct the VMSC to initiate BOR.

4.8.1.1 Terminating Call Handling with BOR

One aspect with BOR is the terminating call handling for the called subscriber. The called party may have a terminating CAMEL subscription. Hence, the HLR sends the T-CSI to the GMSC in the IPLMN. The HLR verifies that the GMSC supports the required CAMEL phase before sending T-CSI. If the terminating call is subject to on-line charging, then the on-line charging may be applied directly in the GMSC in the IPLMN. However, if BOR results in a terminating call leg within the IPLMN (as in Figure 4.80), then the terminating call may be toll-free for the called party. However, this may differ per operator.

If the terminating CAMEL service involves the use of announcements, then the HPLMN operator should ensure that the GMSC in the IPLMN has the required announcements installed. The availability of appropriate announcements should be part of the BOR agreement between operators.

4.8.1.2 Call Forwarding with BOR

The terminating call in combination with BOR may lead to early call forwarding, late call forwarding or ORLCF. Operators may decide to prevent call forwarding in combination with BOR, since this may lead to uncommon call cases. For example, if the call in Figure 4.80 leads to early call forwarding, then there will be a forwarding call leg from the GMSC to, for example, the voice mail box in the HPLMN of the called party. The called party may be charged for this international call forwarding leg.

4.8.2 Immediate Service Termination

Immediate service termination (IST) is specified in GSM TS 02.32 [7] and GSM TS 03.35 [33]. IST is a network feature that enables an operator to immediately terminate ongoing services from a subscriber. Removing subscriber data from the MSC/VLR where a subscriber is registered is not sufficient for this purpose; an ongoing call is not terminated by removing subscriber data from the VLR.

The forced termination of an ongoing call of a subscriber may result from a stimulus from the FIGS; refer to Section 4.3.7 for more information on FIGS. For the network to be able to terminate the ongoing call of a subscriber, there will be a CAMEL service in control of the respective calls of the subscriber. The CAMEL service may use the release call operation to release the call when required.

4.8.3 Operator-determined Barring

ODB is specified in GSM TS 02.41 [9] and GSM TS 03.15 [30]. ODB is similar to the GSM call barring supplementary service. The operator may apply ODB to bar specific services, such as outgoing calls or incoming calls, for a subscriber. Unlike GSM call barring, ODB settings cannot be changed by the subscriber. ODB may be used, for example, when roaming conditions or network conditions do not permit the use of a specific service. Alternatively, a subscriber's service profile may not allow certain services.

A further distinction between ODB and call barring is that call barring may be specified for individual basic services or basic service groups, whereas ODB may be applied only to the entire group of subscribed basic services.

ODB is divided into different categories; not all categories have an impact on CAMEL services. The relevant ODB categories are:

- *Barring of outgoing calls* – this category is checked during outgoing call establishment after the invocation of CAMEL; if this ODB category prevents the call establishment, then this leads to one of the call establishment failure DPs in the O-BCSM. The sub-category 'barring of all outgoing calls' is checked before CAMEL invocation.
- *Barring of outgoing premium rate calls* – this category is checked during outgoing call establishment after the invocation of CAMEL. Call establishment failure due to this ODB category leads to one of the call establishment failure DPs in the O-BCSM.
- *Operator specific barring* – the behaviour of this category of ODB is not specified; it may be determined by the operator. For example, an operator may disallow specific destinations for certain subscribers.
- *Barring of supplementary services management* – this ODB category bars, amongst others, the use of mobile initiated USSD. As a result, USSD services in the SCP cannot be used by the subscriber.

For the remaining ODB categories, refer to GSM TS 02.41 [9].

4.8.4 High-speed Circuit-switched Data

High-speed circuit-switched data (HSCSD) allows for data rates up to 57.6 kbit/s through the GSM radio access network. The high throughput is achieved by allocating multiple time slots over the air interface between MS and BTS. An HSCSD call may be subject to CAMEL control. The HSCSD call is treated as a bearer service call. The number of time slots allocated to HSCSD calls may change during the call. Such change is not indicated to the CAMEL service. The CAMEL service can therefore only base the charge of the call on the information that is available at call setup. The CDR for the call may, however, contain an indication of the changed data rate.

4.8.5 Multiple Subscriber Profile

The multiple subscriber profile (MSP) GSM supplementary service is a mechanism that enables subscribers to have multiple profile definitions in the HLR. The subscriber has a single SIM, and hence a single IMSI, but may have a different MSISDN per profile. The subscriber may select one profile as the active profile. Each profile may have different settings for the supplementary services.
   Two phases of MSP exist:

- MSP phase 1 – refer to GSM TS 02.97 [25] and GSM TS 03.97 [48]. This phase of MSP makes use of CAMEL phase 2 capability.[46]
- MSP Phase 2 – refer to 3GPP TS 22.097 [67] and 3GPP TS 23.097 [85]. This phase of MSP makes use of CAMEL phase 2 and CAMEL phase 3 capability.

For more information on MSP, see Chapter 5.

---

[46] For MSP phase 1, there are no GSM R97 specifications available; refer to GSM R98 for MSP specifications.

5

CAMEL Phase 3

5.1 General Third-generation Networks

CAMEL phase 3 was introduced in 3GPP release R99 of the third generation ('3G') mobile network. The 3GPP third generation mobile network is the evolution of the second-generation ('2G') GSM network. GSM is the second-generation network standard developed by ETSI. GSM is a European standard, albeit deployed worldwide. The third-generation network from 3GPP, on the other hand, is a true global standard, developed by members of various regional standardization organizations.

The core network architecture of the 3GPP 3G network is similar to the core network architecture of the ETSI GSM network. This means that CAMEL phase 1 and CAMEL phase 2 technology, which is specified for the GSM network, may be used in the 3G network as well. The 3G network that is specified by 3GPP, also referred to as the universal mobile telecommunications system (UMTS), is one of a group of third-generation mobile networks. Table 5.1 presents a list of some of the 2G and 3G networks.

The development of the third-generation network architecture was started by ITU, under the name International Mobile Telephony 2000 (IMT 2000). There is a mutual compatibility between CAMEL phases and network generation. CAMEL phase 1 and CAMEL phase 2 technology are specified for the GSM network, but may also be used in the UMTS network. CAMEL phase 3 and CAMEL phase 4 technology form part of the 3G network architecture, but may also be used in the GSM network.

5.1.1 UMTS Network Architecture

When comparing UMTS with GSM, there are a number of architectural differences. However, the migration from GSM to UMTS allows for phased evolution. When deploying a 3G network, an operator does not need to apply all aspects that are specified for the 3G network. The 3G network may contain a mix of 2G and 3G functionality.

Figures 5.1 and 5.2 present the UMTS network architecture for circuit-switched services and for packet-switched services, respectively. Both the MSC and the SGSN in the 3G network architecture may at the same time control 3G radio access network infrastructure and 2G radio access network infrastructure. Tables 5.2 and 5.3 contain legend for figures 5.1 and 5.2.

The GPRS network infrastructure is introduced in GSM R97 and may therefore be labelled as '2G'. However, the fundamental difference between GSM and GPRS, being *packet-switched media transport* vs *circuit-switched media transport*, has resulted in the classification of '2.5G' for GPRS.

The entities in the 2G and 3G network architecture are grouped in logical parts. See Table 5.4. As is reflected in Figures 5.1 and 5.2, a core network may be a mix of CS nodes (MSC, GMSC) and packet-switched (PS) nodes (SGSN, GGSN). In addition, both the CS infrastructure and the

CAMEL: Intelligent Networks for the GSM, GPRS and UMTS Network   Rogier Noldus
© 2006 John Wiley & Sons, Ltd

Table 5.1 Overview of 2G and 3G network technologies

Network	Generation	Organization	Description
TDMA	2G	ANSI	Also referred to as D-AMPS
GSM	2G	ETSI	Global system for mobile communication
DCS1800	2G	ETSI	Digital cellular system – GSM 1800
PCS	2G	ETSI	Personal communication system – GSM 1900
UMTS	3G	3GPP	Universal mobile telecommunications system
CDMA-2000	3G	3GPP-2	Code division multiple access – 2000. This standard was developed mainly by Qualcomm

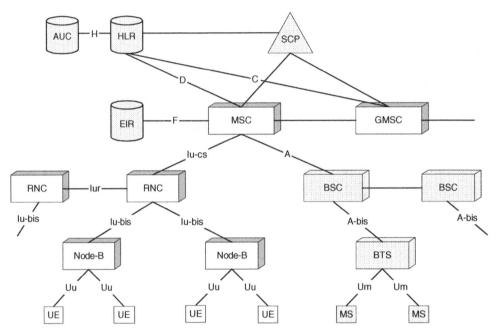

Figure 5.1 UMTS network architecture for CS services

PS infrastructure of the core network may at the same time be controlling a 2G RAN and a 3G UTRAN. CAMEL has control capability for each of the possible combinations:

- CS call or short message through 2G RAN;
- CS call or short message through 3G UTRAN;
- GPRS data connection or short message through 2G RAN;
- GPRS data connection or short message through 3G UTRAN.

5.1.2 2G Cell Planning vs 3G Cell Planning

The cell structure used in the 3G UTRAN is different from the 2G RAN cell structure. Although CAMEL has no direct link with the radio access network, either in GSM or in UMTS, the location information that is reported to the gsmSCF in UMTS is slightly, but significantly, different from the location information in GSM (Figure 5.3). The cells that make up the 3G RAN coverage area are grouped into *service areas*. Figure 5.3 depicts a service area consisting of four cells (*cell 1, cell 2, cell 3, cell 4*). One or more service areas form a location area.

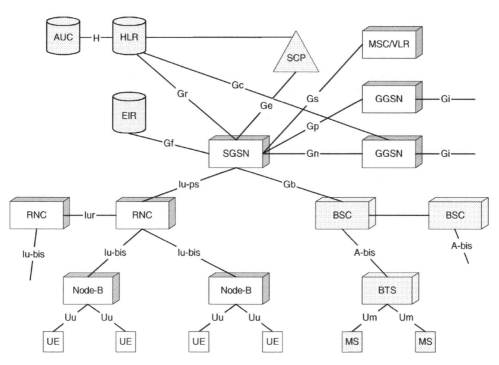

Figure 5.2 UMTS network architecture for PS services

Table 5.2 Legend for Figures 5.1 and 5.2 (entities)

Entity	Full name	Description
UE	User equipment	The UE is the mobile station in the UMTS network. The UE is the UMTS network equivalent of the MS in GSM network
Node-B	Node-B	The Node-B in the W-CDMA radio network has similar function as the BTS in the TDMA radio network
RNC	Radio network controller	The RNC controls one or more node-Bs in the UMTS radio network
MSC	Mobile services switching centre	Switching node for CS services
GMSC	Gateway mobile services switching centre	Gateway switching node for CS services
SGSN	Serving GPRS service node	The SGSN is the node through which GPRS subscribers are attached to the packet-switched network
GGSN	Gateway GPRS service node	The GGSN forms the gateway between the packet-switched PLMN and the IP network

The other entities in Figures 5.1 and 5.2 are the 2G entities that are described in Chapter 1

5.1.3 Location Information

The location information that a 3G RAN reports to the MSC is slightly different from the location information that a 2G RAN reports to the MSC. Consequently, an SCP receives different location information when controlling a call in a 3G network than when controlling a call in a 2G network.

Table 5.3   Legend for Figures 5.1 and 5.2 (interfaces)

Interface	Description
Iu	Interface to control the RNC from the CS or PS core network, by MSC or SGSN respectively. Iu is specified in 3GPP TS 25.410 [99]
Iu-cs	Iu-cs is the CS variant of the Iu interface; Iu-cs is used by MSC to control the RNC
Iu-ps	Iu-ps is the PS variant of the Iu interface; Iu-ps is used by SGSN to control the RNC
Iur	Interface between two RNCs; Iur is specified in 3GPP TS 25.420 [100]
Iu-bis	Interface between the RNC and Node-B; Iu-bis is specified in 3GPP TS 25.430 [101]
Uu	WCDMA radio interface between UE and Node-B
Gb	The Gb interface is used by the SGSN to control a BSC
Ge	The Ge interface carries the CAP dialogue between SGSN and gsmSCF
Gn	The Gn interface is used between SGSN and GGSN and is also known as the GTP. The Gn interface is used when SGSN and GGSN are in the same PLMN. Gn is specified in 3GPP TS 29.060 [105]
Gp	The Gn interface is used between SGSN and GGSN (GTP). The Gp interface is used when the GGSN is in another PLMN than the SGSN. Gp is specified in 3GPP TS 29.060 [105]
Gs	The Gs interface between SGSN and MSC is used when the MSC and SGSN share visitor location register. Gs is specified in 3GPP TS 29.016 [104]
Gi	The Gi interface forms the IP interface between GGSN and the PDN

The other interfaces in Figures 5.1 and 5.2 are the 2G interfaces that are described in Chapter 1

Table 5.4   Logical division in UMTS network

System	Description
Base station system (BSS)	The BSS consists of one BSC and one or more BTS. One MSC or SGSN may control several BSSs. The BSSs within the PLMN form the RAN. The radio transmission technology used in the RAN is TDMA
Radio network system (RNS)	The RNS consists of one RNC and one or more Node-Bs. One MSC or SGSN may control several RNSs. The RNSs within the PLMN form the universal terrestrial radio access network (UTRAN). The radio transmission technology used in the UTRAN is wideband code division multiple access (W-CDMA)
Core network (CN)	The core network consists of three groups of components: • *CS-only nodes* – the MSC and GMSC are used for CS telecommunication services. When split-MSC architecture is used, the core network also contains media gateway • *PS-only nodes* – the SGSN and GGSN are used for PS telecommunication services • *common nodes* – the HLR, AUC and EIR are used for both CS services and PS services
Service network (SN)	The service network contains the gsmSCF. The gsmSCF has interfaces with the core network nodes, but not with the BSS or RNS. CAMEL specifies only the gsmSCF in the service network. There may be various other entities in the services network for service provisioning, service data, subscriber data, operation and maintenance, etc.

Table 5.5 lists the location information elements for 2G and 3G networks, as reported to SCP.

Age of location information
This element provides an indication of the reliability of the location information. Every time a location update takes place, i.e. the location register in the MSC/VLR is refreshed, this element is set

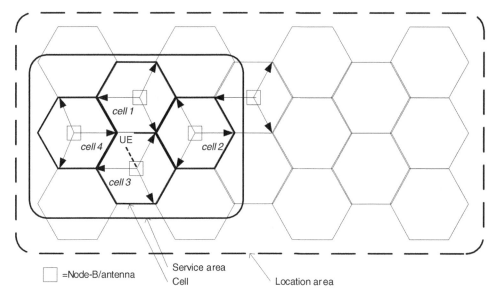

=Node-B/antenna   Service area
                   Cell   Location area

Figure 5.3 W-CDMA cell planning

Table 5.5 Location information elements in 2G and 3G networks

2G location information	3G location information
Age of location information	Age of location information
Mobile country code	Mobile country code
Mobile network code	Mobile network code
Location area code	Location area code
Cell Global Id	Service area identifier; SAI present
Routing area code	Routing area code
Geographical information	Geographical information
	Geodetic information[a]
VLR number	VLR number
	MSC number[a]
Location number	Location number
	Local service area identity[a]
	Current location retrieved[a]

[a] These elements may also be reported in a 2G network when CAMEL phase 3 is supported

to 0 and will start counting. When this element is reported for mobile originated call establishment, there is radio contact between the UE and the MSC and hence the element will be 0.

Mobile country code (MCC)
The MCC is country code that is assigned by ITU-T to the country to which the PLMN belongs.

Mobile network code (MNC)
The MNC identifies the specific PLMN to which the MSC belongs. The MNC values are defined by the national regulators.

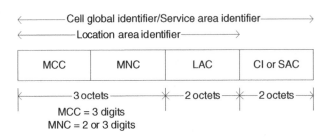

Figure 5.4   Structure of SAI, CGI and SAI

Location area code (LAC)

The LAC identifies a group of cells (the location area) within a VLR service area. A subscriber may move within a location area without performing a location update in the VLR. As a result, when the VLR is read with, for example, ATI, the reported cell Id or service area Id may not be the current one.

Cell Global Id (CGI)

The CGI identifies the cell where the subscriber was registered at the time of the most recent radio contact between UE and MSC. The cell Id is reported from a 2G RAN, but not from a 3G RAN.

Service Area Code (SAC)

The SAC identifies the service area where the subscriber was registered at the time of the most recent radio contact between UE and MSC. The service area is reported from a 3G RAN, but not from a 2G RAN. The MCC, MNC, LAC and CI or SAC are reported in a single information element. In a GSM R97 (or earlier) compliant MSC, the reported parameter is called 'cellIdOrLAI'; in a 3GPP R99 (or later) compliant MSC, the reported parameter is called 'cellGlobalIdOrServiceAreaIdOrLAI'. The structure is as depicted in Figure 5.4.

An MSC may report location area identifier (LAI), consisting of MCC, MNC and LAC. Alternatively, the MSC may report the cell global identifier (CGI) or service area identifier (SAI). The CGI is the combination of MCC, MNC, LAC and CI; the SAI is the combination of MCC, MNC, LAC and SAC. Since CGI and SAI have identical coding, the receiver of the location information may not be able to deduce whether the served subscriber is registered in a 2G network (CGI is reported) or registered in a 3G network (SAI is reported). The following two methods may be applied:

(1) The operator ensures that there is no overlap between the set of CI codes and the set of SAC codes; the reported value for CI/SAC indicates whether the subscriber is registered in 2G or 3G network.
(2) The MSC that reports the location information includes the parameter 'SAI Present'. The presence of this parameter indicates that the subscriber is registered in 3G access network.

Routing Area Code

The routing area code (RAC) is used to identify a location in the 2G or 3G access network for PS connectivity. The RAC forms part of the routing area identifier (RAI). The structure of the RAI is reflected in Figure 5.5.

Geographical Information

The geographical information represents the location of the subscriber in geographical format. It is an operator's option whether the MSC includes the geographical information in the reported location information. The MSC may derive the geographical information from the CGI or SAI. The geographical information is defined in 3GPP TS 23.032 [76] and takes the form of an ellipsoid point with uncertainty circle. The ellipsoid point is expressed in longitude and latitude; the uncertainty

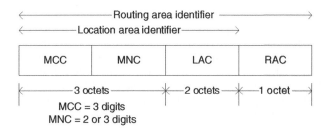

Figure 5.5   Structure of the routing area identifier

represents a radius, expressed in meters. This notation is formally specified in the World Geodetic System 1984 (WGS 84).[1] The ellipsoid point may represent the position of the BTS or Node-B/antenna and the uncertainty circle may reflect the radius of the cell.

Geodetic Information
The reporting of geodetic information to the SCP is introduced in CAMEL phase 3. The geodetic information contains the same information as the geographical information with the following additional information elements:

- *Location presentation restricted indicator (LPRI)* – this element indicates whether the geodetic information of the calling subscriber may be presented to the called party.
- *Screening indicator (SI)* – this element indicates whether the geodetic information for this call was provided by the user or by the network.
- *Confidence* – this element indicates the confidence that the subscriber is indeed in the location defined by the location description in the geodetic information.

There are two cases where the geodetic information is reported to the SCP, in the CAP protocol.

(1) *MO call* – for the MO call, the geodetic information relates to the calling subscriber and is derived from the location information in the VLR. CAMEL does not specify the setting of LPRI in this case. The MSC could contain an option where the LPRI is set per subscriber or per subscriber group. The SI should be set to 'network provided' for the MO call case.
(2) *MT call* – for the MT call, the geodetic information relates to the called subscriber and is obtained by means of the MAP provide subscriber information (PSI) message. An MSC could apply different settings to LPRI when the geodetic information is reported in MAP PSI than when the geodetic information is used during MO call establishment. The ISUP information flow that arrives at the GMSC for an MT call may carry calling geodetic information. This geodetic information is, however, not reported to the SCP, unlike the location number from the same ISUP information flow.

VLR Number
This element contains the number of the VLR, in accordance with the E.164 number plan. For an MO call, the MSC reports this element to the SCP in the CAP IDP operation. For an MT call, the VLR number is provided from the HLR to the GMSC, which reports it to the SCP, in CAP IDP.

MSC Number
This element is used for location services.

---

[1] WGS 84 is defined in Military Standard WGS84 Metric MIL-STD-2401: 'Military Standard Department of Defence World Geodetic System (WGS)'.

Location Number
The MSC may derive a location number (LN) from the LAI, CGI or SAI for this subscriber. For MO calls, the LN of the calling subscriber is reported to the SCP. For MT calls, the LN of the called subscriber is reported to the SCP. In addition, if the ISUP information flow that arrives at the GMSC for the MT call contains LN, then that LN is also reported to the SCP. This second instance of LN relates to the calling subscriber, not to the called CAMEL subscriber.

The allocation of LN to ranges of CGI, LAI or SAI is operator-specific. The LN is commonly used to express the location of a subscriber as an E.164 number that fits in with the operator's number plan. The LN may have the following international format:

$$LN = Country\ Code + National\ Destination\ Code + Subscriber\ Number$$

whereby the national destination code is the network code of the operator and the subscriber number is defined by the operator. The LN may also be reported in national format, e.g.

$$LN = National\ Destination\ Code + Subscriber\ Number$$

Owing to the international usage of CAMEL, it is not recommended to use a LN in national format. The LN in CAP IDP has often led to confusion. The reason is that for, CAP IDP, the LN is defined in two places:

(1) on the main level of the argument of CAP IDP;
(2) inside the location information data type.

The LN on the main level of the argument of CAP IDP relates to the location of the *calling subscriber*. For MO calls, the calling subscriber is the served CAMEL subscriber; for these calls, the LN is provided by the serving MSC. For MT and MF call cases, the location number is retrieved from the incoming ISUP signalling.[2] The location number inside the location information is used only for the MT call case. It relates to the location of the called party. For a mobile terminating call, both instances of the location number may be present in CAP IDP.

Local Service Area Identity
This element is used for *support of localized service area* (SoLSA). SoLSA is a feature that enables an operator to define the localized service area (LSA) to a subscriber. When a subscriber is in her LSA, she will have exclusive service availability. SoLSA is specified for GSM/EDGE access only. Refer to 3GPP TS 43.073 [127].

Current Location Retrieved
This element is used for *active location retrieval*.

5.1.4 Split-MSC Architecture

Prior to 3GPP Rel-4, the (G)MSC is considered an integrated entity, handling both signalling (subscriber mobility, call control, supplementary service control, etc.) and media traffic (speech, data). This is depicted in Figure 5.6. The interfaces between MS and MSC (Um, A-bis, A) carry both the signal messages and the media. When the MSC establishes the call and routes the call to the next node in the network, the call-related signalling and the media are transported through the SS7 network to the required destination. Both the ISUP signalling and the media use the transport

---

[2] When the geodetic information was introduced in CAMEL phase 3, it was defined only within location information. It is therefore not possible in a mobile terminating call to report the geodetic information relating to the calling party.

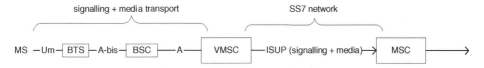

Figure 5.6   Signalling with integrated MSC

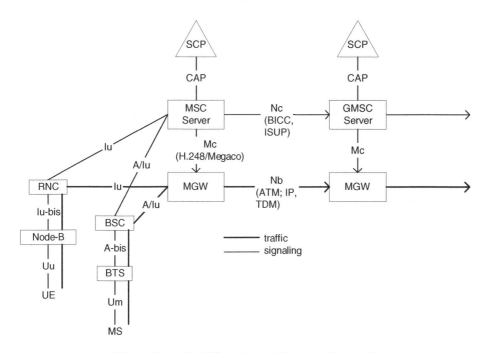

Figure 5.7   Split-MSC architecture ('layered architecture')

capabilities of the MTP layer in the SS7 communication stack; see Chapter 2 for signalling in the GSM network.

A novel aspect in 3GPP Rel-4 is the 'split MSC architecture'. As the name of the feature implies, it entails that the MSC or GMSC is split up into two functional parts: (1) MSC Server/GMSC Server; and (2) media gateway. The split architecture is depicted in Figure 5.7.

The split-MSC architecture is mainly described in the following 3GPP specifications:

- 3GPP TS 23.205 [90]; *bearer-independent circuit-switched core network*; this specification describes the network architecture;
- 3GPP TS 29.205 [107]; *application of Q.1900 series to bearer independent circuit-switched core network architecture; stage 3*; this specification describes the bearer independent call control protocol;
- 3GPP TS 29.232 [110]; *media gateway controller (MGC) – Media gateway (MGW) interface; stage 3*; this specification describes the H.248/Megaco protocol;
- 3GPP TS 29.414 [112]; *core network Nb data transport and transport signalling*; this specification describes the use of IP and ATM for data transport between MGWs;
- 3GPP TS 29.415 [113]; *core network Nb interface user plane protocols*; this specification describes the user part for the Nb protocol between the MGWs.

The MSC Server acts as an MSC for all subscriber registration, subscriber mobility, call management, supplementary service management, etc. The traffic to and from the UE or MS does, however, not run through the MSC server. The traffic is tapped off from the BSC (2G network) or RNC (3G network) and is routed to a MGW. The MSC server controls the MGW through the Mc interface. The embodiment of the Mc interface is the H.248/Megaco (media gateway control), normally referred to as Megaco. Megaco is based on ITU H.248 [130], which is a generic MGW protocol. Megaco is jointly developed by 3GPP and the Internet Engineering Task Force (IETF).

Between MSC servers in a network, signalling is carried through the Nc interface. The Nc interface is embodied by the bearer-independent call control (BICC) protocol. BICC, which is specified in ITU Q.1901 [144], may be regarded as the signalling part of ISUP. The mapping between CAP and ISUP, that is specified from CAMEL phase 1 onwards, is equally valid for the split-MSC architecture. That means, for example, that relevant parameters from CAP connect (CON) are copied on BICC initial address message (IAM).

The traffic bearers between the MGWs, the Nb interface, may be based on real-time transport protocol (RTP), IETF RFC[3] 1889 [166], or user datagram protocol (UDP), IETF RFC 768 [165]. RTP is also used as transport protocol in the IP multimedia subsystem (IMS) network. For an integrated MSC architecture, the media transport uses time division multiplex (TDM). Media transport using TDM is bound by circuits of 64 kbit/s and is less efficient than the transport capabilities of ATM and IP, depending on the type of data.

The split-MSC architecture has no impact on CAMEL call control capability. A CAMEL service may be unaware that a call is handled by an MSC service instead of an integrated MSC. The gsmSSF resides in the MSC server. CAP operations may control the BICC signalling or may result in traffic connection to a specialized resource function (SRF) in the MGW. Certain UMTS features may, however, be used only in combination with BICC. One such feature is service change and UDI/RDI fallback (SCUDIF). SCUDIF has interaction with CAMEL phase 4 (see Chapter 6).

5.1.5 CAMEL Phase 3 Features

Table 5.6 contains an overview of the functionality that is introduced in CAMEL phase 3.

The CAMEL phase 3 functionality is specified in the following 3GPP R99 specifications:

- 3GPP TS 22.078 [66] – CAMEL; service description; stage 1;
- 3GPP TS 23.078 [83] – CAMEL; stage 2;
- 3GPP TS 29.078 [106] – CAMEL; CAMEL application part (CAP).

All 3GPP specifications referred to in the present section are 3GPP R99 specifications, unless otherwise indicated.

5.2 Call Control

5.2.1 Subscribed Dialled Services

Subscribed dialled services (SDS) is a method in CAMEL phase 3 to trigger an additional IN service during MO or MF call establishment. The following subscribed services may hence be invoked during call set up: subscribed service at DP collected info; and subscribed dialled service at DP analysed info. This method uses an enhancement to the O-BCSM: the DP analysed info is added in CAMEL phase 3. Figure 5.8 shows the new DP in the O-BCSM for CAMEL phase 3. The invocation of a subscribed dialled service is based on the presence of the CAMEL subscription element D-CSI. D-CSI, just like O-CSI, is contained in HLR and may be sent to VLR at registration and to GMSC during terminating call handling.

---

[3] IETF standards are referred to as 'requests for comments' (RFC).

Table 5.6 CAMEL phase 3 features

		Feature	Section
1		Call control	5.2
	a	Subscribed dialled services	5.2.1
	b	Serving network-based dialled services	5.2.2
	c	CAMEL control of mobile terminated calls in VMSC	5.2.3
	d	CAMEL service invocation at call failure	5.2.4
	e	Service interaction control	5.2.5
	f	Call gapping	5.2.6
	g	Support of long forwarded-to numbers	5.2.7
	h	On-line charging enhancements	5.2.8
	i	CAMEL control of multiple subscriber profile	5.2.9
	j	General enhancements to CAP	5.2.10
2		CAMEL control of GPRS	5.3
3		CAMEL control of MO-SMS	5.4
4		Mobility management	5.5
5		CAMEL interaction with location services	5.6
6		Active location retrieval	5.7
7		Subscription data control	5.8
8		Enhancement to USSD	5.9

5.2.1.1 Service Invocation

When a call is established, the call flow follows the O-BCSM. At first, a CAMEL service at DP collected info may be invoked, as in CAMEL phase 1 and CAMEL phase 2. This service, the subscribed service, is dependent on availability of O-CSI and fulfilment of trigger criteria of O-CSI.

When processing at DP collected info is completed, the BCSM transits to the state DP analysed info. If the subscriber has D-CSI, then the MSC/gsmSSF may invoke a subscribed dialled service. The invocation of the subscribed dialled service is based on the dialled number for this call. As a result, triggering for the subscribed dialled service may be influenced by the subscribed service (O-CSI). The gsmSSF compares the dialled number with the set of numbers stored in D-CSI. If a match is found, then the gsmSSF invokes the dialled service (Figure 5.9).

Dialled services may be used for MO calls and for MF calls. For MO calls, D-CSI is used if it is present in VLR. For early forwarded calls, D-CSI is used if it is present in MAP SRI-Res and call forwarding takes place. For late forwarded calls, D-CSI is used if it is present in VLR.

When call forwarding is induced by the SCP, by sending CAP connect (with modified destination number) to the GMSC during MT call handling, the SCP has the capability to determine whether O-CSI will be applied for the forwarded leg. The O-CSI, if received in MAP SRI-Res, is applied for the MF leg in GMSC only when the CAP connect that led to this MF leg contains 'O-CSI applicable'. Such a capability, to determine the applicability of O-CSI for the MF call, does not exist for D-CSI. Hence, if D-CSI is available in the GMSC (i.e. received in SRI-Res), it will always be used for the MF call.

If the subscribed service (O-CSI) has modified the called party number, then the modified called party number is used for the triggering check at DP analysed info. The comparison of the dialled number with the set of numbers contained in D-CSI is fully specified by CAMEL.[4] This is required to guarantee international consistency in the handling of D-CSI in MSCs from different vendors.

---

[4] The O-BCSM in INAP CS1 also contains the DP analysed info. The exact criteria for BCSM transition to DP analysed info are, however, not specified in INAP CS1.

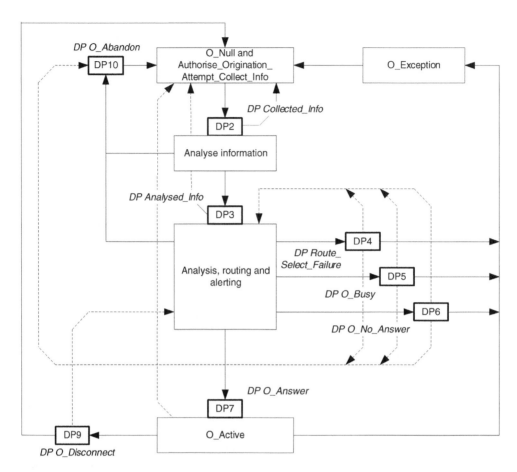

Figure 5.8 O-BCSM for CAMEL phase 3. Reproduced from 3GPP TS 23.078 v3.19.0, figure 4.3, by permission of ETSI

The subscribed dialled service uses CAP v3 (or CAP v4 in CAMEL Phase 4). If the subscribed dialled service is used in a call in combination with a subscribed service, then the subscribed service may use any CAP version. That means that, within one MSC, O-CSI and D-CSI for a call may use different CAP versions. In this way, an operator may introduce subscribed dialled services without having to upgrade the already operational subscribed service to CAMEL phase 3.

5.2.1.2 Contents of D-CSI

D-CSI consists of a list of up to 10 numbers; each number has a CAMEL service associated with it. Table 5.7 shows an example of D-CSI.

The CAMEL service that is invoked at DP analysed info depends on the dialled number. This is different from the trigger criteria that are contained in O-CSI; although O-CSI may contain a set of 10 numbers, there is only one CAMEL service associated with O-CSI.

D-CSI contains one CAMEL capability handling parameter only. That implies that all subscribed services contained in D-CSI use the same CAMEL phase. If the operator wishes to use CAMEL phase 4 for a subscribed service, then all subscribed services contained in D-CSI will be upgraded

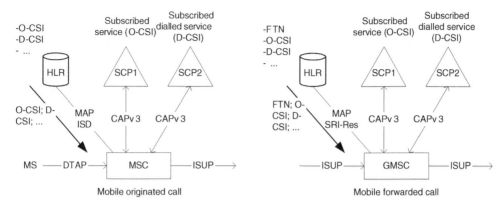

Figure 5.9 Subscribed dialled service for MO and for MT call

Table 5.7 Example contents of D-CSI

	Number	Service key	gsmSCF address	Default call handling
1	0800 2143	125	+27 83 887766	Continue
2	0800	126	+27 83 887766	Continue
3	0900 8754	127	+27 83 887767	Release
4	0900	128	+27 83 887767	Release

CAMEL capability handling = CAMEL phase 3.

to CAMEL phase 4. Alternatively, the operator defines multiple D-CSIs: a CAMEL phase 3 D-CSI and a CAMEL phase 4 D-CSI. However, the HLR may send not more than one D-CSI to VLR or GMSC for a subscriber.

When D-CSI is checked, the gsmSSF compares the dialled number with each number in D-CSI, until a match is found. If overlap exists between two or more numbers in D-CSI, then this may affect the outcome of the D-CSI check and, as a result, a different service may be invoked. The example D-CSI structure in Table 5.8 contains overlap between 0800 2143 and 0800. If the subscriber dials 0800 2144, then different dialled services could be invoked. Compare the two practical cases in Table 5.8.

The operator shall therefore place the numbers in D-CS in the correct order[5].

Table 5.8 D-CSI number check

Dialled number	Sequence of checking	Result
0800 2143	**0800 2143** 0800	entry 1 (SK125; +2783887766; continue)
0800 2143	**0800** 0800 2143	entry 2 (SK126; +2783887766; continue)

---

[5] The order of the numbers in the trigger criteria for O-CSI is not relevant. The reason is that O-CSI specifies a single IN service only for DP collected info.

5.2.1.3 Dialogue Capabilities for D-CSI

A rule that CAMEL has inherited form ETSI CS1 is 'single point of control'. This rule entails that, for a single BCSM instance, at the most one IN service may be in control. When the subscribed dialled service is invoked at DP analysed info, a subscribed service may already be active; this subscribed service was invoked at DP collected info. To guarantee that at most one IN service is in control, the dialled service has a *short dialogue*. The short dialogue has the following characteristics, compared with a *long dialogue*: (1) the short dialogue is not allowed to arm detection points; and (2) the short dialogue is not allowed to request reports. The following CAP operations may therefore not be used by the dialled service: request report BCSM event (RRB), apply charging (ACH), call information request (CIRq).

Pre-arranged end rules specify that, when the SCP gives control of a call back to the MSC/gsmSSF and there are no armed DPs or pending reports, the CAMEL relationship will be closed. Hence, when the dialled service gives control of the call back to the MSC/gsmSSF, by sending CAP continue (CUE), CAP connect (CON) or CAP continue with argument (CWA), the dialled service will terminate. The following capabilities remain available for the dialled service:

(1) user interaction (connect to resource, establish temporary connection, play announcement, etc.);
(2) charging (send charging information, furnish charging information);
(3) call gap;
(4) call termination (release call);
(5) call continuation, possibly with modified call information (continue, connect, continue with argument).

The dialled service may use CAP operations reset timer (RT) and activity test (AT) when required. The CAP v3 application context for the dialled service is the same as the CAP v3 application context for the subscribed service. The dialled service should therefore take care not to use capability that is not allowed for the dialled service, such as sending CAP ACH.

Owing to the fact that the subscribed dialled service can not arm any detection points, the abandon event from the calling party can not be notified to the subscribed dialled service. Therefore, if the abandon event occurs during user interaction, then this leads to CAP dialogue abort (TC_Abort) for the subscribed dialled service.

5.2.1.4 User Interaction

The user interaction capability for the dialled service is the same as for the subscribed service. While the subscribed service is waiting for call answer, the dialled service may apply user interaction (Figure 5.10). The dialled service in Figure 5.10 applies user interaction with through-connect. The

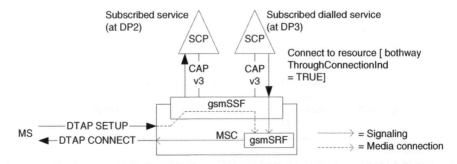

Figure 5.10   User interaction in dialled service with through-connect

through-connection has the effect that an answer message, in the form of a DTAP CONNECT, is sent to the calling party. This answer message does not result in an answer notification to the subscribed service.

5.2.1.5 Charging

The charging capabilities that the subscribed dialled service may use are limited to CAP send charging information (SCI) and CAP furnish charging information (FCI).

• When the subscribed dialled service uses FCI, the free format data carried in FCI is placed in a special information field in the call detail record CDR for this call. Refer to 3GPP TS 32.005 [114] for this information field in the various CDRs.
• The use of SCI by the subscribed dialled service is mutually exclusive with the use of SCI by the subscribed service.

If a CRN is allocated at DP collected info, for the O-CSI service, then this CRN is also used at DP analysed info, for the D-CSI service. Otherwise, if no CRN has been allocated, because the subscriber does not have O-CSI, then a new CRN is allocated for the D-CSI service.

5.2.1.6 Interaction between O-CSI and D-CSI

When both a subscribed service and a subscribed dialled service are invoked, then these two services may interact with one another (Figure 5.11). The following information flows may be distinguished in Figure 5.11:

(1) This flow contains, amongst others, data that is received from the calling party, such as called party number.
(2) This flow contains call-related data, possibly modified or enhanced by the subscribed service. This data flow is used for the subscribed dialled service. The subscribed dialled service does not know for certain parameters whether they are modified by the subscribed service, e.g. called party number, additional calling party number or calling party category. The subscribed dialled service does not know at all whether there was a subscribed service invoked for this call.
(3) This flow contains call-related data, possibly modified or enhanced by the subscribed dialled service. The subscribed service does not know whether any data is modified by the subscribed dialled service. One exception is destination number, which may be reported in the answer notification for the subscribed service.

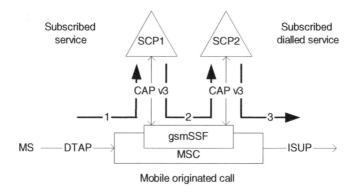

Mobile originated call

Figure 5.11 Interaction between subscribed service and subscribed dialled service

A call release initiated by the subscribed dialled service (i.e. CAP Release Call) will be notified to the subscribed service by means of a call establishment failure DP. The failure event that occurs depends on the cause value that is used by the subscribed dialled service in the release call instruction.

5.2.1.7 Triggering

Subscribed dialled services provide a means to trigger a CAMEL service for *specific* dialled numbers, i.e. the numbers contained in D-CSI. CAMEL does not specify a *wild card number* for D-CSI. Therefore, it is not possible to define a D-CSI that results in *unconditional triggering*, regardless of the dialled number. D-CSI contains a maximum of 10 entries. To accomplish triggering for all numbers, all entries are needed to define the numbers starting with 0, 1, 2 up to 9, but a number starting with 4 may be dialled with different nature of address, e.g. 45634 (unknown format) or +46 634... (international format). Hence, D-CSI remains restricted to triggering for selected numbers.

5.2.1.8 ORLCF

For optimal routing at late call forwarding (ORLCF) calls, D-CSI is used in the GMSC if D-CSI is present in MAP resume call handling (RCH). The inclusion of D-CSI in MAP RCH requires MAP v4.[6] Unlike for O-CSI, the VMSC does not perform a triggering check on D-CSI before sending MAP RCH. The rationale is that the destination number which is used for the D-CSI trigger check may be modified by a subscribed service invoked with O-CSI.

5.2.2 Serving Network-based Dialled Services

A further enhancement to outgoing call handling in CAMEL phase 3 is the *serving network-based dialled services* (also referred to as *network dialled services*). Network dialled services represents a CAMEL service that may be invoked for an MO call or MF call, based on criteria that are defined by the *serving network* operator. This is a deviation from the CAMEL principle that all CAMEL services are subscribed services, which are executed in the HPLMN of the served subscriber.

The trigger criteria used for network dialled services are known as *N-CSI*. A comparison between D-CSI and N-CSI is contained in Table 5.9. The network dialled service is invoked from the same detection point in the O-BCSM as the subscribed dialled service, that is, DP analysed info. The sequence of invoking the three CAMEL services from within the O-BCSM is:

(1) the subscribed service (O-CSI) is invoked from DP collected info;
(2) the subscribed dialled service (D-CSI) is invoked from DP analysed info;
(3) the network dialled service (N-CSI) is invoked from DP analysed info.

Table 5.9   Comparison between D-CSI and N-CSI

Subscribed dialled services (D-CSI)	Network dialled services (N-CSI)
Service is subscribed. D-CSI service is provided to selected subscribers only	Service is not subscribed; the serving network operator may determine to which subscribers N-CSI service is provided
D-CSI is stored in HLR and distributed to MSC and GMSC	N-CSI is not stored in HLR; N-CSI is provisioned in MSC and GMSC only
HPLMN operator is in control of D-CSI service	VPLMN (VMSC) or IPLMN operator (GMSC) is in control of D-CSI service
Contents of D-CSI and triggering check for subscribed dialled services are standardized	Contents of N-CSI and triggering check for network dialled services are not standardized

[6] MAP RCH is the only MAP message that uses MAP v4.

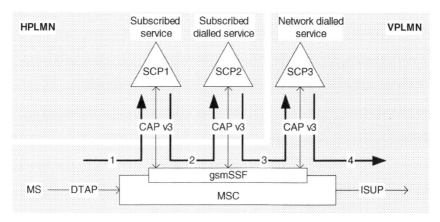

Figure 5.12   Multiple service invocation for MO call

Figure 5.12 gives a graphical representation of the multiple service invocation. In Figure 5.12, the subscribed service may provide modified call data to the gsmSSF (2). The gsmSSF uses the modified call data to perform a triggering check for the subscribed dialled service and to trigger the subscribed dialled service if criteria are fulfilled. The subscribed dialled service may provide modified call data to the gsmSSF (3), after which the gsmSSF uses the modified call data to perform a triggering check for the network dialled service and to trigger the network dialled service if criteria are fulfilled. The network dialled service may provide modified call data to the gsmSSF (4). The call is now routed out of the logical MSC, towards its destination.

The triggering of network dialled service may be based on a combination of various call related parameters, such as: called party number; calling party number (calling party's MSISDN); calling party category; calling party IMSI. The capability of the network dialled service is equal to the capability of the subscribed dialled service. That implies that the network dialled service uses CAP v3 (or CAP v4) and shall be a short dialogue. The same interaction rules apply w.r.t., for example, the use of CAP SCI and the use of through-connect for user interaction.

Example use cases of network dialled service include:

- *Carrier selection* – for every long-distance call, the network dialled service determines the long-distance preferred carrier, depending on destination, day, time, etc. If this carrier selection involves user input (e.g. *'enter 1 for Sprint, enter 2 for MCI'*), then the network dialled service needs to establish a through-connection between the calling party and a gsmSRF internal in the MSC.
- *Freephone and premium rate* – call starting with 0800 or 0900 trigger a network dialled service. The service translates the dialled number into the corresponding business number.
- *Service control* – special numbers may be used to allow the user to change her service settings, e.g. settings related to the calling party's GSM supplementary services. This network dialled service applies user interaction to obtain input from the subscriber. If the triggering of this service is based on the calling party number only, then this has the effect that subscribers from other networks may also call this service. In this case, the service deduces that the calling party belongs to a different network and will terminate the service. In this example use case, the network dialled service is in essence the *destination* of this call. Once the service has completed its processing, it releases the connection. There will not be an ISUP connection established.

5.2.3 CAMEL Control of Mobile Terminated Calls in VMSC

CAMEL phase 3 introduces additional control capability for terminating calls, in the form of the *'VT-CSI'*. VT-CSI is comparable with the T-CSI. CAMEL control of the terminating call in the

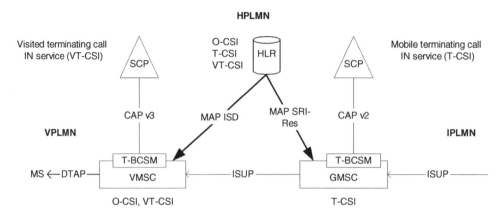

Figure 5.13   VT-CSI CAMEL service triggering

VMSC is referred to as the 'VT call case'. Whereas T-CSI is used for terminating call control from the GMSC, VT-CSI is used for terminating call control from the VMSC (Figure 5.13).

VT-CSI forms part of the CAMEL subscription information that is sent from HLR to VLR during location update. VT-CSI has the same structure as T-CSI, including trigger criteria. When a terminating call arrives at the VMSC (i.e. the GMSC has established the roaming leg with the VMSC), then the MSC contacts the VLR for this terminating call. This terminating call handling process in the VMSC is based on the mobile station roaming number (MSRN) that is received in the ISUP IAM for this call.

If the subscriber has VT-CSI in the VLR, then the MSC invokes a CAMEL service for this terminating call. The invocation of the VT-CSI CAMEL service and the further handling of this CAMEL service is based on the same T-BCSM as is used in GMSC for terminating calls. CAMEL service invocation takes place only if trigger criteria are fulfilled. For T-CSI and VT-CSI, the trigger criteria associated with DP terminating attempt authorized consist of a set of basic service (group) codes. For a T-CSI, the basic service triggering check is performed in HLR. For VT-CSI, this triggering check is performed in VLR.

Since VT-CSI is introduced in CAMEL phase 3, it uses CAP v3 (or CAP v4 in CAMEL phase 4). A subscriber may have T-CSI and VT-CSI from different CAMEL phases. For example, a subscriber may have CAMEL phase 2 T-CSI and CAMEL phase 3 VT-CSI. Hence, when an operator introduces a VT-CSI service, an operational CAMEL phase 2 T-CSI service does not need to be upgraded to CAMEL phase 3.

In broad lines, the capability for a VT-CSI service is the same as the capability of a T-CSI service; however, the VT-CSI service has some unique functionality. An overview of VT-CSI service capability is listed in Table 5.10.

CAMEL phase 4 also introduces a number of features that may be used in a VT-CSI service. When a call is routed to the VMSC, call routing has been completed already. That means that

Table 5.10   VT-CSI capability (not exhaustive)

Terminating AoC	CAMEL control of AoC requires a CAMEL relation between the SCP and the VMSC of the called subscriber
Service interaction	The SCP may control the GSM supplementary services that may be used by the called subscriber
Late call forwarding	The SCP may receive a notification of late call forwarding or may force call forwarding

the VT-CSI has no control over the call routing. Practically, the control for the VT-CSI relates to interaction that requires a direct link with the VMSC. Pre-paid for terminating call leg may be done with VT-CSI instead of T-CSI. If pre-paid charging is done purely for the roaming leg between GMSC and VMSC, then using VT-CSI prevents the invocation of an MT call CAMEL service in the case of early call forwarding; the VT call CAMEL service is invoked only when an actual call leg is established with the VMSC. If late call forwarding occurs, then the VT call CAMEL service is notified through the forwarding notification mechanism. The forwarding notification mechanism in VMSC works identically to the forwarding notification mechanism in GMSC: before GSM call forwarding is performed, the VT call CAMEL service is notified, so that it can stop the charging.

In the example in Figure 5.13, two CAMEL services are indicated. These services may act on the call at the same time. However, an operator may have the following implementation:

- when the called subscriber is in the HPLMN, terminating calls will be subject to a VT CAMEL service from the VMSC;
- when the called subscriber is in the VPLMN, terminating calls will be subject to a MT CAMEL service from the GMSC.

The above mechanism requires the following setting in the HLR: VT-CSI – send to VMSC in HPLMN, but not to VMSC in VPLMN; T-CSI – send to GMSC when the called subscriber is in VPLMN, but not when the called subscriber is in HPLMN. Although the capability to configure the HLR as described above is not defined in CAMEL, this capability is commonly available in HLRs. A rationale for the above mechanism may be that an operator may have CAMEL phase 3 capability in the HPLMN and hence can use VT-CSI. When the subscriber is roaming, she may register in a CAMEL phase 2 MSC and, in that case, terminating call handling is done in GMSC in HPLMN.

5.2.3.1 Charging

The CRN and GMSC address (GMSCA) that are reported to the SCP are the CRN and GMSCA from the GMSC where the terminating call handling is performed. CRN and GMSC are transported via MAP from GMSC to HLR (in MAP SRI) and from HLR to VLR (in MAP PRN); Figure 5.14. The combination of CRN and GMSCA may be used to correlate the CDRs that may be created for the terminating call. The GMSCA may provide an indication of the location of the GMSC and hence may be used to determine the charge for the roaming leg.

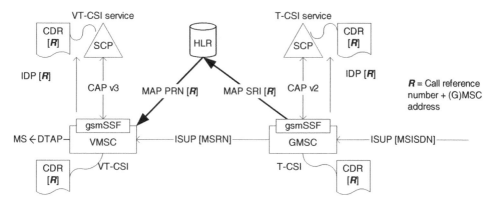

Figure 5.14 CDR correlation for terminating calls

The location information that is reported to the SCP in CAP IDP for the VT-CSI service is still the *stored location information* for the called subscriber. The reason is that CAP IDP is sent to the SCP before alerting of the called party takes place. The location information in the VLR is updated after paging response.

5.2.3.2 Late Call Forwarding

When late call forwarding occurs in the VMSC, the VT-CSI CAMEL service will receive a forwarding notification. If the SCP responds with CAP CUE, then the forwarding continues. The forwarding may be subject to ORLCF. In that case, the roaming leg between GMSC and VMSC is released. The VT-CSI CAMEL service is also released. The SCP may induce late call forwarding in the VMSC by sending CAP CON to the gsmSSF in the VMSC. This SCP-induced late call forwarding may not be subject to ORLCF. The reason is that the 'forwarding reason' in MAP RCH cannot indicate that the forwarding is caused by the SCP.

5.2.4 CAMEL Service Invocation at Call Failure

CAMEL phase 3 defines additional TDPs for the O-BCSM and T-BCSM. Table 5.11 lists the available TDPs for CAMEL phases 1–3. The DPs route select failure, T_Busy and T_No_Answer relate to call establishment failure. When the O-BCSM or T-BCSM makes a state transition to one of these DPs, a CAMEL service may be invoked. These DPs may be used as *event* detection points within an already active CAMEL service. However, by using them as *trigger* detection points, the invocation of a CAMEL service may be restricted to call establishment failure cases. In cases where the CAMEL service needs to act only in these failure cases, arming route select failure, T_Busy and T_No_Answer as TDP may prevent unnecessary CAMEL service invocation; the service is invoked only when necessary. Once the service is invoked, the SCP may generate a follow-on call.

The O-CSI for one subscriber may contain both TDP Collected_Info and TDP route select failure. Although route select failure (and T_Busy and T_No_Answer) may be armed as TDP, possible EDP arming for these DPs takes precedence. This implies that no CAMEL service triggering will take place if a CAMEL service is active already. The precedence rules are:

IF CAMEL service is active THEN
    report DP as EDP, if dynamically armed
ELSE
    report DP as TDP, if statically armed and trigger conditions are fulfilled.

Table 5.11 Overview of TDPs in CAMEL phases 1–3

Trigger detection point	CAMEL phase		
	Phase 1	Phase 2	Phase 3
O-BCSM			
Collected information	√	√	√
Analysed info			√
Route select failure			√
T-BCSM			
Terminating attempt authorized	√	√	√
T_Busy			√
T_No_Answer			√

The definition of TDP route select failure applies to O-CSI for MO calls and to MF calls (in GMSC and in VMSC). The definition of TDP T_Busy and T_No_Answer applies to T-CSI for MT calls and to VT-CSI for VT calls.

Note: when CAMEL phase 3 was developed, O_Busy and O_No_Answer were also proposed as TDPs. However, the O_Busy and O_No_Answer events pertain to conditions related to the called party, rather than conditions related to the calling/forwarding party (i.e. the served CAMEL subscriber). For that reason, it was decided that O_Busy and O_No_Answer should not be defined as TDP.

5.2.4.1 Trigger Criteria

Route select failure, T_Busy and T_No_Answer, when defined as TDP, may have trigger criteria associated with them. These criteria, which may be defined per TDP, consist of a list of up to five ISUP cause values.[7] When the trigger criteria are present, triggering will take place only when the ISUP cause value related to the call establishment failure matches any of the cause values in the trigger criteria. See Figure 5.15 for an example. The trigger criteria for route select failure, and the other DPs enable the operator to refine the CAMEL triggering even further.[8]

5.2.4.2 Call Forwarding

The triggering of the CAMEL service for call establishment failure may be used in combination with call forwarding notification. When call forwarding occurs, possibly in combination with ORLCF, then an internal signal is provided to gsmSSF, enabling the gsmSSF to inform the SCP about the pending call forwarding. In this case, the gsmSSF may trigger a CAMEL service, if the related TDP (e.g. T_Busy) is armed as TDP and no CAMEL service is active at that moment. CAP IDP will in that case contain the gsm-ForwardingPending parameter.

5.2.5 Service Interaction Control

Service interaction control is a mechanism that enables a CAMEL service to control which GSM supplementary services may be used by a subscriber. A subscriber may subscribe to a range of GSM supplementary services and then, per call, the CAMEL service indicates to the gsmSSF which services may be used in that call. This service interaction control is done primarily with the

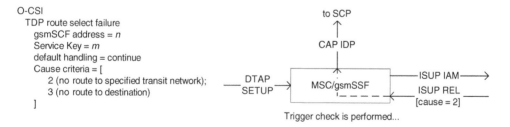

Figure 5.15 Triggering check for TDP route select failure

---

[7] Analogous to the basic service trigger criterion for DP collected info and DP terminating attempt authorized, it could have been considered for CAMEL to define *cause group codes* as a permissible cause value in the trigger list, e.g. cause = 0000 0000 for 'all causes in Class (000): normal event'.

[8] Defining criteria for TDP T_No_Answer is possible, but does not make sense. The only ISUP cause value that may lead to DP T_No_Answer is 'No answer from user (user alerted)' (ISUP cause = 19).

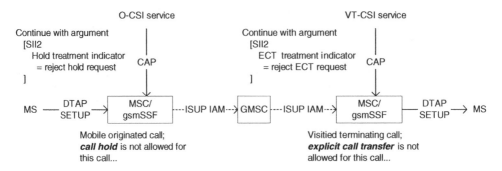

Figure 5.16 Local service interaction control

parameter '*Service Interaction Indicators 2*' (SII2).[9] ETSI CS1 (ETS 300 374-1) defines the service interaction indicators (SII) parameter. The usage of the SII in ETSI CS1 is, however, undefined. ETSI CS2 (EN 301 140-1 [159]) defines the SII2 parameter with usage similar to SII2 in CAMEL phase 3.

For MO and MF calls, the service interaction control may be used both in the subscribed service (O-CSI) and in the dialled service (D-CSI, N-CSI). There are three categories of service interaction control: (1) local service interaction control; (2) forward service interaction control; and (3) backward service interaction control.

5.2.5.1 Local Service Interaction Control

Figure 5.16 reflects the usage of local service interaction control. Local service interaction control entails that the CAMEL service indicates to the gsmSSF that a particular GSM supplementary service may not be used during that call. In the two example cases in Figure 5.16, the subscriber has a subscription to call waiting, call hold and explicit call transfer. The CAMEL service, however, disallows call hold (in the MO call) and explicit call transfer (in the VT call). The following GSM supplementary services can be controlled in this manner:

- *Call hold* – the CAMEL service can indicate that the call may not be placed on hold.
- *Call waiting* – the CAMEL service can indicate that, during this call, no call waiting can be applied, i.e. no second call may be offered to the subscriber.
- *Explicit call transfer* – the CAMEL service can indicate that explicit call transfer may not be applied to this call.

The local service interaction control pertains to GSM supplementary services that involve user control at the VMSC. Therefore, this type of service interaction control applies to MO calls and VT calls only.

5.2.5.2 Forward Service Interaction Control

Figure 5.17 provides an example of *forward* service interaction control. Forward service interaction control entails that a CAMEL service may control whether particular GSM supplementary service(s) may be applied to the call at a further point in the call chain. This control is accomplished by placing applicable parameters in the ISUP signalling flow in the forward direction, during call establishment. These parameters are provided to the gsmSSF in CAP CON or CAP CWA and are mapped to the

---

[9] SII2 is also used in CAMEL phase 2. However, SII2 in CAMEL phase 2 is used to create a through-connection during user interaction.

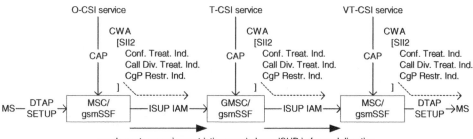

Conf. Treat. Ind.   = conference treatment indicator
Call Div. Treat. Ind. = call diversion treatment indicator
CgP Restr. Ind.   = calling party restriction indicator

Figure 5.17   Forward service interaction control

corresponding parameter in ISUP IAM (see dashed arrow in Figure 5.17). The Annex to 3GPP TS 29.078 [106] specifies the details of this mapping. The following GSM supplementary services can be controlled in this manner:

- *Multi-party* – the CAMEL service can indicate that the call may not become part of a multi-party call;
- *Call forwarding* – the CAMEL service can indicate that the call may not be subject to GSM call forwarding. This restriction applies to all call forwarding categories and to call deflection. When the GMSC processes a terminating call with the call forwarding restriction, the GMSC includes this restriction in the MAP SRI to HLR, so the HLR can suppress call forwarding for this call.[10]
- *Calling line identification presentation* – the CAMEL service can indicate that the calling line identity (CLI) of the calling subscriber may not be presented to the called party. If the CLI needs to be suppressed, then this is done by setting the *Address presentation* in the header of the CLI to 'presentation restricted'.[11] This restriction is applied to both the calling party number and the additional calling party number.

The forward service interaction control may be applied to MO calls, MF calls, MT calls and VT calls.[12] In Figure 5.17, the example call runs from MSC to GMSC and VMSC. Since the forward service interaction indicators are applied in the *forward* direction, these indicators, when used in MO call control, have no effect on service availability for the calling party. For example, if the CAMEL service includes the conference treatment indicator in CAP CON when controlling an MO call, the calling subscriber may invoke MPTY, but the called party may not invoke MPTY.

Owing to the nature of the forward service interaction, the service restriction applies per outgoing call leg. If a follow-on call is created, the CAMEL service may provide a different set of forward service interaction indicators.

5.2.5.3 Backward Service Interaction Control

Figure 5.18 provides an example of *backward* service interaction control. Backward service interaction control entails that a CAMEL service may control whether particular GSM supplementary

---

[10] Another commonly applied method to suppress call forwarding for a call is setting the *redirection counter* in the *redirection information* parameter in ISUP IAM to its maximum value.

[11] The actual CLI is not removed from the signalling. One of the reasons is that the CLI is still needed for called subscribers who subscribe to CLIR override. The CLI is also needed for lawful intercept.

[12] In CAMEL phase 3, the CLI restriction may be applied in MO calls only; in CAMEL phase 4, the CLI restriction may be applied to other call types as well.

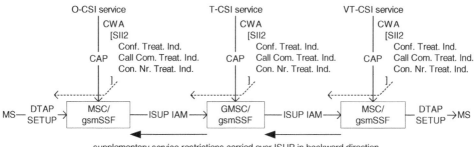

Figure 5.18   Backward service interaction control

service(s) may be applied to the call at an *earlier* point in the call chain. This control is accomplished by placing applicable parameters in the ISUP signalling flow in the backward direction, at call acceptance or at call answer. These parameters are provided to the gsmSSF in CAP CON or CAP CWA and are mapped to the corresponding parameter in ISUP ACM, CPG, ANM, CON or REL (see dashed arrow in Figure 5.18). The Annex to 3GPP TS 29.078 [106] specifies the details of this mapping. The following GSM supplementary services can be controlled in this manner:

- *Multi-party* – the CAMEL service can indicate that the call may not become part of a multi-party call. This restriction is included in ISUP ACM, CPG, ANM or CON. An MPTY request from the calling party will be rejected in this case.
- *Call completion on busy subscriber* – the CAMEL service can indicate that CCBS may not be applied to this call. This restriction is included in ISUP REL.[13] A CCBS request from the calling party will be rejected in this case.
- *Connected number treatment indicator* – the CAMEL service can indicate that the connected line identity (COL) of the connected party may not be presented to the calling party. This restriction is included in ISUP ANM or CON. The *address presentation in* COL will be set to 'presentation restricted' in this case. In order to obtain the COL of the connected party presented to the calling party, the ISUP IAM for the outgoing call shall contain the parameter 'Optional forward call indicators' with '*Connected line identity request indicator*' = *requested*. If the CAMEL service restricts the presentation of the COL to the calling party, then that does not affect this request-for-COL indicator in ISUP IAM.[14]

The backward service interaction control may be applied to MO, MF, MT and VT calls. Since the backward service interaction indicators are applied in the *backward* direction, these indicators, when used in VT call control, have no effect on service availability for the called party. For example, if the CAMEL service includes the conference treatment indicator in CAP CWA when controlling a VT call, the called subscriber may invoke MPTY, but the calling party may not invoke MPTY.

---

[13] A 'CCBS possible' parameter may be present in the diagnostics field of the ISUP cause in ISUP REL. The SCP may prevent CCBS by instructing the gsmSSF to remove that parameter from the diagnostics field.

[14] One of the reasons for not affecting the request-for-COL indicator in ISUP IAM is that the actual COL may still be required for a subscriber who has a COLR-override category. Refer to 3GPP TS 23.081 for more information on the COLR override.

Owing to the nature of the backward service interaction, the service restriction applies for the calling party for the entire call. If a follow-on call is created, then the service restriction for the calling party remains applicable.

A CAMEL service may use any local service interaction control, forward service interaction control or backward service interaction control and then release the control relationship with the gsmSSF. An example is CAP IDP followed by CAP CWA. In that case, the local and backward service interaction control remains applicable for the remainder of the call, even without the further CAMEL control.

5.2.6 Call Gapping

Call gapping is a mechanism that enables an SCP to reduce the number of CAMEL service invocations from a specific MSC/gsmSSF. The CAP call gap (CG) operation may be used at any moment during an active CAMEL dialogue. When the call gap instruction is issued to a gsmSSF, the *call gapping*, i.e. the reduction of CAMEL service invocations, applies to that gsmSSF for subsequent calls. The call gap conditions are set in the gsmSSF for a particular duration (up to 24 h), but may subsequently be explicitly removed by the SCP. A call gap instruction has no effect on the CAMEL dialogue in which the call gap instruction is provided. See Figure 5.19 for a graphical representation.

Call gap may be applied due to near-overload condition in the SCP. When the near-overload condition occurs in the SCP, the SCP waits for every MSC in its network to initiate a CAMEL dialogue. As soon as an MSC initiates a CAMEL dialogue, the SCP sends the call gap instruction to that MSC. Call gapping is now active in that MSC and prevents CAMEL service invocation for selective CAMEL services.[15] The call gap criteria apply to an entire MSC. That means that, when a CAMEL service that is invoked with O-CSI applies call gap, this may also affect CAMEL service invocation related to D-CSI, VT-CSI, etc., in this MSC. However, proper use of call gap criteria (see below) may ensure that call gap is applied to selected services only. Call gap may be used within an operator's own PLMN only. Here, the gsmSSF compares the address of the gsmSCF that sends call gap with the address of the MSC in which it (i.e. the gsmSSF) resides. The gsmSCF address is in this case the address from the CSI. The details of this check are slightly implementation-dependent. The gsmSSF address and the MSC address are both ISDN address strings containing an E.164 number in international format (Figure 5.20).

One operator may have more than one NDC allocated to it. Hence, even when the SCP that issues the call gap instruction belongs to the same PLMN as the MSC, the gsmSCF address may have an NDC value that differs from the NDC value of the MSC. Hence, the gsmSSF will know (i.e. will have adequate configuration for) which NDC values belong to its own PLMN. Such NDC

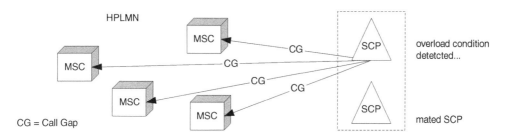

Figure 5.19   Call gapping (in HPLMN) due to an overload condition

---

[15] CAMEL phase 4 includes the capability for the SCP to initiate a CAMEL dialogue with a particular MSC. However, the capability of the SCP-initiated CAMEL dialogue does not include the call gap operation.

Figure 5.20   Structure of international E.164 number

configuration, for the call gap acceptance check, may also be used when a single SCP would be serving more than one PLMN.

The SCP may define criteria for the call gapping to be applied. This implies that, for a particular MSC/gsmSSF, only specific calls will be subject to call gapping. The criteria that may be provided by the SCP consist of a combination of the following elements:

- *Called address* – the gsmSSF applies call gapping if the called number matches the called address criterion.
- *Service* – the gsmSSF applies call gapping if the service key from the applicable CSI matches the service criterion.
- *Calling address* – the gsmSSF applies call gapping if the calling party number (if available) matches the calling address criterion.
- *SCF Id* – the gsmSSF applies call gapping if the gsmSCF address from the applicable CSI matches the SCF Id criterion. If the gsmSCF address of the CSI leads to the selection of one of a group of SCPs (dynamic load sharing), then the call gapping applies to all SCPs of this group.

When the gsmSSF applies call gap for a CAMEL service, then the call for which the CAMEL service had to be invoked may continue or may be released. The action to be taken by the gsmSSF is determined by the DCH parameter in the CSI. When the DCH indicates 'continue', then the gsmSSF allows the call to continue and indicates in CAP IDP that the call gap check has been performed for this call (*Call Gap Encountered*). When the DCH indicates 'release', the gsmSSF applies the gap treatment to the call. Gap treatment forms part of the call gap instruction. The gap treatment may consist of either one of:

- *Announcement* – this parameter identifies an announcement or tone to be played to the calling party. Call gap is used only within an operator's own PLMN, hence the CAMEL service should know which announcement code to use. When the playing of the announcement or tone is complete, the gsmSSF uses a default ISUP cause value in the ISUP release in the backwards direction.
- *Release cause* – this parameter is used in the ISUP release in the backwards direction. The use of cause value and location information is identical to the use of these parameters in the release call operation (see Chapter 4). In addition, the release cause may include diagnostics information. This is further explained in Section 5.2.10.

When the gsmSSF applies DCH, this is indicated in the CDR for the call. With the introduction of Call Gap, DCH has a dual meaning: (1) DCH is used in the case of CAP signalling error; and (2) DCH is used in the case of non-invocation of CAMEL service due to call gap. Billing systems may use the DCH parameter to post-process a CDR when the CAMEL service for that call cannot be invoked. This mechanism may therefore also be applied to calls that were subject to call gap. Placing DCH in the CDR when DCH is caused by call gap is an implementer's option.

If more than one CAMEL service is invoked for one call, then for each CAMEL service invocation, the GAP criteria will be checked (Figure 5.21). In the call case in Figure 5.21, two CAMEL services apply for the call: a subscribed service, based on the presence of O-CSI, and a subscribed

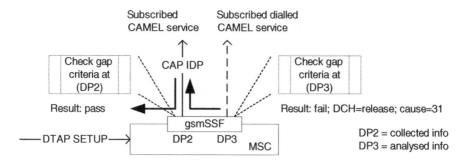

Figure 5.21   Call gap in multi-CAMEL service call case

dialled service, based on the presence of D-CSI. Although a single set of call gap criteria may be active in the gsmSSF, the outcome for the two CAMEL service invocations may differ. This is due to the fact that different call parameters apply for the two CAMEL services; called party number, service key and gsmSCF address may be different. At DP analysed info, the trigger check for the subscribed dialled service is performed. The result of this check may be a service key, gsmSCF address and default call handling. The trigger check is then followed by the call gap check. In the example of Figure 5.21, the call gap check fails and the call is released with ISUP cause 31 ('normal unspecified'). This release is reported to the subscribed service, which may release the call or take other action.

5.2.7 Support of Long Forwarded-to numbers

As from 3GPP R99, a forwarded-to number (FTN) may have a length of up to 28 digits. In earlier GSM versions, the maximum number of digits was 15.[16] The increased length of the FTN is not required for call routing, but for carrying service selection information; the service selection information is 'padded to the number'. This method is useful for forwarding to a PABX with service selection capability, for example.

The encoding of the FTN is as follows:

GSM R97 encoding of forwardedToNumber
    ISDN-AddressString ::= AddressString (SIZE (1..9))
3GPP R99 encoding of longForwardedToNumber,
    FtN-AddressString ::= AddressString (SIZE (1..15))

Using the long FTN in a GSM network requires that the signalling system (e.g. ISUP) in that network is capable of transporting a number of this length. The long FTN is sent from HLR to VLR or GMSC in same the manner as for 'normal FTN', provided that VLR and GMSC support the long FTN.

The long FTN has the following impact on CAMEL.

5.2.7.1 Initial DP

The called party number (CdPN) in CAP IDP has increased length. The encoding of the CdPN in CAP is as follows:

---

[16] Although the encoding of the forwarded-to number allows for 16 digits, 3GPP TS 23.082 specifies a maximum number of 15 digits.

GSM R97 (CAMEL phase 2) encoding of forwardedToNumber,
  CalledPartyNumber ::= OCTET STRING (SIZE (3..12))
3GPP R99 (CAMEL phase 3) encoding of forwardedToNumber,
  CalledPartyNumber ::= OCTET STRING (SIZE (2..18))

The encoding of the contents of the CdPN in CAP is defined in ITU-T Q.763 [137]; two octets are used for header information. For CAMEL phase 2, the CdPN may contain a maximum of 20 address digits; in CAMEL phase 3, the CdPN may contain 32 address digits. It is not foreseen that a call to a long FTN will be subject to further call forwarding. The length of the redirecting party Id is therefore not increased.

When a call to an FTN containing more than 20 digits is controlled by a CAMEL phase 2 service, then the MSC has to truncate the FTN when building the CAP IDP.[17] The effect of this truncation depends on the significance of the digits that are removed in CAP IDP. The call to the long FTN may be subject to further CAMEL service invocation due to D-CSI or N-CSI. Since D-CSI and N-CSI services use CAP v3 (or CAP v4), these services will receive the complete long FTN.

An operator may use the long FTN for the purpose of conveying service selection data to the gsmSCF. When call forwarding takes place, the forwarded call invokes a CAMEL service. The long FTN is reported to the gsmSCF. The FTN may be structured as follows:

$$FTN = \langle address\ digits\rangle\langle service\ selection\ digits\rangle$$

The CAMEL service uses the service selection digits as input to its service processing and then removes the service selection digits from the FTN. The service then sends CAP CON with a destination routing address, containing only the original address digits.

5.2.7.2 Connect

The CAP connect operation may contain a destination routing address that can contain a long FTN. The encoding of the destination routing address is as follows:

DestinationRoutingAddress {PARAMETERS-BOUND : bound} ::= SEQUENCE
SIZE(1) OF
  CalledPartyNumber {bound}

Hence, the gsmSCF may supply a destination routing address containing up to 32 address digits.

5.2.7.3 Optimal Routing at Late Call Forwarding

When ORLCF is applied, the long FTN may be supplied by the HLR to the GMSC during the late call forwarding handling. (Figure 5.22). The method depicted in Figure 5.22 is referred to as *forwarding interrogation.*

The forwarding interrogation from Figure 5.22 enables the operator to send a 'normal FTN' to the VLR. This normal FTN may be a default FTN. When call forwarding occurs from the VLR and ORLCF is applied, the GMSC will receive the normal FTN in the MAP RCH information flow. The GMSC contacts the HLR using MAP send routing information (SRI); the HLR responds by supplying the long FTN to the GMSC. The long FTN overrides the default FTN. The call is now forwarded to the forwarding destination by means of the long FTN. The GMSC uses the forwarding interrogation only when the MAP SRI that contained the MSRN, also contains the 'forwarding interrogation required' parameter.

---

[17] CAMEL does not formally specify this truncation, but is unlikely that an MSC would reject the call in this case. This truncation should in any case apply to CAP only, not to the further call handling over ISUP.

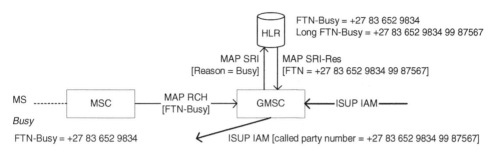

Figure 5.22 Forwarding interrogation during ORLCF handling

If CAMEL handling applies in the GMSC for the MF call, then the long FTN is reported to the SCP. Here as well, if a CAMEL phase 2 service is used for the MF call, then the long FTN in CAP IDP will be truncated.

5.2.7.4 Call Deflection

A subscriber may deflect an incoming call to a destination that has address length in access of 15 digits, a 'long deflected-to-number' (DTN). This requires adequate support in the UE and in the MSC. The further handling of call deflection to a long DTN is similar to the handling call forwarding to a long FTN. The deflected call may be subject to ORLCF and may be subject to CAMEL control. The forwarding reason that is used in MAP RCH for call deflection is 'unconditional'. The GMSC may perform forwarding interrogation during ORLCF handling for forwarding reasons busy, no answer and not reachable. Hence, the GMSC will not perform forwarding interrogation for ORLCF for a deflected call. For further details on the forwarding interrogation mechanism, refer to 3GPP TS 23.082 [84].

5.2.8 On-line Charging Enhancements

CAMEL phase 3 contains a number of enhancements related to on-line charging control. These enhancements relate to certain CAP operations.

5.2.8.1 Free Format Data

The following two enhancements related to CAP FCI are introduced in CAMEL phase 3.

Increased Free Format Data Size

A CAMEL service may place up to 160 octets of free format data in the CDR that is generated in the MSC, as opposed to 40 octets of free format data for CAMEL phase 2. The 160 octets of free format data may be provided for the incoming leg and for each outgoing leg. The limit of 160 octets per CAP FCI allows for the inclusion of one or more extension fields in CAP FCI even when Blue SCCP is used. See Chapter 2 for details related to SCCP message length.

Data Append

The CAMEL service may append free format data to the free format data that was already provided to the gsmSSF. In CAMEL phase 2, free format data always overwrites the free format data that was previously provided for a call leg. See Figure 5.23 for the append mechanism.

The data append mechanism is backwards-compatible with FCI in CAMEL phase 2. The CAMEL service may send multiple CAP FCI operations during a call for one specific leg. If CAP FCI does not include the *append indicator*, then the free format data carried in CAP FCI overwrites the free

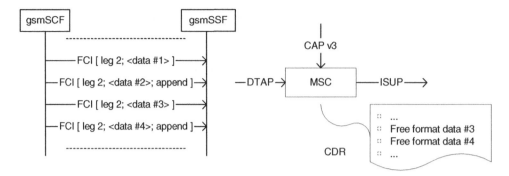

Figure 5.23   Free format data appending

format data that was provided to the gsmSSF for the indicated leg in this call. If, however, CAP FCI contains the *append indicator*, then the free format data carried in CAP FCI is appended to the free format data that was provided to the gsmSSF for the indicated leg in this call.

In the example in Figure 5.23, the gsmSCF provides free format data 1, which is appended with format data 2 by the next CAP FCI. The following CAP FCI replaces data 1 + data 2 with data 3. Finally, the last CAP FCI appends data 3 with data 4. The free format data that is placed in the CDR at the end of this call leg is data 3 + data 4.

One practical example is an on-line charging service that uses CAP FCI to indicate in the CDR the rate of the call, based on information received in CAP IDP. Then, at answer, the CAMEL service receives in the answer notification an indication of the destination address, which may differ from the dialled number (see Section 5.2.8.3 on deferred charging calculation). The CAMEL service may re-calculate the rate of the call and overwrite the free format data in the CDR with the re-calculated charge. Then, at the end, CAMEL service uses the FCI-append mechanism to indicate in the CDR how much was charged for the call.

The total amount of free format data for one specific call leg, including the appended data, should not exceed 160 octets. The reason for this limitation is that the free format data needs to be placed in a transfer account procedure (TAP) file, along with other data from the CDR. The TAP file may need to be sent to a charging system in another PLMN. A too large free format data field might complicate the handling of the TAP file.

5.2.8.2 Charging Continuity Monitoring

The on-line charging mechanism uses the ACH and ACR operations. During a call for which on-line charging is used, sequences of CAP ACH (instruction) and CAP ACR (response) are used. When the gsmSSF sends CAP ACR, it allows the call to continue, but expects the next CAP ACH. Chapter 4 explains that, for CAMEL phase 2, the call continues in the case that the next CAP ACH does not arrive for any reason. To prevent a call continuing in the case of signalling failure, CAMEL phase 3 introduces an ACH-monitoring mechanism. When the gsmSSF sends CAP ACR, it starts the Tccd timer; Tccd is stopped when the next ACH arrives. If the next ACH does not arrive, then Tccd expires and results in the release of the call. The maximum duration for Tccd is defined per operator, but should not exceed 20 s. Figure 5.24 gives a graphical representation of this feature.

The ACH-monitoring mechanism may be used during a through-connected call and during user interaction. When Tccd expiry occurs, the call is always released; the default call handling parameter is not used in that case. A further result is that there is no special indication in the CDR about the forced call release. It might have been useful if the call release due to Tccd expiry was reflected

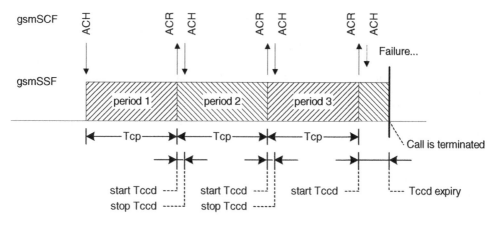

Figure 5.24 Charging continuity monitoring

in the CDR; that would enable CDR processing systems to post-calculate the correct charge of the call and apply post-charging for the unpaid portion of this call.

5.2.8.3 Deferred Charging Calculation

Description of charging mechanisms so far have assumed that all the information that is needed for calculation of the cost of the call is known at the start of the CAMEL service. For an MO call specifically, the above implies that, at DP collected info, the required information is available. This may, however, not be the case for all call scenarios. The number that was dialled by the calling subscriber may be subject to number translation by subsequent processes in the MSC. The call will in that case be established to a different destination than indicated by the dialled number. In another call scenario, the subscriber dials a short code number; a dialled service translates the number to the appropriate destination. (Figure 5.25).

In the example in Figure 5.25, the subscriber dials dest. #1; the subscribed CAMEL service (O-CSI) uses dest. #1 for an initial credit check. This destination address for the call is subsequently modified by the D-CSI service and the N-CSI service, respectively. The B-number analysis that

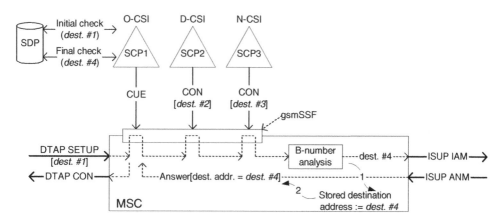

Figure 5.25 Reporting of destination address at answer

follows the N-CSI service also modifies the destination number. The subscribed CAMEL service is, however, aware only of the number that was reported to it, i.e. dest. #1. The final destination number that is used to populate the ISUP IAM, dest. #4, is stored in the MSC (see 1 in Figure 5.25). When the answer event is reported to the subscribed CAMEL service, the stored destination number (dest. #4) is included in the answer notification (see 2 in Figure 5.25). The subscribed CAMEL service uses this number for a final credit check.

Any change to the destination of the call after sending of the ISUP IAM, such as call forwarding, does not affect the reported destination number. Hence, it is not the *connected number* in ISUP ANM that is reported to the subscribed CAMEL service.

The reporting back of the destination number of the call is also used when the MSC receives ISUP REL. The internal release message that is reported to the subscribed CAMEL service contains the stored destination number. If the ISUP REL results in the triggering of a route select failure CAMEL service (see Section 5.2.4), then the stored destination number is used as the called party number for CAP IDP.[18]

5.2.8.4 CAMEL Warning Tone

In CAMEL phase 2, the playing of the warning tone is possible only in combination with the forced call release. CAMEL phase 3 decouples the playing of the warning tone and the forced call release due to call period expiry (Tcp expiry). Therefore, the warning tone may be played in the following call termination methods: (1) gsmSCF-enforced call release (i.e. CAP Release Call); and (2) forced call release (i.e. Tcp expiry with Release If Duration Exceeded = True).

A service may want to use gsmSCF-enforced instead of forced call release, so it can place free format data in the CDR prior to releasing the call. This free format data may contain an indication of the reason for the call release (e.g. low-credit). The CAMEL phase 3 service may include the warning tone instruction in CAP ACH related to the last call period, but not the release if duration exceeded parameter. The gsmSSF plays the warning tone 30 s prior to Tcp expiry, but will not force-release the call at the end of the call period. When the gsmSSF has sent the CAP ACR to the gsmSCF, the gsmSCF may use CAP FCI, followed by CAP RC.[19] In fact, the gsmSCF may instruct the playing of the warning tone multiple times. See Figure 5.26 for an example sequence flow.

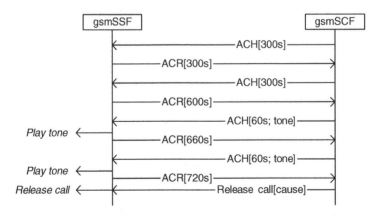

Figure 5.26   Multiple CAMEL warning tones

---

[18] It would have been useful if the stored destination number were included not only in *TDP* route select failure, but also in *EDP* route select failure.

[19] Should, for whatever reason, this release call instruction not arrive, then the Tccd timer will prevent uncharged call continuation.

5.2.9 Multiple Subscriber Profile

The MSP GSM supplementary service is a mechanism that enables subscribers to have multiple profile definitions in the HLR. The subscriber has a single SIM, and hence a single IMSI, but may have a different MSISDN per profile. The subscriber may select a profile as the active profile. Each profile may have different settings for, amongst others, supplementary services.

Two phases of MSP exist:

- MSP phase 1; refer GSM TS 02.97 [25] and GSM TS 03.97 [48]. This phase of MSP makes use of CAMEL Phase 2 capability.
- MSP phase 2; refer 3GPP TS 22.097 [67] and 3GPP TS 23.097 [85]. This phase of MSP makes use of CAMEL phase 2 and CAMEL phase 3 capability.

The present section describes MSP phase 2 only. One of the principles of MSP is that a subscriber always has an active profile; the profile determines what capabilities the subscriber has at any moment. Although MSP is provisioned in the HLR as *supplementary service*, MSP itself does not provide a service; it controls which *other services* are available at any moment.

MSP profiles are kept in two places:

- *HLR* – the HLR contains between one and four profiles; at any moment, one profile is the *active* profile;
- *SCP* – the SCP contains between one and four profiles; at any moment, one profile is the *active* profile.[20]

The profiles in HLR and SCP are provisioned by the operator. The settings in the HLR related to individual supplementary services may be changed by the subscriber, in accordance with the procedures that are specified in 3GPP TS 22.030 [62]. An operator may prohibit the changing of the supplementary services settings by MSP subscribers; the MSP subscriber can change between profiles, but not modify the profiles. Each profile in HLR or SCP may contain an indication about the availability of the following supplementary services and network services: call barring; operator-determined barring; call hold; explicit call transfer; multi-party call; completion of calls to busy subscriber; call waiting; and calling line identification restriction. A profile may also include certain charging characteristics, e.g. different charging may apply for a call that is established with the corporate profile than for a call that is established with a private profile. MSP applies to both circuit switched calls and MO-SMS. For MO-SMS, only a subset of the above-listed services is applicable.

5.2.9.1 Using MSP

When a subscriber registers in a network, the level of CAMEL support in that network determines how MSP is applied to that subscriber whilst registered in that network.

The Network Supports CAMEL Phase 3 or Higher

The HLR will not send the service settings from the active profile to the VLR. Instead, the HLR sends for all subscribed services the state 'active, operational' to the VLR. In addition, the HLR sends CAMEL phase 2 and CAMEL phase 3 subscription data to the VLR. Outgoing, forwarded and terminating calls from this subscriber will result in the invocation of a CAMEL phase 3 MSP service. The MSP service uses the currently active MSP profile stored in the SCP. The profile indicates, amongst others, which supplementary service may be used by the subscriber. The SCP may use the *service interaction indicators 2* parameter in CAP to apply the service restrictions for this call.

---

[20] Although MSP specifies a maximum of four profiles per subscriber, CAMEL does not prohibit the provisioning of more than four profiles in the SCP.

One example is that the HLR sends subscription data for ECT and MPTY. However, when the subscriber establishes a call, the MSP CAMEL service determines that, according to the currently active profile in the SCP, ECT and MPTY are not allowed for this subscriber. The permission to use ECT and MPTY may also depend on dynamic factors such as location and time. The SCP prevents the use of ECT and MPTY by sending CAP CON or CAP CWA[21] to the MSC; the CON or CUE contains SII2 with an indication that ECT and MPTY are not allowed.

The Network does not Support CAMEL Phase 3 or Higher
The HLR will send the active profile to the VLR. The active profile contains settings for the different services. Calls from this subscriber will not lead to the invocation of the MSP service.

5.2.9.2 Setting the Active Profile

The subscriber may use call-related or call-unrelated methods to select a particular profile in HLR or SCP as active profile.

HLR Profile Selection
A subscriber may use the following digit string to select the active MSP profile (see 3GPP TS 22.030 [62]):

$$* \ 59n \ \# \ SEND; n \ \text{identifies the profile}; 1 \leq n \leq 4.$$

SCP Profile Selection
MSP specifies that USSD will be used to select the active profile. MSP does not specify the exact USSD string that will be used for selecting the active profile, but at least the following information should be supplied: USSD service code – this will indicate *MSP selection*; and profile indicator. As an operator option, the USSD string may also include a password.

Although it is not explicitly mentioned in the MSP specification, an operator may use call- or SMS-related mechanisms to select the profile in the SCP. Examples include:

- *Dialled services* – a dialled service, based on D-CSI, may be activated by dialling a specific number, e.g. 0800 59*n*, where *n* identifies the profile. The dialled service sets the identified profile as active profile and then releases the call.
- *User interaction* – a CAMEL service may be invoked, e.g. a dialled service, that uses user interaction to prompt the subscriber to enter the desired profile on the terminal.

MSP does not specify any coordination between the MSP profile in HLR and the MSP profile in SCP. Hence, an MSP subscriber may change the active profile in SCP, but not in HLR. An operator may take special provision that these two profiles are aligned.

5.2.9.3 Dynamic Profile Selection

When the subscriber is registered in a CAMEL phase 3 network, then the selection of profile to be applied to that call may be done in two ways: (1) the SCP uses the currently active profile; (2) the profile to be used is obtained from call data. This differs per call case.

MO Calls
The subscriber may include in the dialled number an explicit indication of the MSP profile. The user may use the following digit string to explicitly select the MSP profile:

$$\langle \text{Directory Number} \rangle * 59n \ \# \ SEND; n \ \text{identifies the profile};$$

---

[21] 3GPP TS 23.097 specifies the use of the connect only. However, the CAMEL MSP service may also use continue with argument.

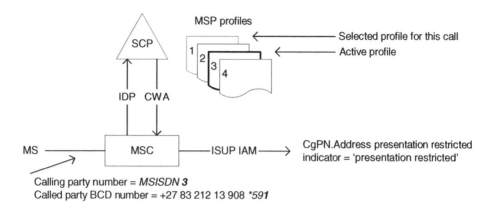

Figure 5.27   Explicit profile selection for MO call

Hence, the digits '*59*n* #' are added to the dialled number. A subscriber may not know whether she is registered in a CAMEL phase 3 network or not. If the network supports CAMEL phase 3, then the SCP will use the suffix to select the profile and remove the suffix from the dialled number. If the network does not support CAMEL phase 3, then call processing continues with the suffix. The serving MSC may remove the suffix from the number or retain it.

As explained earlier, each MSP profile in the HLR may have a different MSISDN. The MSISDN that is sent to the VLR may therefore differ per VLR registration process. When the subscriber initiates a call, the currently registered MSISDN is reported to the SCP. The SCP may use this MSISDN as an indication of the active MSP profile. In Figure 5.27, the active profile is profile 3, but the subscriber selects profile 1 for this call. For this profile, the calling line identity is set to 'restricted'.

MT Calls

Each MSP profile may include a different MSISDN. Hence, an MSP subscriber may be called with different numbers. The number that is used to call the MSP subscriber is used to select the profile that the SCP will use for that call.

For call control from the GMSC, based on the reception of T-CSI from HLR, the called MSISDN is available in the ISUP signalling that arrives at the GMSC; that MSISDN is reported to the SCP in the CAMEL service invocation. For call control from the VMSC, based on the reception of VT-CSI from VLR, the called MSISDN is not available in the ISUP signalling that arrives at the VMSC; that ISUP signalling contains the MSRN. The called MSISDN may, however, be included in the signalling flow from HLR to VLR during call establishment, i.e. in MAP PRN. The VMSC may then use this called MSISDN in the CAMEL service invocation. If the HLR does not provide the called MSISDN to the VLR, then the VMSC uses the registered MSISDN (in VLR) in the CAMEL service invocation.[22]

When the SCP applies a specific profile for the MT call, the SCP may indicate this with a designated alerting pattern. (See also Chapter 4 for the alerting pattern.) In the example in Figure 5.28, the subscriber has MSISDN-2 registered in the VLR. The calling party uses the MSISDN of profile 4 to call the MSP subscriber. Hence, profile 4 is used for controlling the call to the B-party. One practical use case may be that receiving a call addressed to this MSISDN implies that this terminating call is charged against the private account. The alert signal informs the user about the nature of the call, i.e. private call.

---

[22] The called party number in MAP PRN is transported in *external signal info*.

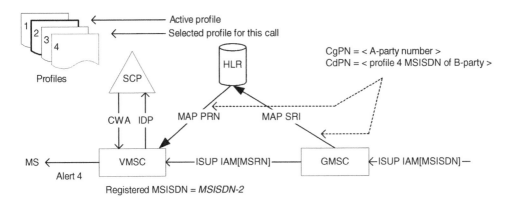

Figure 5.28   Implicit profile selection for MT call

The MSP specifications do not specify a mechanism to explicitly select an MSP profile for the MT call. However, the MSP profile to be used for an MT call could also be selected with the MSP suffix that is also used to select the MSP profile when setting up an MO call. The MSP profile selection for MT calls might e.g. be '*60n', where n indicates the MSP profile. The calling party may dial +27 83 212 13 908 *601. The *601 suffix does not affect the routing of the call and may be ignored by the HLR during MT call processing. The suffix is included in the CAMEL service invocation from the VMSC for the called party. The SCP may use this indication to select the profile for the MT call, irrespective of which MSISDN was used to call the subscriber.

The calling party might use a combination of MSP profile selection for the MO call (for the MSP service of the calling party) and MSP profile selection for the MT call (for the MSP service of the called party). Such a dial string may be +27 83 212 13 908 *591 *601. Explicit MSP profile selection for MT calls is, however, not standardized and would be a proprietary implementation.

Depending on the support of CAMEL phase 3 in HPLMN and in VPLMN, it may occur that, for one call, different MSP profiles are applied in the GMSC, through the T-CSI service, and in the VMSC, through the VT-CSI service.

MF Calls
For MF calls, the same explicit MSP selection mechanism applies in theory. That is to say, the subscriber may register an FTN including the MSP selection suffix (*59n #). However, an HLR may not accept the registration of such an FTN.[23]

Instead of using a suffix to the FTN for selecting the MSP profile for the MF call, the SCP should apply the same profile to the MF call as to the MT call. Presuming that a CAMEL service was invoked for the MT call, the SCP has selected a profile for the MT call. The invocation of the CAMEL service for the MF call will include the same combination of CRN + MSCA as the invocation of the CAMEL service for the MT call. Hence, the SCP, when receiving the service invocation for the MF call, could use the CRN +MSCA to check which profile was already selected for that call.

MO SMS
For MO SMS, the subscriber may select the profile by adding the MSP suffix to the destination address of the SMS. The SCP will remove the MSP suffix from the destination address. If the network does not support CAMEL phase 3 and the subscriber uses the MSP suffix for the SMS, then the SMS routing may continue with the MSP suffix in the destination address.

---

[23] In the case of call deflection or SCP-induced call forwarding, it may be possible for the DTN or FTN to include the MSP suffix.

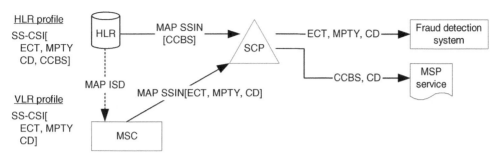

Figure 5.29 CCBS notifications

MT SMS

For MT SMS, no profile selection mechanism is defined by MSP.

5.2.9.4 CCBS Notifications

For the purpose of MSP, the SSIN feature is enhanced. This feature was introduced in CAMEL phase 2 and enables the SCP (or fraud detection system) to receive a notification when certain GSM supplementary services are invoked. These services include ECT, MPTY and CD.

CAMEL phase 3 adds CCBS to the list of GSM supplementary services for which a notification may be sent. The MSP service also uses the CD notifications. The CCBS notifications for CCBS are sent from HLR, not from VLR, as is the case for ECT, MPTY and CD (Figure 5.29).

CCBS state changes are reported to the HLR. The HLR then uses SS-CSI to send a notification to the SCP, which may use the notifications for the MSP service. The CCBS notification also includes the current CCBS state in the HLR.

In the example in Figure 5.29, the SSI notifications from VLR and HLR are sent to the SCP, which forwards the notifications to the fraud detection system and MSP service. The relaying of the SSI notifications through the SCP is required, because SS-CSI has a single gsmSCF address only. Furthermore, the invocation notifications related to CD are used for both fraud detection and MSP.

5.2.10 Other Enhancements to CAP

The following further enhancements to CAP are introduced in CAMEL phase 3.

5.2.10.1 Cause Diagnostics

In CAMEL phase 3, the cause parameter, when included in the CAP information flow, may contain the *diagnostics*. The *diagnostics* is part of the ISUP cause element, which is included in ISUP Release.[24] The ISUP cause is defined in ITU-T Q.850 [142]. It contains additional information related to the reason for the release of the call. Not all cause values may have diagnostics associated with it; see the examples in Table 5.12. The cause parameter may be used in CAP in two directions:

- *From gsmSSF to gsmSCF* – the diagnostics are copied from the ISUP release message. If the call release resulted from MSC-internal action, then the MSC may include the diagnostics information

---

[24] The cause is also included in some other ISUP messages, but those ISUP messages are not reported to the gsmSCF (e.g. call progress), or the ISUP message is reported, but the cause is not included in the report to gsmSCF (e.g. address complete in CAMEL Phase 4).

Table 5.12 Cause diagnostics

Cause value	Meaning	Diagnostics
3	No route to destination	Condition
16	Normal call clearing	Condition
17	User busy	CCBS indicator
21	Call rejected	Call rejected condition

in the internal ISUP release message. The reporting of the cause diagnostics to the gsmSCF enables the CAMEL service to analyse the reason for call release and decide on further action.

• *From gsmSCF to gsmSSF* – the gsmSCF may include diagnostics in the cause parameter in the CAP release call operation and in the call gap operation. When the gsmSCF receives a disconnect or busy notification, for example, then the gsmSCF may copy the diagnostics that it receives from the gsmSSF to the diagnostics in CAP release call.

5.2.10.2 Enhancement to User Interaction

When PA or PC is used, then the gsmSCF may order the playing of a concatenation of 16 pre-recorded announcements. For CAMEL phase 2, a maximum of five pre-recorded announcements may be ordered with a single PA or PC operation. This increase in number of messages per PA or PC operation enhances the flexibility of the user interaction.

5.2.10.3 Called Party Number Length

In CAMEL phase 3, the CdPN may contain a minimum of 0 digits, i.e. the number consists of a number frame only. In certain markets (e.g. North America), specific operator numbers may be identified with a designated nature of address value, without the need for further digits. A calling subscriber does not have the ability to dial such a number; however, number translation in the MSC may yield such a number. The resulting operator number, not containing any digits, may be reported in the answer notification to the gsmSCF. Examples of such operator numbers may be found in ANSI ISUP, T1.113–1995 [160].

The encoding of a CdPN in CAP also allows for a CdPN that does not contain any digits. ANSI ISUP, T1.113–1995 [160] does not specify any nature of address for CgPN related to an operator number without digits.

5.3 CAMEL Control of GPRS

General packet radio system (GPRS) was introduced in GSM R97. GPRS allows for the establishment of PS data connections via the GSM networks. GPRS uses the same RAN as GSM, but a different core network infrastructure. CAMEL control of GPRS is introduced in 3GPP R99.

The core network functionality of GPRS is mainly specified in the following 3GPP specifications:

3GPP TS 22.060 [64]	GPRS, Service description, stage 1
3GPP TS 23.060 [81]	GPRS, Service description, stage 2
3GPP TS 29.060 [105]	GPRS, GPRS Tunneling Protocol (GTP) across the Gn and Gp interface

5.3.1 Network Architecture

Figure 5.30 gives a simplified representation of the GPRS network. Two variants exist for the GPRS network, in as far as the access technology is concerned:

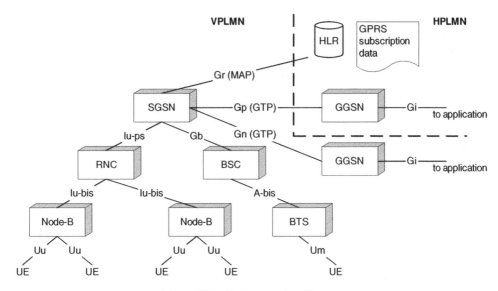

Figure 5.30 GPRS network architecture

- *TDMA GPRS* – the TDMA GPRS uses the TDMA GSM radio access network. The radio connection between the subscriber and the core network runs via the BTS and the BSC. The data transmission for the GPRS connection uses allocated time slots in the TDMA radio access. A GPRS data link may use one or more time slots. This GPRS variant is often referred to as '2.5G'.
- *W-CDMA GPRS* – the W-CDMA GPRS uses the W-CDMA radio access network. The radio connection between the subscriber and the core network runs via the Node-B and the radio network controller (RNC).

The concepts HPLMN and VPLMN are also used for GPRS. A GPRS subscriber is permanently registered in the HLR of a particular PLMN; this PLMN is her HPLMN. The GPRS subscriber may at any moment be *attached* to an SGSN in the VPLMN. Being attached to an SGSN may be compared with being registered with an MSC. The attachment to the SGSN may occur through the TDMA RAN (via BTS and BSC) or through the W-CDMA RAN (via Node-B and RNC).

5.3.1.1 PDP Context
The access control protocol for GPRS is DTAP. DTAP is also used for CS call control over the RAN. DTAP is specified in 3GPP TS 24.008 [94]. The protocol between the SGSN and the GGSN is GTP. In UMTS, GTP may also be used between SGSN and the RNC. GTP has a dual purpose:

(1) *GPRS control* – this includes attaching to SGSN, creating and terminating a PDP context (PDPc), routing area update (RAU), etc. This part of GTP is known as GTP-C.
(2) *Payload transfer* – this part of GTP is used for transferring the payload through a *tunnel*. The tunnel is the logical data connection that is established between the SGSN and GGSN (2G RAN) or between RNC and GGSN (3G UTRAN). This part of GTP is known as GTP-U.

Each tunnel through which GTP-U messages may be transferred is associated with a logical control path controlled by GTP-C. One GTP-C logical control path may control one or more tunnels (Figure 5.31).

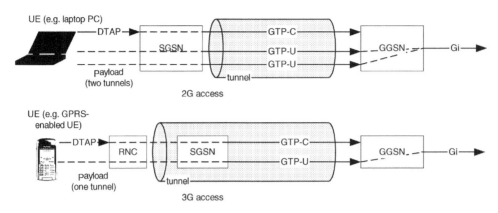

Figure 5.31   GTP between SGSN and GGSN

Table 5.13   PDP context element description

Element	Description
End user address	This element contains the address that will be used by the PDN to address the application in the UE that has established this PDP context. Various addressing protocols may be used, such as IP v4, IP v6 and point-to-point protocol (PPP)
Quality of service	The quality of service (QoS) defines various quantities of the PDP context, such as: • maximum bit rate for uplink (UE to network) • maximum bit rate for downlink (network to UE) • guaranteed bit rate for uplink • guaranteed bit rate for downlink • peak throughput Refer also to 3GPP TS 24.008 [94] for a definition of QoS
Access point name	The access point name (APN) defines the access into the PDN. The APN contains a domain name, in accordance with the domain name server (DNS) naming rules. When the subscriber establishes a PDP context, she may provide an APN to the SGSN
Charging Id	The charging Id is a number that is allocated by the GGSN, when a PDP context is activated by that GGSN. The charging Id is used together with the GGSN address to uniquely identify a PDP context
GGSN address	The GGSN address identifies the GGSN through which the PDP context is established
PDP initiation type	The PDP initiation type indicates whether the PDP context is established under initiative of the user or the PDP context is established under instruction of the network ('network-initiated PDP context')

Hence, the PDP context is the data channel between the user and the GGSN. The GTP-U messages that are transferred through the PDP context contain the T-PDUs, which are the protocol data units used for payload transfer. T-PDU may contain IP packets (or other protocol is used). At the GGSN, the IP packets (or other) from the T-PDU are sent to the PDN.

When a PDP context is established, various elements may be supplied by the initiator of the PDP context, i.e. by the UE. The SGSN and GGSN may also determine certain elements for the PDP context. The PDP context elements that are relevant for CAMEL are listed in Table 5.13. These

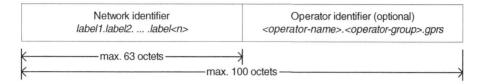

Figure 5.32 Structure of APN

elements are reported to the gsmSCF during the CAMEL service invocation. For more details on the PDP context, refer to 3GPP TS 29.060 [105].

APN Structure

The APN is composed of a network identifier (NI) and optionally an operator identifier (OI); see Figure 5.32. The NI is used to select a service, whereas the OI is used to select an operator network.

The NI consists of one or more labels. If the NI contains two or more labels, then the NI will comply with the rules for internet domain names. The NI identifies the place in the internet domain to which the data connection will be established. The OI, if present, always consists of three labels: a label identifying the operator name; a label identifying the operator group or country; and the label 'gprs'.

A default OI may be used and will have the structure 'mnc'<MNC> · 'mcc'<MCC> ·gprs. The MNC and MCC represent the mobile network code and mobile country code respectively; MNC and MCC are derived from the SIM card, universal subscriber identity module (USIM) or IMS subscriber identity module (ISIM) of the GPRS subscriber.[25] An example of default OI is 'mnc002.mcc655.gprs', which identifies MTN, South Africa. For more details on the APN structure, refer to 3GPP TS 23.003 [73].

5.3.1.2 CAMEL Control

The CAMEL control of GPRS occurs through a CAMEL dialogue between the SCP and the SGSN; see Figure 5.33. The CAMEL control of GPRS runs strictly between SCP and SGSN. Here, the SGSN has a gprsSSF, which may be compared with the gsmSSF in the MSC and GMSC. The gprsSSF forms the link between the GPRS handling in the SGSN and the gsmSCF. Since CAMEL control of GPRS is introduced in CAMEL phase 3, the control protocol between the gprsSSF and gsmSCF is CAP v3. The CAP v3 that is used for GPRS control is different from the CAP v3 that is used for call control.

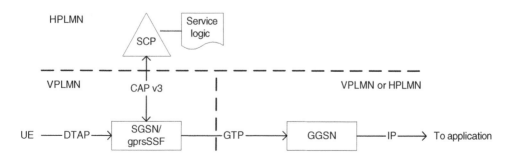

Figure 5.33 CAMEL control in the GPRS network

---

[25] The SIM card may contain a subscriber identity module (SIM), a USIM or an IMS subscriber identity module (ISIM).

The establishment of the CAMEL relationship between gsmSCF and SGSN is dependent on the availability of CAMEL subscription data element GPRS-CSI, which may be received from HLR. This implies that GPRS CAMEL services are always *subscribed services*; there is no serving network-based CAMEL control, comparable to the N-CSI service for CS call control.

5.3.1.3 Location Update

When a GPRS subscriber attempts to register with an SGSN, the SGSN contacts the HLR of that subscriber. This *GPRS location update* procedure has similarities with the location update procedure from an MSC. The SGSN uses the IMSI of the GPRS subscriber to derive the HLR address. The SGSN then uses MAP signalling to contact the HLR and to request subscription data for this subscriber. When the attachment procedure is successful, the subscriber is *attached* to the SGSN. The subscriber may now start activities such as establishing a PDP context or sending an SMS. In addition, the subscriber may receive SMSs and the network may perform network-initiated PDP context establishment.

During the location update procedure, the SGSN and HLR may perform CAMEL capability negotiation, similar to the CAMEL capability negotiation between MSC and HLR (during registration at the MSC) and between GMSC and HLR (during MT call handling); see Figure 5.34.

An operator may administer different IMSI ranges in the SGSN. For each IMSI range, the SGSN contains an indication of the offered CAMEL phase(s). When a subscriber attaches to the SGSN, the IMSI of this subscriber is used to determine which CAMEL phase(s) will be offered to that subscriber. In this manner, the SGSN may offer CAMEL capability to selected subscriber groups.[26] For the SGSN, the lowest CAMEL phase that may be offered to a subscriber is CAMEL phase 3; there is no CAMEL phase 1 and CAMEL phase 2 capability defined for the SGSN. When the SGSN sends the MAP location update message to the HLR, the SGSN includes the supported CAMEL phases for this subscriber. The supported CAMEL phases may be one of: no CAMEL support; CAMEL phase 3; or CAMEL phase 3 + CAMEL phase 4 (support of CAMEL phase 4 requires the SGSN to comply with 3GPP Rel-5 or later). If the SGSN supports the required CAMEL phase for this subscriber, then the HLR may send the CAMEL data to the SGSN. The subscriber is now registered as CAMEL subscriber in the SGSN.

5.3.2 Subscription Data

The GPRS-CSI, just like O-CSI and some other CSIs, consists of one or more TDP. Table 5.14 gives a breakdown of the contents of GPRS-CSI. Each TDP consists of the elements in Table 5.15.

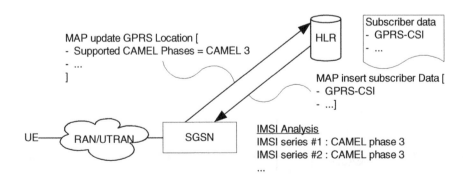

Figure 5.34   GPRS attach procedure

---

[26] The definition of CAMEL support in the SGSN per IMSI series is not specified by CAMEL; this mechanism is vendor-specific.

Table 5.14 Contents of GPRS-CSI

Element	Description
TDP list	List with TDPs. Each TDP in GPRS-CSI represents a TDP in one of the two GPRS state models, where a CAMEL service may be invoked. The individual TDP are discussed in the next sections
CAMEL capability handling	This element indicates the CAP version that will be used for the CAMEL dialogues. For CAMEL control of GPRS, the CAP version will always be CAP v3

Table 5.15 GPRS-CSI TDP definitions

Element	Description
TDP	Indication of the position in a GPRS state model
Service key	Service key to be used for the CAMEL service invocation
gsmSCF address	Address of the SCP to which the service invocation shall be addressed; this address is a GT
Default handling	The default handling indicates what handling the gprsSSF shall take in the case of CAMEL dialogue failure; the handling may be 'continue' or 'release'

GPRS-CSI also contains the elements 'notification to CSE' and 'csi active'. These elements relate to the CAMEL phase 3 feature 'Subscription data control'. This feature is described in Section 5.8.

5.3.3 GPRS State Models

For the purpose of CAMEL control, two state models are defined in GPRS: (1) GPRS session state model; and (2) PDP context state model. For both state models, CAMEL control may be applied. The contents of GPRS-CSI determine for which state models a CAMEL service may be invoked.

5.3.3.1 GPRS Session State Model

The GPRS session state model relates to the attach/detach state of a subscriber. An instance of this state model is created when the subscriber attaches to the SGSN. The state model instance remains active for the entire attachment duration (unless an inter-SGSN routing area update takes place). If GPRS-CSI was received from the HLR during the location update procedure (i.e. during attachment) and GPRS-CSI contains a TDP related to this state model, then a CAMEL service will be invoked. Figure 5.35 reflects the GPRS session state model.

When the GPRS Session state model is instantiated, i.e. the subscriber has attached to the SGSN, a CAMEL service may be invoked, depending on the contents of GPRS-CSI. GPRS-CSI may contain the following TDPs related to the GPRS session state model:

- *Attach* – this TDP indicates that a CAMEL service will be invoked as soon as the subscriber has attached to the SGSN. The CAMEL service that is invoked at this point may remain active for the entire attachment duration.
- *Change of position* – this TDP indicates that a CAMEL service will be invoked when inter-SGSN routing area update occurs (change of position). The use of this TDP is explained further in Section 5.3.7.2.

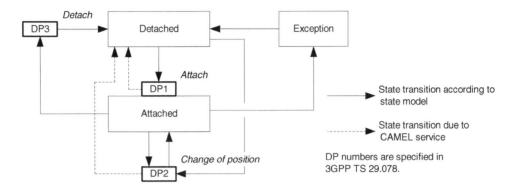

Figure 5.35   GPRS session state model. Reproduced from 3GPP TS 23.078 v3.19.0, Figure 6.3, by permission of ETSI

Table 5.16   Detection points for the GPRS session state model

DP name	Arming mode	Description
Attach	TDP-R	Subscriber attaches to the SGSN
Detach	EDP-N, EDP-R	Subscriber detaches from the SGSN
Change of position session	TDP-R, EDP-N	Subscriber changes position within the SGSN service area or has changed position from another SGSN service area

The GPRS session state model follows many of the rules that apply to the basic call state models (BCSM) that are used for CS call control. An overview of the DPs for the GPRS session state model is contained in Table 5.16.

The use of the change of position DP is explained in Section 5.3.7.2. A CAMEL service that is invoked with TDP attach has full control over the GPRS session and over the PDP contexts that are established during the GPRS attachment duration. The control of the GPRS session and the PDP contexts is done through a single CAMEL relationship between gsmSCF and gprsSSF. Besides the state model for the GPRS session, there is a state model per established PDP context. This implies that one CAMEL relationship may control multiple state model instances: one state model for the GPRS session and one state model per PDP context. This is a distinctive difference from one of the principles of the CAMEL relationships for call control, whereby one CAMEL relationship always relates to one state model instance.

The CAMEL service that is invoked at DP attach may be used to apply 'attach authorization'. The service has the following control capability:

GPRS Session Control

- Allow/disallow GPRS attachment
- Detach the subscriber from the SGSN at any moment
- Apply duration-based charging ('duration' relates here to being GPRS attached)
- Monitor the location of the subscriber during the GPRS attachment duration

PDP Context Control

The CAMEL service that is invoked for the GPRS Session may control each individual PDP context that is established by the subscriber.

5.3.3.2 PDP Context State Model

The PDP context state model relates to the establishment and disconnection of a single PDP context. A subscriber who is attached to an SGSN may establish one or more PDP contexts. The maximum number of PDP contexts that may be established at any moment varies per SGSN vendor. Typically, for GSM access, six to eight PDP contexts may be established; for UMTS access, 10–12 PDP contexts may be established. For each PDP context that is established, an associated state model is instantiated. The state model for a PDP context is presented in Figure 5.36.

When the PDP context state model is instantiated, i.e. the subscriber establishes a PDP context, a CAMEL service may be invoked, depending on the contents of GPRS-CSI. GPRS-CSI may contain the following TDPs related to the PDP Context state model:

- *PDP context establishment* – this TDP indicates that a CAMEL service shall be invoked when the subscriber initiates the establishment of a PDP context. The CAMEL service that is invoked at this point may remain active for the entire PDP context duration.
- *PDP context establishment acknowledgement* – this TDP indicates that a CAMEL service will be invoked when a PDP context is established. The CAMEL service that is invoked at this point may remain active for the remainder of the PDP context duration.
- *Change of position* – this TDP indicates that a CAMEL service will be invoked when inter-SGSN routing area update occurs (change of position).

The DPs for the PDP context state model are listed in Table 5.17. The usage of the change of position DP is explained in Section 5.3.7.2. When scenario 1 (GPRS Session control) is used, all PDP context-related DPs may be armed as EDP.

When a CAMEL service is invoked at PDP context establishment, it may apply the following control to this PDP context:

- allow PDP context establishment without modifying data
- allow PDP context establishment with SCP-supplied APN
- disallow PDP context establishment

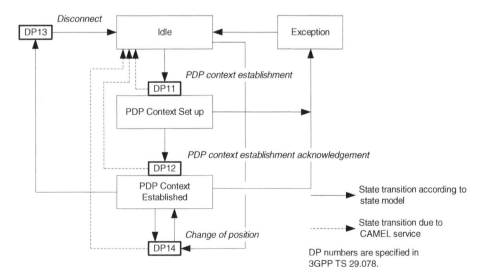

Figure 5.36   PDP context state model. Reproduced from 3GPP TS v3.19.0, Figure 6.4, by permission of ETSI

Table 5.17 Detection points for the PDP context state model

DP name	Arming mode (scenario 1)	Arming mode (scenario 2)	Description
PDP context establishment	EDP-N, EDP-R	TDP-R	The subscriber, who is attached to the SGSN, establishes a PDP context
PDP context establishment acknowledgement	EDP-N, EDP-R	TDP-R, EDP-N, EDP-R	The SGSN has received confirmation from the GGSN that the PDP context is established. The PDP context may now be used for data transfer
PDP context disconnection	EDP-N, EDP-R	EDP-N, EDP-R	The PDP context is disconnected
Intra-SGSN change of position context	EDP-N	EDP-N	Subscriber changes position within the SGSN service area
Inter-SGSN change of position context	EDP-R	TDP-R	Subscriber has changed position to another SGSN service area

When a CAMEL service is invoked at PDP context establishment acknowledgement, it may apply the following control to this PDP context:

- allow the acknowledgement of the PDP context; the PDP context may now be used for data transfer
- release the PDP context

It is not possible for the CAMEL service invoked at PDP context establishment acknowledgement to modify the APN.

PDP Context Establishment
The present section describes the process of PDP context establishment by a subscriber and how a CAMEL service may control the PDP context.

When the subscriber establishes a PDP context, the UE sends the *Activate PDP context request* DTAP message to the SGSN. This DTAP message may contain parameters such as APN, requested QoS and requested PDP address. If GPRS-CSI is present in the SGSN for this subscriber and the GPRS-CSI contains TDP PDP context establishment, then the PDP context establishment results in the invocation of a CAMEL service from the SGSN. There are no trigger conditions associated with GPRS-CSI, such as a list of APNs. That means that every time the subscriber establishes a PDP context, a CAMEL service will be invoked. If the CAMEL services decides that for this PDP context no CAMEL control is required, then the service may send CAP CUE-GPRS; PDP context establishment now continues without CAMEL control. The details of the CAMEL service to be invoked are retrieved from GPRS-CSI. Specifically, the gsmSCF address and service key are used to invoke the CAMEL service. It is permissible for a subscriber to establish a PDP context without supplying an APN. In that case, the SGSN would select a default APN for this PDP context.[27]

At this point in the PDP context establishment process, no APN analysis has taken place yet in the SGSN. This implies that the APN as received from the UE is reported to the CAMEL service. If the subscriber did not provide an APN, then the CAMEL service is invoked without reporting an APN to the SCP. The CAMEL service has now gained control over the PDP context. The control

---

[27] Unlike CS call establishment, where a subscriber must always provide a called party BCD number.

over the PDP context runs through a gprsSSF, which is a logical entity that resides in the SGSN and serves as a relay between the PDP context process in the SGSN and the gsmSCF. The gprsSSF may in many aspects be compared with the gsmSSF that resides in the MSC or GMSC. Protocol rules for the gprsSSF are further explained in Section 5.3.3.4.

The interworking between the SGSN and the gprsSSF is reflected in 3GPP TS 23.060 [81]. The GPRS procedures are defined in 3GPP TS 23.060 [81] by means of sequence diagrams. The sequence diagrams for the various procedures contain *CAMEL interaction boxes*. These interaction boxes may be compared with the DPs from the BCSMs form CS call control. One example of such sequence diagram is the PDP context activation procedure for UMTS; see Figure 5.37.

The rectangular boxes 1 and 2 in Figure 5.37 constitute the DPs PDP context establishment and DP PDP context establishment acknowledgement, respectively. It is at these points that the corresponding CAMEL procedure may be invoked. These CAMEL procedures are specified in 3GPP TS 23.078 [83].

The CAMEL service invocation request (i.e. the IDP-GPRS information flow) contains various elements that may be used by the CAMEL service for its internal processing. Unlike CS call control, not all information elements that are needed for PDP context handling are available. Some of the required elements are reported only at PDP context establishment acknowledgement.

If the CAMEL service allows the PDP context establishment to continue, then it sends a CAP CUE-GPRS or CAP CON-GPRS operation to the gprsSSF:

- *Continue GPRS* – this operation indicates to the SGSN that PDP context establishment will continue with the information that was available in the SGSN just prior to CAMEL service invocation.
- *Connect GPRS* – this operation indicates to the SGSN that PDP context establishment will continue, using the APN that is included in CAP CON-GPRS. If an APN was already provided by the UE, then the APN that is provided by the CAMEL service will overwrite the UE-provided APN. Both the NI part and the OI part of the APN will be overwritten.

When the CAMEL service has instructed the gprsSSF to continue the PDP context establishment, the SGSN runs the 'APN and GGSN selection' procedure. This procedure is specified in 3GPP

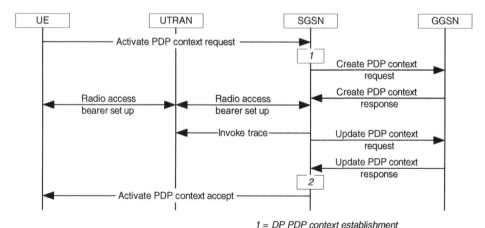

1 = DP PDP context establishment
2 = DP PDP context establishment acknowledgement

Figure 5.37 PDP context activation procedure for UMTS. Reproduced from 3GPP TS 23.060 v3.14.0, figure 63, by permission of ETSI

TS 23.060 [81], Annex A. The main tasks of this procedure include: validating APN; selecting default APN, if no APN is available; deriving GGSN address to be used for this PDP context; and determining the quality of service for this PDP context. If neither the subscriber nor the CAMEL service had provided an APN, then the SGSN selects a default APN in the following way: if a default APN exists for this subscriber, then this default APN is used; else, a default APN for this SGSN is used.

The APN and GGSN selection procedure uses *PDP context subscription records*. These records, maximum 50 per subscriber, are permanently stored in the HLR and are sent to the SGSN during location update. A PDP context subscription record contains a definition of a PDP context that may be established by the subscriber. The APN that is provided by the subscriber or by the CAMEL service is used to select a corresponding PDP context subscription record. When a corresponding record is found, data is used from this record for the PDP context establishment. Table 5.18 lists the elements that are contained in the PDP context subscription record.

The outcome of the APN and GGSN selection procedure is used to contact the GGSN. The SGSN sends the create PDP context request message to the GGSN. If the GGSN accepts the PDP context establishment, then it responds with create PDP context response. This response message has the effect that the SGSN triggers the DP PDP context establishment acknowledgement. This is reflected in Figures 5.38 and 5.39.

When the PDP context establishment is reported to the gsmSCF, not all information related to the PDP context may be available. For example: the user may not have supplied an APN; the user may not have supplied an end user address; or the negotiated quality of service may not be known. These elements may be required for charging the PDP context. When the PDP context activation is confirmed by the GGSN, the missing elements are included in the response; these elements are included in the PDP context establishment acknowledgement notification. The CAMEL service may now perform the charging analysis and apply adequate charging for this PDP context.

An operator has the option to invoke the CAMEL service at PDP context establishment acknowledgement instead of at PDP context establishment. This is determined by the contents of GPRS-CSI. The advantages and disadvantages of either method are listed in Table 5.19.

PDP Context Disconnection
The disconnection of the PDP context is reported to the gsmSCF through the DP PDP context disconnection. Figure 5.40 contains an example of CAMEL signal sequence including the PDP context disconnection.

Table 5.18 PDP context subscription record

Information element	Description
PDP type	The PDP type to be used for this PDP context
PDP address	The PDP address to be used for this PDP context
Access point name	The APN for which this PDP context subscription record applies. This APN consists of the network identifier only
QoS profile subscribed	The subscribed QoS is the maximum QoS that may be applied for this PDP context
VPLMN address allowed	This element indicates whether the subscriber may establish a PDP context using an APN that belongs to the VPLMN domain or only a PDP context using an APN that belongs to the HPLMN
PDP context charging characteristics	This element contains characteristics for the PDP context related to charging. When CAMEL is used for on-line charging for this PDPc, then this element will have the value 'prepaid service'. However, this element is not used by the SGSN to determine whether CAMEL shall be invoked. 3GPP TS 32.015 [115] specifies the permissible values for this element

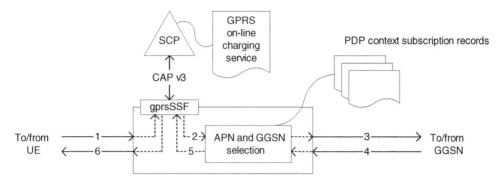

Figure 5.38 PDP Context activation with CAMEL control. (1) The subscriber establishes the PDP context. An APN and other information elements may be included in the request. The gprsSSF sends CAP IDP-GPRS to the gsmSCF. (2) The gsmSCF sends CAP CUE-GPRS to the gprsSSF. PDP context establishment continues, which includes the execution of procedure APN and GGSN selection. (3) The SGSN has derived the APN (if not yet available) and the address of the GGSN. A PDP context activation request is sent to the GGSN. (4) The GGSN validates the PDP context activation request and returns an acknowledgement to the SGSN. The acknowledgement includes the selected APN, negotiated quality of service, charging Id and end user address. (5) The gprsSSF reports the PDP context activation to the gsmSCF, by sending CAP event report GPRS with event PDP context establishment acknowledgement. (6) The gsmSCF confirms the activation notification to the gprsSSF, by sending CAP CUE-GPRS. The PDP context is now active and data transfer between UE and addressed application may take place.

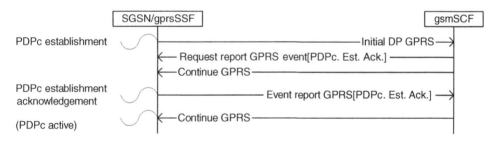

Figure 5.39 CAP signalling sequence for PDP context activation

Table 5.19 Comparison of triggering methods

TDP	Description
PDP context establishment	Triggering at DP PDP context establishment will be used when the CAMEL service wants to be able to modify or supply the APN. This method will also be used when the CAMEL service wants to be able to bar PDP context establishment, e.g. when the user has reached her spending limit
PDP context establishment acknowledgement	Triggering at DP PDP context establishment acknowledgement results in less signalling. The CAMEL service is invoked only when the PDP context is established successfully. All the data that is needed for charging the PDP context is reported to the gsmSCF

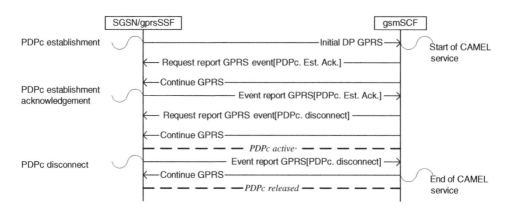

Figure 5.40   PDP context activation and disconnection

The CAMEL service may arm the PDP context disconnect DP at PDP context establishment or at PDP context establishment acknowledgement. When the PDPc disconnect is reported to the gsmSCF, an indication is included about the entity that caused the PDPc disconnection. The following entities may cause the disconnection of the PDPc: UE, SGSN, HLR or GGSN. The disconnect by the HLR may be related to operator requirement, i.e. the HLR instructs the SGSN to release all active PDPcs. The disconnect by the GGSN may be related to charging activities in the GGSN. The addressed application would normally not initiate the PDPc disconnect.

The release of a PDP context before the PDP context has reached the active phase may be caused by a failure in the SGSN or GGSN in PDP context establishment. These events include, but are not limited to:

- APN and GGSN selection cannot find a matching PDP context subscription record;
- the requested APN is not allowed;
- no GGSN address can be determined from the APN;
- the GGSN disallows the PDP context establishment.

The PDP context state model does not have PDP establishment failure DPs, comparable to the O-BCSM and T-BCSM for CS call control. Neither does the PDP context state model include an abandon DP. The behaviour of the gprsSSF in the case of premature PDP context release is as follows:

- if a CAMEL relationship exists and DP PDP context disconnect is armed, then the gprsSSF reports DP PDP context disconnect;
- if a CAMEL relationship exists and DP PDP context disconnect is not armed, then the gprsSSF sends entity released GPRS;
- if the gprsSSF has sent CAP IDP-GPRS, but has not received the first TC_Continue message, then no CAMEL relationship exists yet. If the user abandons the PDP context establishment at that moment, then the gprsSSF aborts the TC dialogue locally.

5.3.3.3 Combined Session and PDP Context Control

The previous sections (5.3.3.1 and 5.3.3.2) have described the two state models that are defined for CAMEL control of GPRS: GPRS session state model and PDP context state model. For both state model types, a CAMEL service may be invoked, depending on the contents of GPRS-CSI

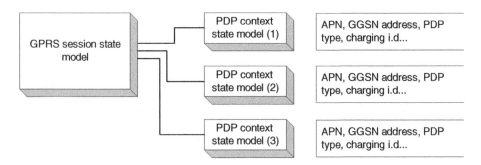

Figure 5.41 Multiple state model instances

for the subscriber. Hence, for one subscriber, there may be multiple state model instances active at a particular moment: a GPRS session state model instance; and zero or more PDP context state model instances. For each PDP context that the subscriber has established, the corresponding state model is instantiated.

Figure 5.41 contains an example of a subscriber who has established three PDP contexts. Each PDP context may have different characteristics, such as APN, GGSN address, PDP type etc. To coordinate the CAMEL control over the various state models that may be active for a subscriber, two control scenarios are defined.

GPRS Control Scenario 1
Scenario 1 entails that a single CAMEL service is invoked to control the GPRS session state model and the individual PDP context state models. This combined service is invoked at DP attach and remains active as long as the subscriber is attached to the SGSN, i.e. until DP detach is reported. From within this combined service, the CAMEL service may arm the PDP context establishment as a *generic event*. That means that, when a PDP context is established, this will be reported to the CAMEL service by means of an event detection point. The PDP context establishment is in this case not reported as TDP but as EDP. Alternatively, the PDP context establishment acknowledgement event may be armed generically. In that case, the acknowledgement of a PDP context from the GGSN is reported to the CAMEL service as EDP. As soon as a PDP context is established and a notification is sent to the CAMEL service, the CAMEL service has gained control over that PDP context. There will be multiple *CAMEL relationships* through a single CAMEL dialogue. The combined CAMEL service can assert full control over the PDP context as described in Section 5.3.3.2. The CAP operations that are exchanged between the gsmSCF and the gprsSSF, related to a specific PDP context, include a *PDP Id*. The PDP Id is allocated by the SGSN and included in the first notification of a PDP context to the gsmSCF. Figure 5.42 gives a graphical representation of this scenario.

Scenario 1 has the advantage that the CAMEL service has control over the attach/detach state of a subscriber. This enables the CAMEL service to control attachment to the SGSN, detach a subscriber from the SGSN at any moment and apply duration based charging for the attachment duration.

GPRS Control Scenario 2
Scenario 2 entails that, for every PDP context the subscriber establishes, a CAMEL service is invoked. A CAMEL service that is invoked for a PDP context may control that PDP context, but has no control over the GPRS attachment. When a subscriber establishes five PDP contexts, for example, then five CAMEL service instances will be invoked. These service instances may exist simultaneously or be invoked at different points in time, depending on when the subscriber establishes the PDP contexts. For scenario 2, the PDP Id is not used in the CAP operations,

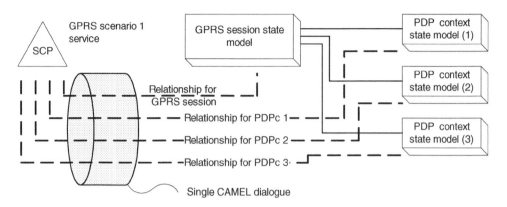

Figure 5.42 Multiple relationships through scenario 1 control

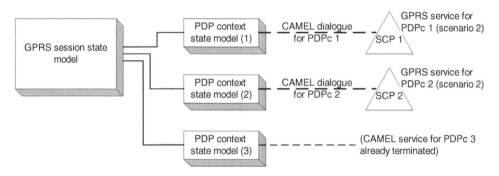

Figure 5.43 Multiple CAMEL dialogues through scenario 2 control

because one CAMEL dialogue relates to exactly one PDP context. In the example in Figure 5.43, three CAMEL services are invoked, one for each PDP context. These services are invoked from the same GPRS-CSI; hence, the same gsmSCF address is used for these services. If dynamic load sharing is used, then the services may be distributed over two SCPs. Even though PDP context 3 in Figure 5.43 is still active, the CAMEL service for PDP context 3 has already terminated, because PDP context 3 was established with a toll-free APN. If the CAMEL service that is invoked for PDP context 3 determines that the PDP context is toll-free, it may send CUE-GPRS to the gprsSSF, without arming any DP. The PDP context establishment continues with CAMEL control. The CAMEL control may be terminated at any moment of the PDP context establishment or during the active phase of the call, allowing the PDP context (establishment) to continue.

Scenario 2 is less complex than scenario 1. This is due to the fact that, in scenario 2 control, the SCP does not need to apply generic DP arming and maintain multiple CAMEL relationships. Practically, if the CAMEL service has no requirement to apply duration-based charging for the attachment duration, then scenario 2 may be the preferred method. Even though there are more CAMEL dialogues used in scenario 2 control, this does not have an adverse effect on the SS7 signalling in the network. This is explained in further detail in Section 5.3.3.4.

Sequential Scenario 1 Control and Scenario 2 Control
Although scenarios 1 and 2 are mutually exclusive at any moment, it is feasible to invoke the two control scenarios in sequence. (Figure 5.44).

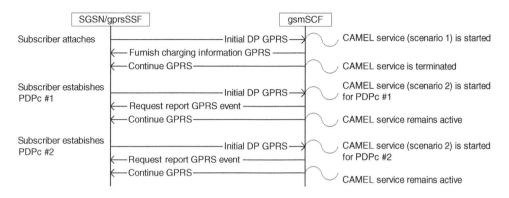

Figure 5.44   Sequential scenario 1 and 2 invocation

The scenario 1 CAMEL service instance in Figure 5.44 does not arm any events or request charging reports. As a result, the scenario 1 CAMEL service terminates when it sends CUE-GPRS. The purpose of this scenario 1 service may be, for example, access control. For that service, it is not necessary for the CAMEL service to remain active after attachment.

After the scenario 1 CAMEL service instance has terminated, the subscriber is attached and may establish PDP contexts. When the subscriber establishes a PDP context, there is no scenario 1 service instance active for this subscriber (any more). Hence, the SGSN may check GPRS-CSI to determine whether a scenario 2 service instance may be invoked for this PDP context. If GPRS-CSI contains TDP PDP context establishment, then a scenario 2 CAMEL service instance is invoked for this PDP context. If a second PDP context is established, another scenario 2 service instance may be invoked. The scenario 2 service instances use a different TDP definition in GPRS-CSI than the scenario 1 service instance; hence different gsmSCF addresses may be used for the scenario service instance and the scenario 2 service instances. The scenario 2 service instances may therefore run on a different SCP than the scenario 1 service instance.

The rules for reporting a PDP context related event to a scenario 1 or to a scenario 2 service are as follows:

```
IF scenario 1 service active for this subscriber THEN
    IF event armed as EDP THEN
        report event as EDP to scenario 1 service;
ELSE IF scenario 2 service active for this PDP context of this subscriber THEN
    IF event armed as EDP THEN
        report event as EDP to scenario 2 service;
ELSE IF event contained in GPRS-CSI THEN
    invoke scenario 2 service logic instance.
```

5.3.3.4 gprsSSF and CAMEL Dialogue Rules

The dialogue rules for the CAMEL dialogue between gprsSSF and gsmSCF are similar to the dialogue rules that apply for the CAMEL call control. One exception to this is the TCAP segmentation, which is explained in Section 5.3.3.5.

gprsSSF Finite State Machine

The gprsSSF is instantiated when a CAMEL service is invoked. The gprsSSF contains a FSM. The state of the gprsSSF FSM is determined by events that take place in the SGSN and by the signalling

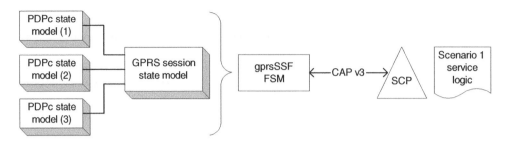

Figure 5.45   gprsSSF FSM for scenario 1 control

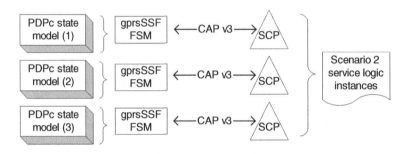

Figure 5.46   gprsSSF FSM for scenario 2 control

between gsmSCF and gprsSSF. The state of the gprsSSF FSM indicates the type of control that the gsmSCF may assert on the state models that are under control by this gprsSSF. This is reflected in Figures 5.45 and 5.46.

In the example in Figure 5.45, one gprsSSF FSM is controlling multiple state models: one GPRS session state model and multiple PDP context state models. In the example in Figure 5.46, each gprsSSF FSM is controlling exactly one PDP context state model.

Table 5.20 lists the permissible gprsSSF FSM states. Figure 5.47 represents how the gprsSSF FSM may transit between the permissible states.

CAMEL Relationship

The CAMEL dialogue between gsmSCF and gprsSSF is described by means of the type of CAMEL *relationship* it represents. As long as the CAMEL dialogue between gsmSCF and gprsSSF is active, there exist one or more relationships. For CS call control, a CAMEL dialogue has exactly one

Table 5.20   gprsSSF FSM states

gprsSSF FSM state	Description
Idle	The gprsSSF is currently not invoked
Wait_for_Request	The gprsSSF is invoked and is waiting for the initial service event (e.g. PDP context establishment)
Waiting_for_Instructions	The gprsSSF has established a relationship with the gsmSCF and is waiting for instructions from the gsmSCF
Monitoring	The gprsSSF is active and is monitoring a GPRS session or PDP context. There is a relationship with the gsmSCF

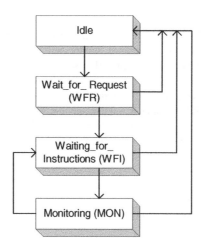

Figure 5.47   gprsSSF FSM state transitions

CAMEL relationship associated with it. For CAMEL control of GPRS, there may be multiple relationships per CAMEL dialogue, depending on the control scenario.

- *Scenario 1* – the gsmSCF has a CAMEL relationship with the GPRS session state model and with each individual PDP context;
- *Scenario 2* – the gsmSCF has a CAMEL relationship with the PDP context that is controlled by this scenario 2 CAMEL service.

The following types of CAMEL relationship are defined:

- *Control relationship* – a control relationship entails that for the GPRS session or for the PDP context, at least one DP is armed as EDP-R or the gprsSSF is processing a TDP-R. A control relationship is needed to release the GPRS session or a PDP context, to arm events for a state model or to apply charging control on the GPRS session or a PDP context.
- *Monitor relationship* – a monitor relationship entails that for the GPRS session or for the PDP context, at least one DP is armed as EDP-N or a charging report is pending. A monitor relationship allows a CAMEL service to monitor the state transitions of a state model, but does not allow the CAMEL service to assert control over that state model, such as forcing detach or releasing a PDP context.
- *Transparent relationship* – the transparent relationship is used in scenario 1 only. This type of relationship entails that a scenario 1 CAMEL service has an established relationship with a PDP context, but has released the relationship with that PDP context, prior to the termination of that PDP context. The CAMEL service is, however, still active and has at least one control or monitor relationship with the gprsSSF. Service logic designers should take care when relinquishing the relationship with an individual PDP context state model in scenario 1, prior to the release of that PDP context. Without a relationship, the CAMEL service will not know when the PDP context is released.

When the last remaining CAMEL relationship with the gprsSSF is released, the CAMEL dialogue is also released.

Tssf

The Tssf timer, which resides in the gprsSSF, is the timer that monitors the CAMEL dialogue when the gprsSSF FSM is in the WFI state. As long as the gprsSSF FSM is in the WFI state, the Tssf is running. There is one Tssf associated with a gprsSSF instance. The rules for the Tssf are as follows:

- Tssf is started when a CAMEL service is started. At this point, Tssf is loaded with a default value. The default value is SGSN/gprsSSF vendor-specific and may have a value between 1 and 20 s. The gprsSSF FSM has made a transition to the state WFI.
- Tssf is stopped when gprsSSF FSM transits from WFI to MON or from WFI to idle.
- Tssf is started with the default value and restarted when the gprsSSF FSM transits from MON to WFI.
- When the gprsSSF FSM is in state WFI, the gsmSCF may use the reset timer GPRS (RT-GPRS) operation to reload Tssf with a service-specific value, which is carried in RT-GPRS. Tssf is restarted after the reload thereof. The Tssf value that is carried in RT-GPRS is valid as long as the gprsSSF FSM is in state WFI.
- When a CAMEL operation is transferred between gprsSSF and gsmSCF whilst the gprsSSF FSM is in state WFI, Tssf is reloaded with the last used timer value and is restarted. The last used timer value may be the default Tssf value or the value that was received with RT-GPRS.
- When Tssf expires, the gprsSSF assumes that a communication failure has occurred between gsmSCF and gprsSSF. The gprsSSF aborts the CAMEL dialogue and applies default transaction handling to the GPRS session or PDP context. The default handling ('continue' or 'release') to be used is obtained from the TDP definition in GPRS-CSI that was used to establish the CAMEL service.

Pre-arranged End

The pre-arranged end rules for GPRS control are identical to the pre-arranged end rules for CS call control. The GPRS CAMEL relationship is retained as long as one or both of the following conditions are fulfilled: there is at least one DP armed in the GPRS session state model or in the PDP context state model; and there is at least one charging report pending.

5.3.3.5 TCAP Segmentation

An aspect of the CAP signalling that is very specific to CAMEL control of GPRS is *TCAP segmentation*. For CS call control, a CAMEL dialogue may remain active for the entire call duration. As long as the CAMEL dialogue for the call is active, a TCAP dialogue is allocated to this CAMEL dialogue.

One aspect of GPRS is that a GPRS session (period of being attached to an SGSN) and a PDP context may have a long duration. A subscriber may be attached to an SGSN for days or weeks. Likewise, a PDP context may be active for an equally long duration. The rationale for having a PDP context active for an indefinite long duration is the 'always connected' concept. A subscriber may use a PDP context for internet connection, for corporate LAN connection or for IMS registration. If charging is based on the volume of the transferred data, then it is convenient to keep a PDP context active. A CAMEL service may be active for the entire duration of a PDP context. When an SGSN or SCP needs to handle a large number of simultaneously active PDP contexts, this may lead to too large a number of TCAP dialogues active simultaneously. Although TCAP allows for 2^{32} simultaneous TCAP identifiers (see ITU-T Q.773 [141]), the number of active TCAP dialogues may still be too large for an SGSN or SCP. To overcome this dilemma, CAMEL phase 3 has introduced the TCAP segmentation mechanism; the TCAP segmentation in CAMEL is used only for GPRS control; it is not used for call control or SMS control. Refer to Figure 5.48 for a graphical representation. The CAP abbreviations of the CAP operation names used in Figure 5.48 are listed in Section 5.3.4.

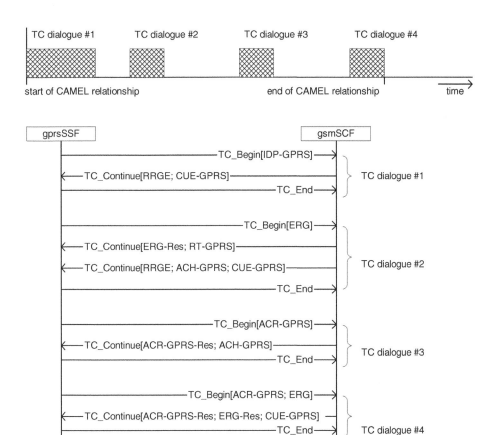

Figure 5.48 TCAP segmentation for CAMEL dialogue

The principle of TCAP segmentation entails that, when a CAMEL service is started, a TCAP dialogue is opened between gprsSSF and gsmSCF. This dialogue remains open as long as the gprsSSF FSM is in state WFI. The rationale is that, as long as the gprsSSF FSM is in state WFI, the gprsSSF expects a CAP operation from the gsmSCF, in order for the gprsSSF to transit to the state monitoring or idle. When the gprsSSF receives a CAP operation that results in a gprsSSF FSM state transition to monitoring or idle, it closes the TCAP dialogue. The resources that were occupied by the TCAP dialogue are now released and may be used for other TC-Users. When an event occurs that requires the transfer of CAP operations between the gprsSSF and the gsmSCF, a new TCAP dialogue is opened. The required CAP operations may now be transferred. When the transfer of CAP operations resulting from this event is complete, the gprsSSF closes this TCAP dialogue. This opening and closing of TCAP dialogues continues until the *CAMEL dialogue* terminates. In the example in Figure 5.48, the gprsSSF reports apply charging report GPRS (ACR-GPRS) and event report GPRS (ERG; PDP context disconnect). The gsmSCF responds with CAP CUE on the PDP context disconnect notification. Pre-arranged end rules specify that the CAMEL dialogue for this scenario 2 service is now closed.

From the TCAP point of view, the opening and closing of the TCAP dialogues follows the normal routines (except for the reporting of the gsmSCF address to the TC-User, as will be explained below). A TCAP dialogue is opened with a TC_Begin primitive; TC_Continue is used for the

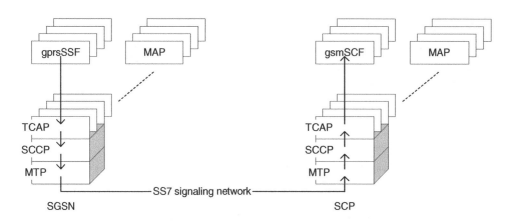

Figure 5.49   Allocation of gprsSSF and gsmSCF entity to TCAP dialogue

transfer of CAP components (operations and errors) through an open TCAP dialogue; TC_End is used to close a TCAP dialogue.

GPRS Reference Number

The TCAP dialogue between two TC-Users, such as gprsSSF and gsmSCF, take care that TCAP messages exchanged between SGSN and SCP are delivered to the correct application in this SGSN or SCP. In the example in Figure 5.49, the SGSN has several gprsSSF instances active, as well as a number of MAP applications, e.g. with the HLR. The SCP has several gsmSCF instances active and also a number of MAP applications, e.g. for any time GPRS interrogation (see Chapter 6). TCAP uses the TCAP transaction identifier to couple an incoming TCAP message to an application. When a new TCAP dialogue is established, a new application instance is invoked.

When TCAP segmentation is applied, as for CAMEL for GPRS control, the TCAP dialogue between the gprsSSF and the gsmSCF may be released while the gprsSSF and gsmSCF themselves remain active. An incoming TC_Begin message during an active CAMEL dialogue could therefore not be offered to the correct application. Instead, a TC_Begin from the gprsSSF, carrying the Event Report GPRS (PDP Context Establishment Acknowledgement), would start a new gsmSCF application instance.

To overcome the above, a GPRS reference number is used. This information element has the following structure (see 3GPP TS 29.078 [106]):

```
CAP-GPRS-ReferenceNumber   :: = SEQUENCE {
   destinationReference         [0]   Integer4     OPTIONAL,
   originationReference         [1]   Integer4     OPTIONAL}
```

The gprsSSF allocates an *origination reference* when it invokes a CAMEL service in the gsmSCF; it includes this origination reference in the first TCAP message to the gsmSCF. Upon receipt of the first TC_Begin message, the gsmSCF also allocates an origination reference and includes it in the first TCAP response message to the SGSN. The origination reference that the gsmSCF received from the gprsSSF is used as destination reference in the first TC_Continue message that the gsmSCF sends to the gprsSSF. The gprsSSF and the gsmSCF have now exchanged one another's GPRS reference number for the current CAMEL dialogue. From here onwards, the CAMEL dialogue between the gprsSSF and the gsmSCF use the GPRS reference number when a new TCAP dialogue is opened. The GPRS reference number is used to couple a new TCAP dialogue related to an existing CAMEL relationship, to the correct CAMEL service.

The GPRS reference number is carried in the *dialogue portion* of the TCAP message. ITU-T Q.764 [138] specifies which TCAP primitives may contain a dialogue portion.[28] For example, the TC_End messages in Figure 5.48 will not contain the GPRS Reference Number.

The closing of the TCAP dialogue when the gprsSSF FSM is in the 'monitoring' state has the following impact on the operation definitions for CAP v3 for GPRS.

(1) When the gprsSSF is in the monitoring state, both gprsSSF and gsmSCF should be able to start a new TCAP dialogue in order to send a CAP operation. For example, the gprsSSF should be able to send a charging report during monitoring and the gsmSCF should be able to release a PDP context. For this purpose, two ACs are specified:

    (a) One AC is used by the gprsSSF to start a TCAP dialogue with the gsmSCF. This TCAP dialogue may be opened with the following CAP operations: IDP-GPRS, event report GPRS, ACR-GPRS and entity released GPRS.

    (b) The other AC is used by the gsmSCF to start a TCAP dialogue with the gprsSSF. This TCAP dialogue may be opened with the following CAP operations: ACH-GPRS, cancel GPRS, FCI-GPRS, release GPRS, request report GPRS event and SCI-GPRS.

(2) The CAP operation activity test GPRS will always open a new TCAP dialogue. The GPRS reference number is used to verify the existence of a CAMEL relationship.

(3) In CS call control, there is a strict relation between CAMEL dialogue closure and TCAP dialogue closure. When the CAMEL dialogue closes (e.g. due to pre-arranged), the TCAP dialogue is also closed and vice versa. With CAMEL control of GPRS, there are separate rules for closing the TCAP dialogue and closing the CAMEL relationship:

    (a) the gprsSSF closes the TCAP dialogue when the gprsSSF FSM transits to monitoring state or idle and the gprsSSF does not expect a charging operation from the gsmSCF;

    (b) The gprsSSF and gsmSCF may close the CAMEL relationship with pre-arranged end.

Retrieving the gsmSCF GT from TCAP

The TCAP segmentation places a requirement on the TCAP part of the SS7 stack in the SGSN, in order to support SCP dynamic load sharing. Load sharing for CS call control is explained in Chapter 2. When the SGSN establishes a CAMEL relationship, it uses the gsmSCF address from GPRS-CSI to address the SCP for the required CAMEL service. The gsmSCF address has the form of a GT; the GT will be translated in the HPLMN of the served CAMEL subscriber, into the SPC of one of the SCPs associated with this GT. When the SCP responds to the gprsSSF, it includes its own GT in the response message; the TCAP layer in the SGSN will store this GT and use it in the remainder of the TCAP dialogue, to address this SCP. However, during the lifetime of the CAMEL dialogue, the TCAP dialogue may be terminated. This has the effect that the GT of this SCP is discarded.[29]

Using the gsmSCF address from the GPRS-CSI, which was used to start the CAMEL service, may result in addressing the wrong SCP, due to dynamic load sharing. For this reason, CAMEL specifies that the GT of the SCP, as received in the first TC_Continue message, is reported to the gprsSSF; the gprsSSF will store this GT. When a TCAP dialogue is opened for an existing CAMEL dialogue, the gprsSSF will use the stored gsmSCF address. CAMEL also allows the gsmSCF address from GPRS-CSI to be used for opening a TCAP dialogue for an existing CAMEL dialogue. However, that method cannot be used in combination with dynamic load sharing. That method should therefore not be used when inbound roaming subscribers are served with CAMEL capability, as the VPLMN operator may not know whether the served subscriber's HPLMN uses dynamic load sharing for the GPRS SCP.

---

[28] The TCAP dialogue portion is supported by Q.764 1994 onwards. The dialogue portion is also used to carry the CAP application context.

[29] The ITU-T TCAP recommendation prescribes that the GT of the responding entity, as received in the first TC_Continue message, should be stored in the TCAP layer and not be provided to the TC-User.

An issue that is closely related to the gsmSCF address to be used for opening a TCAP dialogue in an existing CAMEL dialogue, is the signalling link selection (SLS). The SLS is a mechanism that is used by SCCP; the SLS is used to control the sequence of delivery of successive SCCP messages addressed to the same entity, such as a gsmSCF or gprsSSF. SLS is defined for class 1 SCCP protocol; refer to ITU-T Q.714 [136]. During an active TCAP dialogue, in-sequence delivery of successive CAP components may be guaranteed using the same SLS value when addressing the remote party. However, when a TCAP dialogue is terminated, the SLS value that was used for that TCAP dialogue is no longer known. The new TCAP dialogue within the CAMEL dialogue may use a different SLS value.

A rare condition, resulting from using different SLS values in successive TCAP dialogues, may be the following (Figure 5.50). The gprsSSF sends a TC_End message to the gsmSCF to close TCAP dialogue n. That TCAP dialogue was used for the transfer of charging operations. A fraction of a second later, a PDP context disconnect occurs, which is reported in a new TCAP dialogue $n + 1$. If TCAP dialogue $n + 1$ uses a different SLS value than TCAP dialogue n, then under rare conditions, the TC_Begin related to TCAP dialogue $n + 1$ may arrive at the SCP prior to the arrival of the TC_End related to TCAP dialogue n.

Using the same SLS values for successive TCAP dialogue within one CAMEL dialogue may resolve the above dilemma. However, that method has not been explored by 3GPP. Nevertheless, the chance of the above dilemma occurring is probably very small and may occur only in networks that use satellite communication.

TC Guard Timer

Previous sections explain that, when the gprsSSF FSM transits to the monitoring or idle state, it closes the TCAP dialogue. When the gprsSSF has sent an apply charging report GPRS operation, the gprsSSF FSM state remains in monitoring. In that case, the gprsSSF *would* also close the TCAP dialogue. However, normal charging behaviour for CAMEL control of GPRS is that, when the gprsSSF sends apply charging report GPRS for an active PDP context, the gsmSCF responds with apply charging GPRS, for the next charging period or charging volume. Hence, after the gprsSSF has closed the TCAP dialogue that is used for sending apply charging report GPRS, the gsmSCF opens the next TCAP dialogue for sending apply charging GPRS.

To prevent this unnecessary closing and opening of TCAP dialogues in quick succession, the *TC guard timer* is introduced. When the gprsSSF has sent CAP ACR-GPRS, as in the above example scenario, the gprsSSF keeps the TCAP dialogue open until the expected CAP ACH-GPRS for this PDP context has arrived. This method prevents unnecessary signalling. This sequence of events is reflected in Figure 5.51.

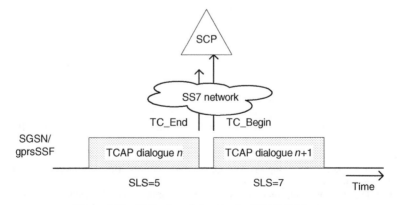

Figure 5.50   SLS values for successive TCAP dialogues

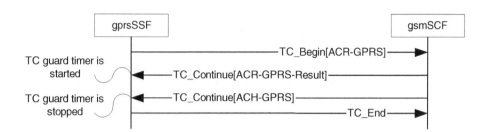

Figure 5.51 Usage of the TC guard timer

In the example flow in Figure 5.51, the gsmSCF sends the ACR-GPRS result component and the following ACH-GPRS operation component in separate TC_Continue messages. The gsmSCF may, however, combine these two TCAP components in a single TC_Continue message. However, the ACR-GPRS Result component is governed by the operation timer associated with the ACR-GPRS operation, T_{acrg} (see 3GPP TS 29.078 [106], section 8, for the operation timers). Before the gsmSCF can send the next ACH-GPRS operation, it may have to contact a charging database. Hence, to prevent operation timer expiry in the gprsSSF, the gsmSCF may first send TC_Continue containing ACR-GPRS result, then contact the charging database and then send TC_Continue containing the next ACH-GPRS.

If no new CAP ACH-GPRS arrives at the gprsSSF within a specific time after the gprsSSF has sent ACR-GPRS, then the gprsSSF may close the CAMEL relationship. This functionality may be compared with the Tccd timer in gsmSSF. However, this functionality on the gprsSSF is optional, whereas the use of the Tccd timer in gsmSSF is mandatory. This timer in gprsSSF should have a value that is longer than the TC guard timer.

5.3.4 CAP v3 Operations for GPRS

The present section provides an overview of the information flows used for GPRS control. CAMEL phase 3 has dedicated sets of CAP v3 operations for the three main functionalities: call control, GPRS control and MO SMS control. Table 5.21 lists the CAP v3 operations that are used for GPRS control.

For the contents of the individual operations, see 3GPP TS 23.078 [83], sections 6.6.1 and 6.6.2. The ASN.1 syntax of each operation and its argument is specified in 3GPP TS 29.078 [106], section 5. The reader should further be aware that most parameters in the CAP operations for GPRS are syntactically marked 'optional'. The semantic rules of presence of each parameter in an operation are specified in the information flow descriptions in 3GPP TS 23.078 [83].

5.3.5 On-line Charging for GPRS

The on-line charging mechanism for GPRS uses mechanisms that are similar to the on-line charging mechanism that is used in call control. A main parallel is the use of the CAP operations that are used for this purpose:

- *Apply charging GPRS* – this operation is used by the gsmSCF to define a charging threshold in the gprsSSF.
- *Apply charging report GPRS* – this operation is used by the gprsSSF to report a charging event to the gsmSCF. In principle, each charging threshold will lead to the subsequent reporting of a charging event, when that event occurs, unless the cancel GPRS operation is used to remove the charging thresholds from the gprsSSF.

Table 5.21 Overview of CAP v3 operations for GPRS control

CAP Operation	Description
Activity test GPRS (AT-GPRS)	This operation is used between gsmSCF and gprsSSF to check for the existence of the CAMEL relationship. This operation always opens a new TCAP dialogue. It uses the GPRS reference number to identify the CAMEL relationship
Apply charging GPRS (ACH-GPRS)	This operation is used to set a charging threshold in the gprsSSF. It is explained in more detail in Section 5.3.5
Apply charging report GPRS (ACR-GPRS)	This operation is used to report the occurrence of a charging event. It is explained in more detail in Section 5.3.5
Cancel GPRS (CAN-GPRS)	CAN-GPRS may be used to cancel the outstanding charging reports and to disarm event detection points. The gsmSCF may use this operation when it wishes to terminate an existing CAMEL relationship
Connect GPRS (CON-GPRS)	CON-GPRS is used in PDP context establishment to supply an APN
Continue GPRS (CUE-GPRS)	When the gprsSSF has contacted the gsmSCF for instructions, CUE-GPRS is used to continue GPRS processing at that point in the state model where processing was interrupted
Entity released GPRS (ER-GPRS)	This operation is used to inform the gsmSCF about the release of the GPRS session or a PDP context. It is used only when the related event is not armed for reporting. Rationale of this operation is that the gsmSCF may be controlling multiple state models. One state model may terminate, e.g. due the release of a PDP context
Event report GPRS (ERG)	ERG is used to inform the gsmSCF about the occurrence of an event
Furnish charging information GPRS (FCI-GPRS)	FCI-GPRS is used to place free-format data in a CDR. FCI-GPRS may be used for the GPRS session and for each individual PDP context. See Section 5.3.5 for more details
Initial DP GPRS (IDP-GPRS)	IDP-GPRS is used to start the CAMEL relationship, i.e. to start the CAMEL service
Release GPRS (REL-GPRS)	REL-GPRS may be used to detach a subscriber or to release a PDP context. To release the GPRS session or a PDP context, a control relationship is needed with the session or the PDP context
Request report GPRS event (RRGE)	The gsmSCF may use RRGE to arm events for reporting or to disarm an event. The event arming or disarming may apply to the GPRS session or to an individual PDP context. This operation requires a control relationship with the state model for which the operation applies
Reset timer GPRS (RT-GPRS)	The gsmSCF may use RT-GPRS to restart the Tssf timer. See Section 5.3.3.4 for more details
Send charging information GPRS (SCI-GPRS)	SCI-GPRS is used for CAMEL control of advice of charge. See Section 5.3.5 for more information

CAMEL control of GPRS defines three charging methods:[30] (1) duration-based charging for PDP context; (2) volume-based charging for PDP context; and (3) duration-based charging for GPRS session. Refer to 3GPP TS 32.015 [115] for general charging principles for GPRS.

5.3.5.1 Duration-based Charging for PDP context

This charging method may be used in scenario 1 and scenario 2 control. When a PDP context is established and the gsmSCF has gained control over that PDP context, it may use CAP ACH-GPRS

---

[30] For CS call control, there is just one charging threshold defined: call duration.

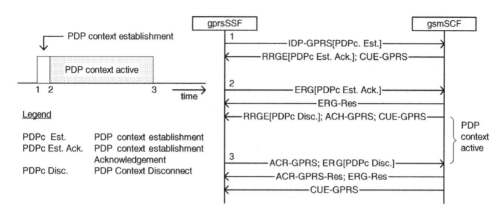

Figure 5.52   Chargeable duration for PDP context

to set a duration threshold. For a PDP context, the duration charging commences when the PDP context becomes active (Figure 5.52).

The gsmSCF may send ACH-GPRS at PDP context establishment or at PDP context establishment acknowledgement. In either case, the duration charging starts no earlier than PDP context establishment acknowledgement. The differences in Table 5.22 apply.

When the PDP context becomes active, the gprsSSF starts duration timing, using the duration threshold received from the gsmSCF. The threshold may have a value between 1 s and 24 h. This implies that, if duration-based charging is applied for a PDP context, a charging report will be generated at least once every 24 h. Every charging report contains the total active duration for the PDP context. Figure 5.53 gives a graphical representation of the duration charging model. (Operation result components are not reflected in the figure.) In this example, the gsmSCF reserves 1 h PDP context duration's worth of credit. As a result, a charging report is generated every hour. The PDP context is terminated by the subscriber after 2.5 h. The final charging report for this PDP context reports the total duration of 2.5 h. When a PDP context exists for more than 24 h, the PDP context duration report uses a 'roll-over counter' to indicate multiples of 24 h periods. The maximum PDP context duration that may be reported through CAMEL is 255 days. The details of the roll-over counter are specified in 3GPP TS 29.078 [106].

Table 5.22   Sending 'apply charging' GPRS to gprsSSF

Sending ACH-GPRS at PDP context establishment	Some parameters that may be needed for charging may not yet be available, such as APN and end user address
	DP PDP context establishment acknowledgement may be armed as EDP-N, resulting in slight increase in CAP signalling efficiency
	If PDP context establishment fails, then the gprsSSF will send an empty charging report to the gsmSCF
Sending ACH-GPRS at PDP context establishment acknowledgement	All parameters that may be needed for charging will be available
	DP PDP context establishment acknowledgement needs to be armed as EDP-R, resulting in a slight increase in CAP signalling (additional CAP CUE-GPRS operation is needed)
	If PDP context establishment fails, then the gprsSSF will not send an empty charging report to the gsmSCF

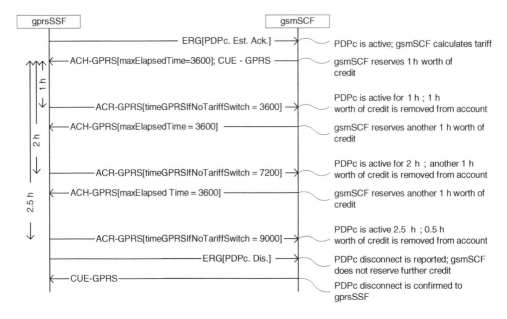

Figure 5.53 PDP context duration charging

Unlike on-line charging for call control, there is no 'release if duration exceeded' mechanism, whereby the gsmSCF can instruct the gsmSSF to release the call when the call period expires. Therefore, when the gsmSCF has determined that the subscriber has reached the maximum permissible PDP context duration and the charging report for the last PDP context has been sent, the gsmSCF will send release GPRS to release the PDP context.

When a subscriber who has an active PDP context receives a CS call, the PDP context may be suspended (on hold).[31] While the PDP context is on hold, the subscriber may answer the CS call. The PDP context is kept active, but no data transfer takes place through this PDP context. After the CS call is terminated, data transfer through the PDP may resume. This suspended PDP context state is, however, not notified to the gprsSSF or gsmSCF. Hence, duration-based charging continues during the suspended state.

5.3.5.2 Volume-based Charging for PDP Context

The volume based charging for the PDP context relates to the total volume of data that is transferred through the PDP context:

$$PDP \text{ context volume} =< \text{volume sent from UE to network} >$$

$$+ < \text{volume received by UE from network} >$$

This charging method may also be used in scenario 1 and scenario 2 control. When the PDP context is established, the gsmSCF determines the price per volume quantity. The price may depend on APN, subscriber location, quality of service, etc. When the PDP context has become active, the PDP context is ready for data transfer. When the volume defined by the threshold is transferred

---

[31] Dual transfer mode (DTM) GSM terminals allow for simultaneous GPRS usage and CS calls. UMTS terminals are always 'multi-RAB', which means that they support simultaneous PS sessions and CS calls.

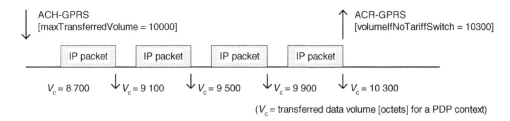

ACH-GPRS
[maxTransferredVolume = 10000]

ACR-GPRS
[volumeIfNoTariffSwitch = 10300]

| IP packet | IP packet | IP packet | IP packet |

$V_c = 8\ 700$ $V_c = 9\ 100$ $V_c = 9\ 500$ $V_c = 9\ 900$ $V_c = 10\ 300$

(V_c = transferred data volume [octets] for a PDP context)

Figure 5.54   Volume report in excess of defined threshold

through this PDP context, a charging report is generated by the gprsSSF. The volume threshold that may be defined by the gsmSCF has a maximum of 4 Giga octets. Hence, if volume-based charging is applied for a PDP context, at least one report will be generated for every transferred 4 Giga octets. The total amount of data that may be transferred through the PDP context may, however, be larger. The maximum PDP context volume that may be reported through CAMEL, using the roll-over counters, is approximately 1 Tera octet. The sequence of CAP operations that is used for PDP context volume-based charging is the same as for duration-based charging, as reflected in Figure 5.53. Likewise, the reasons for sending the volume charging threshold at PDP context establishment or at PDP context establishment acknowledgement are the same as for volume-based charging.

The reported volume for a PDP context may be larger than the defined threshold. This is due to the fact that an SGSN will not update the PDP context volume counter for every individual octet that is transferred through the PDP context. Data is transferred through a PDP context in IP packets. The SGSN updates the volume counter when an IP packet has been transferred. As a result, the volume counter may exceed the threshold, depending on the IP packet size (Figure 5.54). Nevertheless, the actual transferred data volume is reported, so the gsmSCF can charge accordingly. The average size of the IP packet varies per application. For conversational voice, the IP packet size is kept small to maintain small packet transfer latency. For file download, larger packets may be used.

5.3.5.3 Combined Duration and Volume Charging for PDP Context

A CAMEL service may use duration-based charging and volume-based charging for a PDP context at the same time. The gsmSCF sends two charging thresholds to the gprsSSF: a duration threshold and a volume threshold. As a result, the gprsSSF will run both the duration timer and the volume counter when the PDP context becomes active. When the duration threshold is reached, the gprsSSF generates both a duration charging report and a volume charging report, even though the volume threshold was not reached. The gsmSCF will then supply a new threshold for both duration and volume. The gprsSSF keeps the TCAP dialogue open until it has received both thresholds. The same principle applies when the volume threshold is reached: gprsSSF generates both duration charging report and volume charging report. When the PDP context is released, the gprsSSF generates a charging report for duration and a charging report for volume.

5.3.5.4 Duration-based Charging for GPRS Session

This charging threshold may only be used in control scenario 1. When a CAMEL service is started at DP attach, the gsmSCF may use CAP ACH-GPRS to start the duration-based charging. The charging for the GPRS session runs independently of any PDP contexts that are established by the subscriber (Figure 5.55).

In the example in Figure 5.55, the subscriber attaches to the SGSN and establishes three PDP contexts. PDP context 2 and PDP context 3 overlap with PDP context 1. The gsmSCF may use the

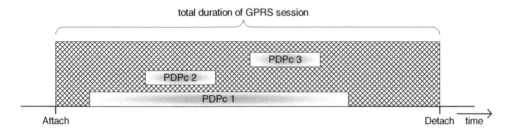

Figure 5.55 Duration-based charging for GPRS session

GPRS charging operations to charge for the total duration of the GPRS session. The detach may be initiated by the subscriber or by the gsmSCF. When the detach occurs, the gprsSSF reports the duration of the GPRS session to the gsmSCF. The gprsSSF may also send intermediate charging reports to the gsmSCF, depending on the charging thresholds set by the gsmSCF.

The duration-based charging for the GPRS session may be combined with duration-based charging and volume-based charging for the PDP contexts. The charging operations for the GPRS session-related charging will run independently of the charging operations for the individual PDP contexts. When the subscriber detaches from the SGSN, the gprsSSF will generate charging reports for the individual PDP contexts, followed by a charging report for the GPRS session.

5.3.5.5 Tariff Zones

When duration-based charging is used, multiple tariff zones may be defined. A tariff zone is defined as a time period in the day or in the week. The gsmSCF sends CAP ACH-GPRS to the gprsSSF and includes a *tariff switch interval* in CAP ACH-GPRS. The tariff switch interval defines an absolute duration, starting from the reception of CAP ACH-GPRS in gprsSSF. When the tariff switch takes place, gprsSSF stores the current charging period duration and records the remainder of the charging period in another counter. When the charging period expires, gprsSSF reports both values to gsmSCF. The gsmSCF may apply high and low tariffs for the respective time zones.

5.3.5.6 Advice of Charge for GPRS

CAMEL has specified the mechanism for the gsmSCF to supply the SGSN with advice of charge (AoC) information for a PDP context. The AoC information is contained in CAP SCI-GPRS. The principles of AoC for GPRS are the same as for CS call control. The gsmSCF supplies the gprsSSF with AoC information at PDP context establishment or at PDP context establishment acknowledgement. The gprsSSF sends the AoC information to the UE when the PDP context has become active.

Although CAMEL control of GPRS includes AoC, the SGSN does not support this feature. The intention of 3GPP, at the time of developing CAMEL phase 3, was to introduce AoC for GPRS at a later stage. CAMEL has specified the AoC control mechanism for the purpose of forward compatibility. The SGSN may signal the support of advice of charge in CAP IDP-GPRS. The support of AoC is indicated in a designated BIT in the parameter SGSNCapabilities. Practically, this BIT will always indicate 'AoC not supported by SGSN'. The gsmSCF will not use CAP SCI-GPRS in that case.

5.3.6 Quality of Service

One of the qualifiers of a PDP context is the quality of service (QoS). A PDP context has three kinds of QoS associated with it:

Requested QoS

The UE includes the Requested QoS in the following DTAP messages:

- *Activate PDP Context Request*, when establishing a PDP context;
- *Activate Secondary PDP Context Request*, when establishing a secondary PDP context;
- *Modify PDP Context Request*, when requesting a change in QoS for an active PDP context.

The requested QoS is reported to the gsmSCF when the PDP context is established. It may be used by the CAMEL service to determine whether the PDP context, with this QoS, may be established. Service designers should take care when using the requested QoS for charging purposes, as the actual QoS for the PDP context may differ from the requested QoS.

Subscribed QoS

The subscribed QoS is contained in the PDP context profile descriptions, which are downloaded from HLR to SGSN during attach. The subscribed QoS is checked in the APN and GGSN selection procedure, which is executed after the PDP context establishment. Therefore, the subscribed QoS is reported in the PDP context establishment acknowledgement notification.

Negotiated QoS

The negotiated QoS is defined during the PDP context establishment process; it is negotiated between SGSN and GGSN. The negotiated QoS is the QoS that is actually used for the PDP context and hence the CAMEL service may use the negotiated QoS for charging purposes. The QoS is reported at PDP context establishment acknowledgement.

The QoS may change during an active PDP context. If the QoS changes, then a charging report will be generated, enabling the CAMEL service to adapt the charging rate for the PDP context (Figure 5.56).

In the example in Figure 5.56, the transmission bit rate changes from 60 to 40 kb/s. When the QoS change occurs, both a duration charging report, if pending, and a volume charging report, if pending, will be generated. The charging report contains the new QoS. Not every change in QoS would have an effect on on-line charging. Therefore, change in QoS for a PDP context will result in a charging report to be sent to gsmSCF only when this QoS is considered to have impact on charging. A QoS parameter such as *guaranteed bit rate for downlink*, for example, may be used for charging. For this parameter, value ranges may be defined; as long as the *guaranteed bit rate for downlink* remains within this range, then there is no impact on charging. 3GPP TS 23.107 [86] specifies details about the PDP context QoS profiles. Exact rules to determine which QoS changes should result in a charging report to be sent to gsmSCF are, however, not specified by CAMEL.

There are no explicit means for the gsmSCF to select a quality of service for the PDP context; CAP CON-GPRS may contain only the APN. One possible mechanism for the gsmSCF to select a QoS is to define two or more PDP context profiles (in HLR) for a specific APN. The profiles would have equal settings except for APN and subscribed QoS. See Table 5.23 for an example.

By providing an APN, the gsmSCF effectively selects a QoS for the PDP context. Both APNs in above example must be defined in DNS and should route to the same application.

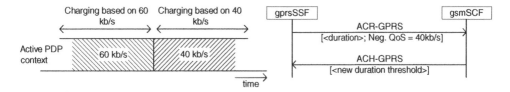

Figure 5.56 Reporting QoS change during active PDP context

Table 5.23   QoS differentiation

PDP context profiles

. . .

Profile 3 (high-speed corporate LAN access)
. . .
APN =
HighSpeedLanAccess.myCompany
QoS = 60 kbyte/s
. . .

Profile 4 (low-speed corporate LAN access)
. . .
APN =
LowSpeedLanAccess.myCompany
QoS = 40 kbyte/s
. . .
. . .

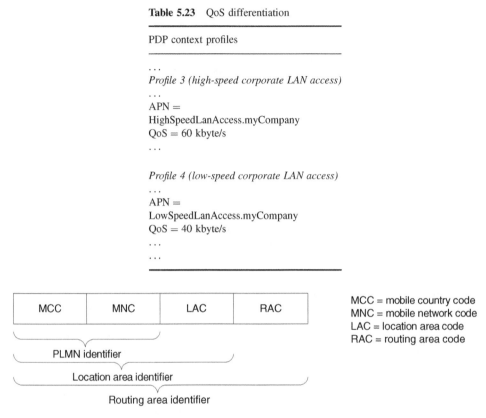

Figure 5.57   Structure of routing area identifier

5.3.7 Routing Area Update

When a subscriber is attached to the SGSN, a routing area update (RAU) may occur. RAU entails that the subscriber moves from one routing area (RA) to another. The RA is an area within a PLMN (Figure 5.57).

The RA consists of one or more cells within the mobile network. The RAU occurs when the subscriber leaves one RA and enters another. A RAU may occur within an SGSN service area (intra-SGSN RAU) or between the service area of one SGSN and the service area of another (inter-SGSN RAU). Detailed RAU procedures are described in 3GPP TS 23.060 [81].

5.3.7.1 Intra-SGSN RAU

The CAMEL service may arm the change of position DP; this arming may be done for the GPRS session state model (scenario 1) or for the PDP context state model (scenario 1 or scenario 2). When the intra-SGSN RAU occurs, the gprsSSF sends a notification to the gsmSCF. The gprsSSF does not generate a charging report as a result of the RAU. Therefore, service designers should take care when applying location-dependent on-line charging. The change in location may be reflected in an adapted charging rate only once the current charging period or volume threshold is reached or a QoS change occurs.

5.3.7.2 Inter-SGSN RAU

The inter-SGSN RAU has specific CAMEL handling associated with it. The inter-SGSN RAU differs from the in-call handover for CS calls. When handover occurs for a CS call, the MSC where the subscriber is registered acts as anchor MSC; all signalling with the UE is relayed via the E-interface, between the anchor MSC and the MSC the subscriber is currently connected to.

When a subscriber who is attached to one SGSN performs an RAU to another SGSN, the subscriber will detach from the one SGSN and attach to the other. This process also includes the termination of the ongoing CAMEL service(s) for that subscriber and the re-initiation of the CAMEL service(s) from the new SGSN. The rationale of this mechanism of deregistration at one SGSN followed by registration at the other SGSN is that a PDP context may have a long duration. Keeping traffic interfaces and control interfaces occupied between two SGSNs for this purpose would be inefficient usage of network resources. The inter-SGSN RAU procedure, including the CAMEL control, is represented in Figure 5.58.

3GPP TS 23.060 [81] describes the inter-SGSN RAU procedures, including the communication between the respective SGSNs and the GGSN. When the subscriber performs a RAU from SGSN (1) to SGSN (2), SGSN (1) will terminate the CAMEL service for the active PDP context. This service termination is done in the normal manner, meaning that the gprsSSF sends the pending charging reports and the PDP context disconnect notification (PDP context disconnect). The disconnect notification contains the *routing area update* parameter, which serves as indication that the disconnect results from inter-SGSN RAU. This indication prompts the SCP to store relevant information related to this PDP context; the information will be needed when the control of this PDP context is resumed form the new CAMEL service. When the SGSN (2) has resumed the handling of the PDP context, then this SGSN may invoke a CAMEL service for this PDP context. The inter-SGSN RAU includes, amongst others, the transfer of the GPRS-CSI from HLR to SGSN (2). If GPRS-CSI for this subscriber contains TDP change of position context, then SGSN (2) invokes a CAMEL service for this PDP context. This CAMEL service may resume the PDP context control. Since this PDP context is not a newly established PDP context, but an existing PDP context, the CAMEL service may want to take cognizance of the history of this PDP context. It may consider the duration of this CDR or the transferred volume; this is required when volume charging of the PDP context depends on the amount of data already transferred. In addition, the new CAMEL service should normally not perform APN validation, since the validation was already performed at PDP context establishment, in SGSN (1).

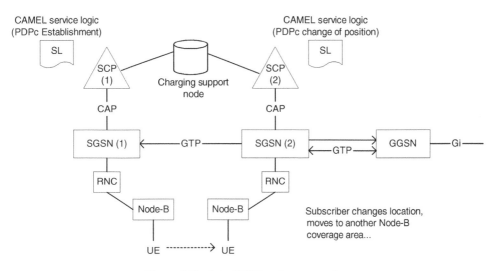

Figure 5.58 Inter-SGSN routing area update

The new CAMEL service, invoked at change of position context, may run in a different SCP than the first CAMEL service, invoked at PDP context establishment or PDP context establishment acknowledgement. This applies even when the same GT is used for the TDP for these respective CAMEL services; dynamic load sharing may have the effect that a GT is translated into one of a set of SPC. However, relevant history information related to the PDP context is stored in the charging support node; the new CAMEL service retrieves this history information for this subscriber. If the subscriber has more than one PDP context active at the time of RAU, then the Charging Id for the PDP context should be used to identify the PDP context in the charging support node.

If the subscriber performs RAU to an SGSN that does not support CAMEL, then there will not be a new CAMEL service invoked for this PDP context. In that case, the stored history information in the charging support node will expire after a predefined time. CAMEL does, however, not define details of this charging support node.

5.3.8 Network-initiated PDP Context Establishment

The PDP context establishment is always performed by the subscriber, i.e. *MO* PDP context establishment. The network may perform charging for this PDP context, using APN, QoS, end user address, etc., as parameters. An application may, however, need to establish a PDP context with a subscriber, i.e. *MT* PDP context establishment. To a large extent, the network-initiated PDP context establishment is a regular PDP context establishment. For network-initiated PDP context establishment, network signalling is used to request the UE to establish a PDP context. The SGSN sends the DTAP message *Request PDP Context Activation* to the UE. This request for PDP context establishment contains parameters that are required by the UE to establish the PDP context. Details of this message are described in 3GPP TS 24.008 [94].

When the UE establishes a PDP context in response to the request from the network, the SGSN includes an indication in CAP IDP-GPRS, *PDP initiation type = network initiated*. The CAMEL service may decide not to charge the subscriber for this PDP context. When a PDP context that is established in this manner is subject to inter-SGSN RAU, then the CAMEL service invocation from the new SGSN will not include this indication; this is due to the fact that the GTP SGSN context response message, sent from old SGSN to new SGSN during inter-SGSN RAU, does not include this indication. However, the CAMEL service that is active before RAU should take care to store this information in the charging support node.

5.3.9 Secondary PDP Context

The secondary PDP context mechanism enables a subscriber to associate a second PDP context with an already existing PDP context. This secondary PDP context uses the same APN and PDP address. The QoS of the secondary PDP context may differ from the QoS of the associated primary PDP context. The establishment of the secondary PDP context may lead to CAMEL service invocation in the same manner as a primary PDP context. A parameter in CAP IDP-GPRS, 'secondary PDP context', indicates that this PDP context is a secondary PDP context. The CAMEL service may skip the APN validation procedure, as this was already performed during the establishment of the associated primary PDP context. The CAMEL service should also take care when modifying the APN, as that may result in failure for this PDP context establishment. If the CAMEL service supplied or modified the APN for the primary PDP context, then the CAMEL service should use the same, modified APN for the secondary PDP context.

The GGSN allocates different charging Id's for the primary PDP context and the associated secondary PDP context. Therefore, if the subscriber has multiple PDP contexts active, then CAMEL service may use APN and end user address to associate the secondary PDP context with a primary PDP context. The CAMEL service may apply a reduced rate for the secondary PDP context,

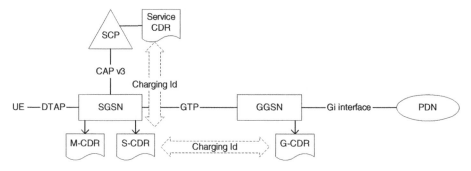

Figure 5.59   Generation of CDRs for GPRS

since such PDP context may be regarded as a means to accomplish higher data throughput for an application for which a PDP context is already active.

The secondary PDP Context may be used to gain a combined, higher-throughput data channel.

5.3.10 Impact on CDRs

The SGSN and GGSN may generate CDR to reflect the subscriber's use of the radio network and core network. Figure 5.59 indicates the CDRs for the GPRS network. The SMS-related CDRs are not reflected in the figure.

The use and contents of CDRs for GPRS are specified in 3GPP TS 32.015 [115].[32] For GPRS, the following CDRs are relevant.

S-CDR

The SGSN may create an S-CDR for each PDP context that is activated in that SGSN. The following CAMEL specific elements may be included in this CDR:

- CAMEL APN – this is the APN that the subscriber provided to the network;
- SCF address – this is the address that is used to invoke the CAMEL service; it is obtained from GPRS-CSI.
- Service key – this is the service key that is used to identify the CAMEL service; it is obtained from GPRS-CSI.
- Default transaction handling – this parameter indicates that default handling has taken place for the PDP context handling; the default transaction handling is described above. This parameter may be used by a CDR post-processing system to decide that the CDR shall still be subject to charging calculation, since the CAMEL service invocation failed or the CAMEL service was terminated prematurely.
- Number of DP encountered – this parameter serves as an indication of the complexity of a CAMEL service; the more DPs are encountered, the more processing needs to be performed by the SGSN. A VPLMN operator may use this parameter to charge the HPLMN operator of the served subscriber for the use of the resource in the VPLMN.
- Level of CAMEL service – this parameter also contains an indication of the complexity of the CAMEL service; the VPLMN operator may use it in the same manner as the number of DP encountered. The parameter may indicate whether on-line charging was used for the PDP context.
- Free-format data – a CAMEL service may place free-format data in the CDR for a PDP context. A CAMEL service uses CAP FCI-GPRS hereto. The use of CAP FCI-GPRS is identical to the use of CAP FCI for call control. The CAMEL service may place up to 160 octets in the CDR.

---

[32] For 3GPP Rel-4 and Rel-5, refer to 3GPP TS 32.215 [118]; for 3GPP Rel-6 onwards, refer to 3GPP TS 32.251 [121].

Although it is not very likely, it is possible that a CAMEL service is invoked at PDP context establishment and another CAMEL service at PDP context establishment acknowledgement. This may occur when GPRS-CSI contains the TDP for both DPs and the CAMEL service that is invoked at PDP context establishment does not retain the CAMEL relationship. In that case, the S-CDR should contain two sets of CAMEL data (except for CAMEL APN). The charging specification (3GPP TS 32.015 [115]) does not cater for this double triggering scenario; the S-CDR may contain at the most one set of CAMEL data.

An SGSN may be configured to generate the S-CDR only when PDP context activation is successful. In that case, no CAMEL data will be recorded for that (failed) PDP context.

M-CDR

The SGSN may generate a *mobility CDR* (M-CDR) for the duration that a subscriber is attached to the SGSN. It may contain the same CAMEL data as the S-CDR, except for the CAMEL APN. When scenario 1 is used, then the CAMEL service may be controlling several PDP contexts for the subscriber. It is unclear from 3GPP TS 32.015 [115] whether any encountered DP for a PDP context that is controlled through this CAMEL dialogue should be reflected in the value of the *number of DP encountered* parameter in the M-CDR or in the value of the *number of DP encountered* parameter in the S-CDR. The same applies to the level of CAMEL service in scenario 1.

G-CDR

The G-CDR is generated in the GGSN per activated PDP context. The impact on the G-CDR by CAMEL is described below.

5.3.10.1 Linking of CDRs

The combination of charging Id and GGSN address may be used for CDR correlation. The following groups of CDRs may be correlated:

- S-CDR – the S-CDR, generated by the SGSN, contains the Charging Id and GGSN address. If the control of a PDP context is resumed by another SGSN, due to inter-SGSN RAU, then the charging Id and GGSN address are transferred to the new SGSN. The S-CDR generated by the new SGSN will therefore contain the same charging Id and GGSN Address.
- G-CDR – the G-CDR, generated by the GGSN, also contains the charging Id and GGSN address.
- Service CDR – the charging Id and the GGSN address are reported to the gsmSCF at PDP context activation. The CAMEL service may place these elements in a service CDR. The generation of a service CDR is optional. If the control of a PDP context is resumed by another gsmSCF, due to inter-SGSN RAU, then the charging Id and GGSN address are transferred to the new gsmSCF. The service CDR generated by the new gsmSCF may therefore contain the same charging Id and GGSN address.

The CDRs generated by SGSN(s), GGSN and gsmSCF(s) may be collected and correlated by CDR processing systems.

5.3.10.2 Transfer of CAMEL Data to GGSN

3GPP Rel-6 specifies the transfer of CAMEL data from SGSN to GGSN. When a PDP context is created, the SGSN sends the GTP message create PDP context request to the GGSN. This GTP message contains a copy of the CAMEL information of the S-CDR of this PDP context. The GGSN places this CAMEL information into the G-CDR for this PDP context. The GGSN performs no further processing on this data.

Rationale of the transfer of CAMEL data form SGSN to GGSN is that it enables operators to perform post-processing on G-CDR only, without having to correlate the G-CDR with the S-CDR. When the GGSN is in the HPLMN, then the G-CDR is more readily available than the S-CDR.

This method of transferring CAMEL data to the GGSN during PDP context activation has some limitations. For example, the elements number of DP encountered and level of CAMEL service may change during the duration of the PDP context. Also, the free format data may be supplied or appended during the PDP context duration.

No method is defined for the transfer of CAMEL data from the SGSN to the GGSN during an active PDP context.

5.3.11 Operator-determined Barring

When the gsmSCF supplies the APN to the SGSN, this APN may still be subject to ODB. ODB for GPRS is introduced in 3GPP Rel-4; refer to 3GPP TS 22.041 [63] (section on 'packet oriented services') for the ODB service requirement description and 3GPP TS 23.015 for technical implementation of ODB. When a subscriber initiates a PDP context, the SGSN checks whether the subscriber is prohibited from establishing a PDP context to this APN. This check takes place during APN and GGSN selection. If the PDP context is barred due to ODB, then this will be notified to the CAMEL service, if the CAMEL service is active. The notification may take the form of event notification (PDP context disconnect) or the entity released GPRS operation. There is no call barring supplementary service defined for PDP context establishment.

5.3.12 GPRS Roaming Scenarios

CAMEL control of GPRS applies to HPLMN usage and VPLMN usage. When a subscriber attaches to an SGSN in a PLMN other than the HPLMN, then the SGSN may or may not support CAMEL phase 3. See Section 5.3.1.3 for GPRS location update procedure. When the SGSN does not support CAMEL phase 3, then the HLR may apply one of the following actions:

(1) disallow attachment to the SGSN – the subscriber may attempt to attach to an SGSN from another operator in that country;
(2) allow attachment to the SGSN;
(3) allow attachment, with reduced set of PDP context profiles – the HPLMN operator may disallow use of a GGSN in VPLMN;
(4) allow attachment, but apply operator determined barring (ODB) for PDP context activation; the subscriber may attach, but cannot establish PDP contexts. As a further option, the ODB setting may entail network-initiated PDP context being allowed, but UE-initiated PDP context being disallowed.

The action to be taken by the HLR in this case is not specified by CAMEL; CAMEL specifies only that the HLR shall not send GPRS-CSI in this case. An HLR vendor may define more actions that may be taken in this case.

The use of GGSN in the HPLMN or GGSN in the VPLMN depends on operator agreement and may be configured in the PDP context profile. Using a GGSN in the VPLMN may result in more efficient traffic routing. On the other hand, using a GGSN in the HPLMN allows the operator to have the GGSN apply validation on the PDP context, for example.

5.3.12.1 PDP Context Charging from the GGSN

If a roaming subscriber establishes a PDP context using the GGSN in the HPLMN, then charging may take place in the GGSN. Figure 5.60 presents an example scenario whereby the GGSN invokes a CAMEL phase 3 service from the GGSN. The protocol used between GGSN and gsmSCF is denoted by CAP v3', meaning that not all the CAP v3 control capability is available in this method. This method is not specified by CAMEL, but is used in some networks. The provisioning of GPRS-CSI in the GGSN is vendor-specific. The GGSN may trigger the CAMEL service when it deduces from the SGSN address in the 'create PDP context request' GTP message that the subscriber is attached to a non-CAMEL phase 3 supporting SGSN.

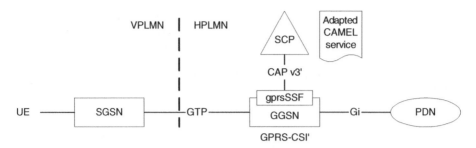

Figure 5.60   CAMEL service invocation from the GGSN

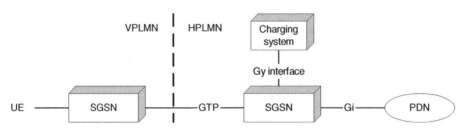

Figure 5.61   PDP context charging via GGSN

Figure 5.61 presents an example scenario whereby the GGSN applies flow-based charging control. Flow-based charging is described in 3GPP TS 23.125 [87]. Here as well, the GGSN may use the flow-based charging when it deduces that the SGSN does not support CAMEL phase 3. The charging system depicted in Figure 5.61 may include an SCP; details of such combination of flow-based charging and CAMEL are described in 3GPP TS 23.125 [87]. The Gy interface is a charging protocol based on remote authentication dial in user service (RADIUS); RADIUS is described in IETF RFC 2865 [168].

5.3.13 Enhanced Data Rates for GSM Evolution

Enhanced data rates for GSM evolution (EDGE) is a technique that increases the data throughput in the GSM TDMA radio access. EDGE may be used in combination with GPRS or may be used for a CS (data) calls. When EDGE is used for GPRS, the PDP contexts that are established may be controlled by CAMEL phase 3 control in the same manner as when standard TDMA is used. However, the QoS may indicate a higher data rate. When EDGE is used for data calls, then one of the GSM bearer services is indicated in the CAMEL service invocation. The CAMEL service may adapt the charge of the call to the data rate. For technical details about EDGE, refer to 3GPP TS 43.051 [125].

5.4 CAMEL Control of MO-SMS

Short Message Service (SMS) exists as from GSM phase 1. Two SMS-related services exist:

- MO-SMS – this service entails the sending of a short message from the UE to a particular destination; this destination may be a subscriber of the same mobile network, a subscriber of another network or a network-based destination.
- MT-SMS – this service entails the receiving of a short message on one's UE.

Table 5.24 Specifications and reports for SMS

Specification	Title
3GPP TS 23.038 [77]	Alphabets and language-specific information. This specification specifies the data coding scheme that is used for SMS
3GPP TR 23.039 [78]	Interface protocols for the connection of short message service centres (SMSCs) to short message entities (SMEs).[33] This report lists the protocols that are used for communication with the SMSC
3GPP TS 23.040 [79]	Technical realization of SMS. This specification describes the overall SMS functionality
3GPP TS 23.042 [80]	Compression algorithm for SMS. This specification describes the Huffman data compression technique that may be applied to SMS
3GPP TS 23.060 [81]	GPRS; service description. This specification describes the SMS functionality for GPRS
3GPP TS 24.011 [95]	Point-to-point (PP) SMS support on mobile radio interface. This specification describes the signalling procedures over the RAN, between UE and MSC and between UE and SGSN, for SMS transfer
3GPP TS 29.002 [103]	Mobile application part. This specification specifies the SMS transfer via MAP messages

The CAMEL control of SMS in 3GPP R99 relates to the MO-SMS only. As from 3GPP Rel-5 onwards, CAMEL control of SMS relates to both MO-SMS and MT-SMS. SMS is specified mainly in the 3GPP specifications and reports as listed in Table 5.24.[33]

The SMS is defined as a GSM basic service. The following basic service (group) codes are allocated to SMS:

Teleservice 20	short message service group
Teleservice 21	short message MT (MT-SMS)
Teleservice 22	short message MO (MO-SMS)
Teleservice 23	cell broadcast service

Refer to 3GPP TS 22.003 [59] for a description of GSM teleservices. The subscription data for SMS is contained in HLR and is transferred to MSC or SGSN. SMS may be subject to supplementary services and operator-determined barring. 3GPP TS 22.004 [60] specifies which supplementary services may be applied to SMS.

5.4.1 Network Architecture

The network architecture for MO-SMS is presented in Figure 5.62. The SMS may be used in the CS and PS domains:

- *CS domain* – the SMS may be sent and received via the MSC where the subscriber is registered; this method of SMS transfer is referred to as 'MSC-based SMS';
- *PS domain* – the SMS may be sent and received via the SGSN that the subscriber is attached to; this method of SMS transfer is referred to as 'SGSN-based SMS'.[34]

---

[33] 3GPP TR 23.039 [78] is a technical *report*, not a technical *specification*.

[34] The terms CS-based SMS and PS-based SMS, as sometimes used, are confusing, as SMS is by definition packet switched.

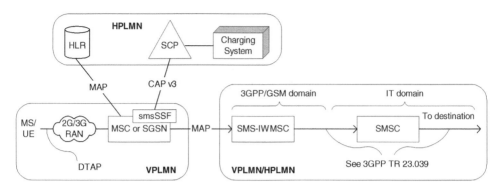

Figure 5.62   Network architecture for CAMEL control of MO-SMS

A subscriber may, depending on subscription data, send SMS via MSC, SGSN or both.[35] When a subscriber submits an SMS,[36] the SMS is transferred from UE to MSC or SGSN. Provided that the subscriber has a subscription to TS22, the MSC uses MAP signalling to transfer the SMS to the SMSC. When the SMS has arrived at the SMSC and is accepted by that SMSC, the SMS submission is considered successful. Delivery of the SMS to the destination is not part of the MO SMS functionality. When the SMS has arrived at the SMSC, the subscriber is charged for the SMS submission.[37]

The SMS-IWMSC has a specific role in the SMS submission. The SMS-IWMSC functions as an interface between the GSM/3GPP domain and the IT domain. The MSC and SGSN are part of the GSM/3GPP domain, whereas the SMSC is part of the IT domain. The interface between MSC or SGSN and SMS-IWMSC is MAP; see 3GPP TS 29.002 [103]. The interface between SMS-IWMSC and SMSC is not defined in GSM/3GPP. A number of industry *de facto* standard protocols are defined for accessing the SMSC; these protocols are listed in 3GPP TR 23.039 [78]. Some of these protocols are: short message peer-to-peer (SMPP); and universal computer protocol (UCP).

Owing to the fact that the protocol between SMS-IWMSC and SMSC is a non-GSM standard protocol, the SMS-IWMSC and SMSC are located in the same country. Although the SMS-IWMSC is defined as a separate functional entity, it may be integrated in one of the MSCs in the operator's network or in the SMSC. Practically, the SMS-IWMSC is often integrated in the SMSC, meaning that MAP signalling is used between the MSC and the combined SMSC and SMS-IWMSC node.

The SMSC address that is used by a subscriber for SMS submission is normally configured in the mobile station. Although the SMSC address may be pre-configured when purchasing a GSM telephone, subscribers may modify the SMSC address. Strictly speaking, the 'SMSC address' that is used for SMS submission and SMS delivery is not the address of the SMSC.

- *MO-SMS* – the 'SMSC address' (destination of the SMS submission) is the address of the SMS-IWMSC between the MSC or SGSN and the SMSC.
- *MT-SMS* – the 'SMSC address' (origin of the SMS delivery) is the address of the SMS-GWMSC between the SMSC and the MSC or SGSN.

The SMS-IWMSC and SMS-GMSC act in this regard as the MAP addressing front-end for the SMSC. The architecture for MT-SMS is described in Chapter 6.

---

[35] GPRS-capable UE may contain user setting for preferred method for submitting SMS: via MSC or via SGSN.

[36] Sending SMS from UE to SMSC is known as 'SMS submission'; sending SMS from SMSC to UE is known as 'SMS delivery'.

[37] SMS charging rules may differ per operator.

5.4.2 CAMEL Control of MO-SMS

The CAMEL control of MO-SMS entails that a CAMEL relationship may be established between the MSC and the gsmSCF or between the SGSN and the gsmSCF. CAMEL control of MO-SMS is a subscribed service, which implies that the subscriber has CAMEL subscription data in the HLR: MO-SMS-CSI.[38] MO-SMS-CSI is sent from HLR to MSC, during registration in the MSC or to the SGSN, on attachment to the SGSN.

5.4.2.1 Location Update

Refer to Figure 5.63; when the subscriber registers in the MSC or attaches at the SGSN, the MSC or SGSN initiates location update procedure. The MSC or SGSN reports to the HLR which CAMEL phases it (the MSC or SGSN) supports. If CAMEL phase 3 is supported, then the HLR may send MO-SMS-CSI to the MSC or SGSN. If CAMEL phase 3 is not supported, then the HLR may take fallback action, such as suppressing TS22 or allowing normal registration without CAMEL. Fallback action may normally be configured per subscriber. CAMEL does not, however, specify the fallback actions. The MAP insert subscriber data message, which forms part of the location update procedure, may also contain, besides the MO-SMS-CSI, a selection of the other CSIs, such as O-CSI, VT-CSI etc. The sending of subscriber data from HLR to MSC or SGSN may also be triggered by a data restoration procedure in the MSC or SGSN; in addition, the HLR may perform stand-alone subscriber data update or deletion.

5.4.2.2 CAMEL Service Invocation

When the subscriber is registered in the MSC and has appropriate subscription data, she may send SMSs. If MO-SMS-CSI is present in the MSC, then the submission of an SMS by that subscriber leads to unconditional triggering of a CAMEL service. The CAMEL service is identified by the contents of MO-SMS-CSI; see Section 5.4.3. Figure 5.64 depicts the process for MSC-based SMS submission; the same architecture applies to SGSN-based SMS submission. The CAP information flow between smsSSF and gsmSCF contains information that is obtained from the SMS header; the content of the SMS, i.e. the text string or data string, is not sent to the gsmSCF. The gsmSCF instructs the smsSSF to continue SMS submission, after which the MSC will send the MAP message MO-ForwardSM to the SMS-IWMSC.

The SMS-Submit message is described in 3GPP TS 23.040 [79]; it is one of the transfer protocol data units (TPDU) used for SMS transfer. Some of the TPDUs relate to MS-to-SMSC data transfer; other TPDUs relate to SMSC-to-MS data transfer. Table 5.25 lists the available TPDUs.

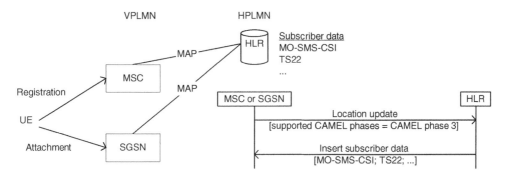

Figure 5.63   Sending CAMEL SMS data at location update

---

[38] This subscription element was initially called SMS-CSI; with the introduction of MT-SMS control in 3GPP Rel-5, SMS-CSI was renamed MO-SMS-CSI. SMS-CSI and MO-SMS-CSI are, however, identical.

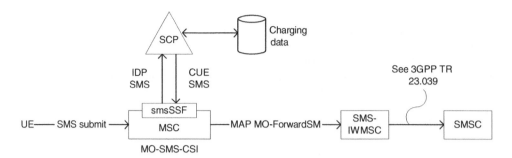

Figure 5.64   SMS submission

Table 5.25   TPDUs for short message service

TPDU type	Direction	Description
SMS-Submit	MS to SMSC	This TPDU contains a short message to be sent to the SMSC
SMS-Command	MS to SMSC	This TPDU is used to send an SMS command from MS to SMSC
SMS-Deliver-Report	MS to SMSC	This TPDU constitutes an acknowledgement or an error indication related to SMS delivery or status report delivery
SMS-Deliver	SMSC to MS	This TPDU is used to transfer a short message from SMSC to MS
SMS-Status-Report	SMSC to MS	This TPDU conveys an SMS status report from SMSC to MS
SMS-Submit-Report	SMSC to MS	This TPDU constitutes an acknowledgement or an error indication related to SMS submission or SMS command

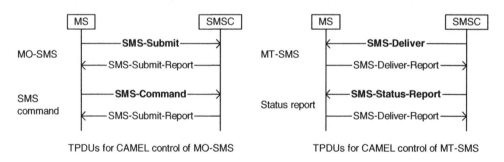

Figure 5.65   TPDUs for CAMEL control

CAMEL control of MO-SMS relates to data transfer from MS to SMSC. The SMS-Submit TPDU and the SMS-Command TPDU may trigger a CAMEL service. The SMS-Deliver-Report TPDU does not constitute data/message transfer, but is used to report the success or failure of the transmission of a previous TPDU. Therefore, the SMS-Deliver-Report TPDU does not trigger a CAMEL service. The relation between these TPDUs is depicted in Figure 5.65. The figure also contains the TPDUs for MT-SMS. If the subscriber has MO-SMS-CSI, then both TPDU SMS-Submit and TDP SMS-Command lead to unconditional triggering of a CAMEL service. One exception is the BAOC category of call barring or ODB. If BAOC is active for TS22 (i.e. for MO-SMS), then both SMS submission and SMS command are barred; as a result, no CAMEL service is invoked.

5.4.3 Subscription Data

The operator may provision each subscriber in the HLR with MO-SMS-CSI, which consists of the elements listed in Table 5.26. In addition, MO-SMS-CSI may contain the elements 'notification-ToCSE' and 'csi-Active'. These elements are described in Section 5.8.

5.4.4 SMS State Model

SMS submission uses a fairly rudimentary state model, compared with call control and GPRS control; see Figure 5.66. The meaning of the Detection Points is described in Table 5.27. When a subscriber sends an SMS that exceeds the permissible length for SMS, then the UE may invite the subscriber to send the SMS as a 'concatenated SMS'. The concatenated SMS is sent as two

Table 5.26   Contents of MO-SMS-CSI

Element	Description
Trigger detection point	This element indicates the point in the SMS state model from where the CAMEL service shall be invoked. For CAMEL control of SMS, this element will always have the value 'sms-CollectedInfo'; see Section 5.4.4 for a description of the state model for MO-SMS
Service key	The service key indicates to the gsmSCF which CAMEL service will be invoked
gsmSCF address	The gsmSCF address is the GT of the gsmSCF where the CAMEL service is located
Default SMS handling	The default SMS handling indicates the action that the smSSF will take in the case of CAMEL dialogue failure; it may have the value 'continue' or 'release'
CAMEL capability handling	This element indicates which version of CAP will be used for the CAMEL service. For CAMEL control of SMS, this element will always have the value 'CAP v3'

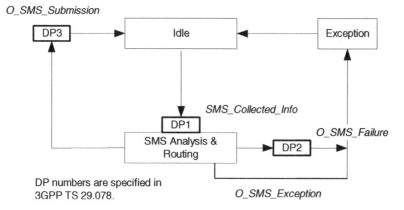

Figure 5.66   State model for CAMEL control of MO-SMS. Reproduced from 3GPP TS 23.078 v3.19.0, Figure 7.3, by permission of ETSI

Table 5.27   Detection points for the MO-SMS state model

DP	Description
SMS_Collected_Info	The subscriber has sent the short message or the SMS command to the MSC or SGSN. There are no conditions prohibiting the SMS or command submission. The CAMEL service is started
O_SMS_Failure	A failure has occurred in the SMS or command submission. The failure may have occurred internally in the MSC or SGSN or may have occurred externally, e.g. in the SMSC
O_SMS_Submission	The short message or SMS command is successfully submitted to the SMSC

or more individual SMs through the network. Each SMS of a concatenated SMS will trigger its own CAMEL service. CAP IDP-SMS does not contain an indication that the SMS is part of a concatenated SMS.

5.4.4.1 CAMEL Dialogue Rules

The CAP v3 protocol that is used for MO-SMS control is different from the CAP v3 that is used for call control. A dedicated application context is used for MO-SMS control.

The arming and disarming rules for the DP in the SMS state model are similar to the rules for CS call control. When the CAMEL SMS service is started, the service logic may arm the O_SMS_Failure and the O_SMS_Submitted DPs. Arming may be done in *notify and continue* mode (EDP-N) or in *interrupt* mode (EDP-R). Arming at least one DP results in retention of the CAMEL relationship after the service has sent CAP CUE-SMS or CAP CON-SMS. When the successful or unsuccessful SMS submission is reported to the gsmSCF, the CAMEL service will terminate.

If the CAMEL service arms neither of the two DPs, then a pre-arranged end occurs as soon the gsmSCF sends CAP CUE-SMS or CAP CON-SMS.

Although the concept of *control relationship* and *monitor relationship* applies for the CAMEL relationship for SMS control, this has no practical meaning. The CAP operations furnish charging information SMS, release SMS and request report SMS event are permissible only when the smsSSF FSM is in the state waiting for instructions.

The smsSSF uses, just like the gsmSSF and the gprsSSF, a Tssf timer. The Tssf timer serves as a guard timer for the smsSSF when the smsSSF FSM is in the state waiting for instruction. The CAMEL service may use the reset time SMS (RT-SMS) operation to reset this timer. A CAMEL service may use the RT-SMS operation in response to the initial DP SMS (IDP-SMS) operation, to 'buy' itself some more response time. A CAMEL service may also use CAP RT-SMS when O_SMS_Failure or O_SMS_Submitted is reported in interrupt mode. However, using CAP RT-SMS at that point in the state model would not be useful. If Tssf expires, then the smsSSF applies Default SMS Handling, which may indicate 'continue' or 'release'.

The pre-arranged end rules used for call control and GPRS control also apply to SMS control. CAP operations other than IDP-SMS may be sent in a TC_Continue message. When the gsmSCF has given control back to the MSC or SGSN and there are no armed detection points, then both smsSSF and gsmSCF may close the TCAP dialogue locally.

5.4.5 Information Flows

Table 5.28 lists the operations that are used for MO-SMS control.

Table 5.28   CAMEL Phase 3 operations for SMS control

CAP operation	Description
Connect SMS (CON-SMS)	This operation may be used by gsmSCF to instruct smsSSF to continue SMS processing, with modified data
Continue SMS (CUE-SMS)	This operation may be used by gsmSCF to instruct smsSSF to continue SMS processing
Event report SMS (ERB-SMS)	This operation is used by smsSSF to inform gsmSCF about the successful or unsuccessful SMS submission
Furnish charging information SMS (FCI-SMS)	The CAMEL service may use FCI-SMS to place free-format data in the SMS CDR
Initial DP SMS (IDP-SMS)	This operation is used to start the CAMEL service
Release SMS (REL-SMS)	The CAMEL service may use REL-SMS to bar the submission of the SMS
Request report SMS Event (RRB-SMS)	RRB-SMS is used to arm the DPs related to successful and unsuccessful SMS submission
Reset timer SMS (RT-SMS)	The gsmSCF may use this operation to reset the Tssf timer in smsSSF

5.4.6 Information Reporting and SMS Steering

When the CAMEL MO-SMS service is started, the smsSSF reports the following information to the gsmSCF.

- *Service key* – this parameter indicates which CAMEL service will be invoked. It is obtained from MO-SMS-CSI.
- *Destination subscriber number* – this is the destination of the SMS, i.e. the intended receiver of the SMS, as entered by the subscriber on the mobile station. If the CAMEL service is invoked for an SMS command, then this parameter identifies the SMS in the SMSC to which the command applies. If the SMS command is not intended for a specific SMS, then this parameter may be absent or may consists of a header only.
- *Calling party number* – this parameter contains the MSISDN of the sender of the SMS.
- *Event type SMS* – this parameter indicates the point in the SMS state model where the service is invoked. It is obtained from MO-SMS-CSI and will always have the value SMS_Collected_Info.
- *IMSI* – this parameter contains the IMSI of the sender of the SMS.
- *Location information* – this information element indicates the location in the GSM access network (for MSC-based SMS) or the GPRS access network (for SGSN-based SMS). This information may be used for charging. If a subscriber sends an SMS whilst roaming, the cost of sending the SMS may be higher.
- *SMSC address* – the SMSC address indicates the SMSC (or SMS-IWMSC, as discussed earlier) to which the SMS shall be sent. The SMSC address is configured in the mobile station of the sender.
- *Time and timezone* – this parameter contains the time and time zone of the MSC or SGSN from where the SMS submission takes place. If the MSC serves an area spanning multiple time zones, then the reported time may differ from the time at the location of the sender.
- *Short message submission info* – this parameter contains various elements related to the SMS-Submit or SMS-Command; see Tables 5.29 and 5.30.[39]

---

[39] The sets of specific elements are a copy of the first octet of SMS-Submit TPDU or SMS-Command TPDU. Not all of the reported elements may be useful for the CAMEL service.

Table 5.29 SMS-Submit specific elements

Data element	Description
Message-Type-Indicator	This element indicates the message type, i.e. SMS-Submit
Reject-Duplicates	This element indicates whether a duplicate SMS will be rejected by the SMSC
Validity-Period-Format	This element indicates whether a validity period is defined for the SMS
Status-Report-Request	This element indicates whether the sender of the SMS has requested a status report
User-Data-Header-Indicator	This element indicates whether the SMS user data contains a header. The user data header may contain a qualifier of the contents of the short message
Reply-Path	This element indicates that a reply path is requested

Table 5.30 SMS-Command-specific elements

Data element	Description
Message-Type-Indicator	This element indicates the message type, i.e. SMS-Command
Status-Report-Request	This element indicates whether the sender of the SMS has requested a status report. The status report may indicate whether the SMS command was executed successfully
User-Data-Header-Indicator	This element indicates whether the SMS user data contains a header

- *Protocol identifier* – the short message may be used in various higher-layer protocols, such as telefax, teletex, videotext, etc. This is indicated by the protocol identifier.
- *Data coding scheme* – this element defines which alphabet is used for the message contents. The alphabets are specified in 3GPP TS 23.038 [77].
- *Validity period* – this element indicates the maximum duration the SMSC will store the short message, if it cannot be delivered to the intended destination due to temporary delivery failure, e.g. because the MS of the intended receiver is currently switched off.
- *SMS reference number* – see Section 5.4.7.
- *MSC address* – see Section 5.4.7.
- *SGSN address* – see Section 5.4.7.

Not all of the above-listed elements are available in the MO-SMS CDR that is generated by the MSC or SGSN. In addition, the MSC or SGSN does not generate a CDR for an SMS command. Therefore, SMS commands cannot be charged by post-processing, whereas CAMEL offers this capability.

When the CAMEL service sends CAP CON-SMS to the smsSSF, it may include the following information elements.

- *SMSC address* – the CAMEL service may force the use of a particular SMSC e.g. pre-paid SMSC or post-paid SMSC. The CAMEL service could also force the use of an SMSC in the VPLMN. Using an SMSC in a PLMN other than the HPLMN would require an agreement between the involved operators.
- *Destination subscriber number* – the CAMEL service may modify or determine the destination of the SMS. This may be useful within the context of a VPN service or when sending an SMS to a short-code destination.
- *Calling party number* – the CAMEL service may change the CPN. The CPN may be set to a VPN number. This CPN has the form of ISDN-Address String; that implies that it does not contain a

presentation indicator. The CAMEL service is not able to restrict the presentation of the CPN to the destination subscriber. Refer to 3GPP TS 22.004 [60]; the CLIP and CLIR supplementary services are not applicable to SMS. Service designers should take care when modifying the CPN in the SMS. The MAP message that conveys the SMS, MO-ForwardSM, does not always contain the IMSI of the sending subscriber. The transport of IMSI in MO-ForwardSM was introduced in 3GPP R98. If the serving MSC does not include the IMSI in the MAP message and the CAMEL service has modified the CPN, then the MAP message may not contain an indication of the sender of the SMS. Such situation may hamper services such as lawful intercept (LI). Refer to 3GPP TS 33.106 [124] for LI requirements.

5.4.7 Charging and Call Detail Records

The MSC or SGSN may generate a CDR during SMS submission; CDRs are specified in 3GPP TS 32.005 [114] (for MSC-based SMS submission) and in 3GPP TS 32.015 [115] (for SGSN-based SMS submission). The applicable CDR for SMS submission is the SMS-MO record (from MSC) and S-SMO-CDR (from SGSN).

5.4.7.1 CDR Impact

The following CAMEL-specific parameters are specified for the SMS-MO record and the S-SMO-CDR record.

- *gsmSCF address* – this is the GT address of the gsmSCF; this element is obtained from the MO-SMS-CSI. Owing to dynamic load sharing, this address may differ from the GT of the gsmSCF handling this particular CAMEL service.
- *Service key* – this is the service key used for the CAMEL service invocation.
- *Default SMS handling* – this element indicates that default handling has taken place for the SMS. The CDR post-processing system may filter out CDRs containing this element, for the purpose of post-charging.
- *Free-format data* – this element contains the free format data provided by the CAMEL service.
- *Calling party number* – this element contains the CAMEL service-provided CPN.
- *Destination subscriber number* – this element contains the destination number provided by the CAMEL service.
- *CAMEL SMSC address* – this element contains the SMSC address provided by the CAMEL service
- *SMS reference number* – the SMS reference number is used for CDR correlation.

5.4.7.2 CDR Correlation

Refer to Figure 5.67 for CDR correlation; when the MSC or SGSN invokes the CAMEL MO-SMS service, it may include an SMS reference number and the MSC address in CAP IDP-SMS. The combination of SMS reference number and MSC address forms a unique identifier of the SMS. The MSC, SGSN and gsmSCF may place this identifier in the respective CDR. A billing gateway (BGW) that is processing CDRs for the operator may use this identifier for correlating the CDRs. If the MSC or SGSN is in another network or in another country than the BGW, then the CDRs may be transported in the form of a TAP file. For TAP file usage and format, refer to GSM PRD TD.57 [162].

The use of the SMS reference number is an optional feature in CAMEL phase 3. For CAMEL phase 4, this feature is a mandatory part of MO-SMS control. The SMS-IWMSC may generate an MO SMS IW CDR. This CDR may be used by the operator of the SMSC for accounting purposes.

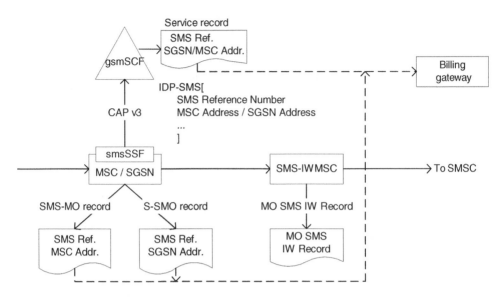

Figure 5.67   CDR correlation for SMS submission

It does not contain an SMS reference and therefore cannot be correlated to the CDRs generated by MSC or SGSN and gsmSCF.

5.4.7.3 Free-format Data

The free-format data capability for CAMEL control of MO-SMS is the same as for call control and GPRS control. The gsmSCF may place and append up to 160 octets in the CDR for the SMS. The CAMEL service may place an indication of the cost in the CDR or an indication that the SMS is sent whilst the subscriber is in her home zone.

5.4.8 Supplementary Services and Operator-determined Barring

Call Barring (CB) is the only GSM supplementary service that is applicable to MO-SMS; refer to 3GPP TS 22.004 [60]. This TS also lists support of private numbering plan (SPNP), CNAP and MSP as supplementary services that relate to SMS. However:

- SPNP is specified in GSM, but is not specified in 3GPP; practically, SPNP may be offered by means of a CAMEL service such as VPN.
- The applicability of CNAP to SMS is for further study.
- MSP does not affect MSC behaviour, but may be offered through a CAMEL service.

CB for SGSN-based SMS is introduced in 3GPP Rel-5. For MO-SMS, *conditional* call barring is applied to the SMSC address, not to the destination subscriber number. The following CB categories may be applied to MO-SMS:

- Barring of all outgoing calls – when BAOC is active for TS22, then the subscriber is barred from sending an SMS. No CAMEL service is invoked in that case.
- Barring of outgoing international calls – if BOIC is active, then CAMEL is invoked first. When CAMEL has returned control to the MSC, BOIC is checked. If BOIC is active, then the subscriber

is barred from using an SMSC in another country. If the CAMEL service modifies the SMSC address, then BOIC is checked against the modified SMSC address.

- Barring of outgoing international calls except to home country – the handling for BOIC-exHC is the same as the handling for BOIC. BOIC-exHC allows an SMSC to be used in the home country of the subscriber or in the visited country. If the CAMEL service modifies the SMSC address, then BOIC-exHC is checked against the modified SMSC address.

CB applies to both SMS-Submit and SMS-Command. The other CB categories, BAIC and barring of incoming calls when roaming (BIC-Roam), do not affect on MO-SMS. ODB has the same effect as CB. ODB may be used to bar the sending of an SMS, either conditionally or unconditionally. When SMS submission or SMS command fails as a result of CB or ODB, the CAMEL service is informed by means of the O_SMS_Failure notification.

5.4.9 Service Examples

The present section contains a number of service examples for successful and unsuccessful SMS submissions. Figure 5.68 depicts successful SMS submission without CAMEL-modified data. The CAMEL service performs a credit check and, upon affirmative response from the charging system, arms the SMS events and allows SMS submission to continue. Actual credit deduction takes place when the CAMEL service has received a notification that the SMS is successfully submitted to the SMSC. The figure also depicts unsuccessful SMS submission. In that case, the credit is released in the charging system; the credit that was reserved at the beginning of the service is now available again for other services.

The use of RT-SMS and the use of FCI-SMS at the start of the service and at SMS submission are optional. In Figure 5.69, the charging system has determined that the SMS may be submitted toll-free. The CAMEL service does not keep the CAMEL service active. Alternatively, the charging system may need to monitor the SMS submission, in order to keep track of the number of successful toll-free SMSs.

5.4.10 International Roaming

When CAMEL control of MO-SMS is used in international environment, then the operator needs to cater for the situation that the subscriber registers in a network that does not support CAMEL phase 3. The HLR of the roaming subscriber determines the fallback action that should be taken during registration in the MSC. This fallback action is operator-specific and is not defined by CAMEL.

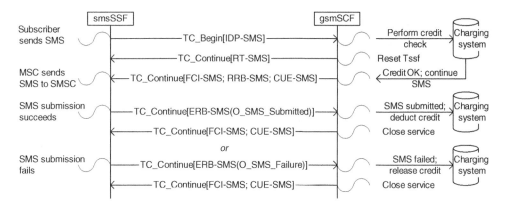

Figure 5.68 Successful and unsuccessful SMS submission

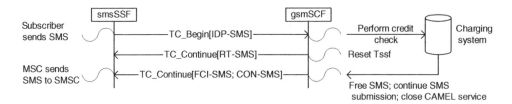

Figure 5.69   Free SMS submission

Furthermore, the fallback action depends on the type of subscriber. Possible actions include, but are not limited to:

- *Normal registration* – the subscriber may register in the MSC and may send short messages. If the subscriber is a pre-paid subscriber, then the charging of the SMSs will be done by other means than CAMEL phase 3.
- *Deny registration* – the HLR disallows registration with this PLMN. The MS will attempt to register with an MSC from another PLMN in that country. This option should be applied when the operator has a CAMEL phase 3 agreement with at least one roaming partner in each country.
- *Restricted registration* – the HLR prevents the sending of short messages by the subscriber by not including basic service TS22 in MAP insert subscriber data message to the VLR.

When the operator applies the option 'normal registration', charging of the SMS may be done in several ways. These include, but are not limited to:

- Post-paid charging – the CDR that is generated by the MSC in the VPLMN is processed by the charging system in the HPLMN of the roaming subscriber and the applicable charge is deducted from the subscriber's pre-paid account. A BGW in the HPLMN will need to know which MO-SMS CDRs need to be forwarded to the charging system, since the SMS may have been charged with CAMEL phase 3 in the VPLMN already. This method has the disadvantage that the transfer of CDRs from the VPLMN to the charging system in the HPLMN may be subject to long delays. In addition, the charging system in the HPLMN has less information available than the CAMEL service. For example, location number and message type will not be available to the charging system.
- CAP v3 service triggering from signalling transfer point; see Figure 5.70.

When the MAP message containing the short message, MO-ForwardSM, is sent from MSC in VPLMN to SMS-IWMSC in HPLMN, it passes through a border STP in the HPLMN. This STP may apply several criteria to deduce that this MAP message contains a short message from a pre-paid subscriber. Figure 5.70 lists the checks that should typically be performed by the STP. These checks may be performed on various levels of the signalling, including SCCP, TCAP and MAP. One of the checks includes an LDAP query to the operator's subscriber database, to verify whether the sender of the SMS is a pre-paid subscriber. LDAP, which stands for lightweight directory access protocol, is a TCP/IP based protocol that is commonly used for database queries. LDAP is defined by the IETF.

If the STP has deduced that the SMS is from a pre-paid subscriber, then the STP invokes an smsSSF instance and triggers the CAMEL phase 3 MO-SMS service, denoted CAP v3' in the figure. This usage of CAP v3 is not standard usage; it would require the gsmSCF address and service key to be configured in the STP, as there is no MO-SMS-CSI available. It is not uncommon for an STP to have the capability for performing the above-described operator-specific logic.

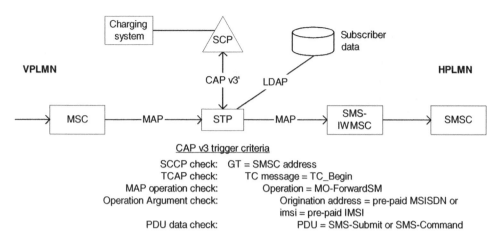

Figure 5.70   CAP v3 triggering for SMS charging, from border STP

When the CAMEL phase 3 service is triggered in this manner, it will not obtain information elements such as location information. Typically, the CAMEL service, when triggered in this manner, should not change any message data, i.e. it should not use CAP CON-SMS.

5.5 Mobility Management

5.5.1 Description

Mobility management is a mechanism that enables an operator to keep track of the location and the attached/detached state of a subscriber. The location of a subscriber is known in the HLR down to the level of VLR address. The HLR requires the VLR address to route incoming calls to the subscriber, to deliver short messages, etc. However, the HLR does not keep track of the location of the subscriber within the VLR service area. Neither does the HLR keep track of the attached/detached state of the subscriber. Hence, the HLR does not keep accurate enough location or state-related information for certain value added services. CAMEL phase 3 offers a solution to this dilemma in the form of mobility management. Figure 5.71 presents the architecture for mobility management.

The mobility management in CAMEL phase 3 relates to CS mobility management only. The CS mobility management relates to registration in the MSC. PS mobility management, related to registration in the SGSN, is introduced in CAMEL phase 4; see Chapter 6.

When a subscriber is not engaged in a call, she may move from one location to another and perform a location update. The location update may be notified to various nodes, depending on the type of location update. The location update types are listed in Table 5.31.

The rightmost column in Table 5.31 indicates whether the mobility management may result in a CAMEL phase 3 mobility management notification to gsmSCF. The location management procedures, including the CAMEL handling, are specified in 3GPP TS 23.012 [74].

The mobility management notifications that may be sent to gsmSCF do not relate to call handling. When a subscriber moves from one location area to another during a call, then there is no GSM location update. Changing location during a call is referred to as 'hand-over'. Hand-over notifications were introduced in CAMEL phase 4.[40] When hand-over occurs, the corresponding

---

[40] Hand-over notifications in CAMEL phase 4 use CAP, not MAP. That feature is a complement to CAMEL phase 3 mobility management.

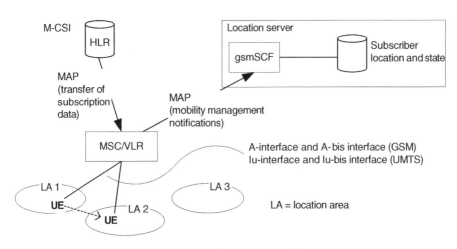

Figure 5.71 Functional architecture for mobility management

Table 5.31 Location update types

Location update type	Handling in GSM/UMTS	Notification to gsmSCF
Location update to other cell, within location area	Not recorded in VLR	No
Location update to other location area, within VLR service area	Location register in VLR is updated; HLR is not informed	Yes
Location update to other MSC/VLR, within same PLMN	Registration in other MSC/VLR; new VLR address is stored in HLR	Yes
Location update to other PLMN	Registration in other MSC/VLR; new VLR address is stored in HLR	Yes
MS attaches to MSC	Subscriber state in VLR is updated; if no subscription data available for this subscriber, then HLR is informed (location update)	Yes
MS detaches from MSC	Subscriber state in VLR is updated; HLR is not informed	Yes
MS is implicitly detached from MSC	Subscriber state in VLR is updated; HLR may be informed	Yes

location update, including the updating of the stored location in the VLR and mobility management notification, takes place after completion of the call.

The mobility management notification to the gsmSCF uses MAP. The notification is sent to the gsmSCF after the related mobility management procedure is completed. The gsmSCF has no control over the mobility management. The mobility management notifications may be sent to another gsmSCF than the gsmSCF that is used for CAMEL service execution. The gsmSCF may relay the mobility management notifications to a database for subscriber location and subscriber state.

5.5.2 Subscription Data

The sending of mobility management notifications is subscription-based. Individual subscribers may have M-CSI in their subscriber record in HLR. The HLR sends M-CSI to the MSC during

Table 5.32 Contents of M-CSI

Element	Description
gsmSCF address	This is the address, in GT format, of the gsmSCF to which the mobility management notifications will be sent
Service key	The service key may be used to select between different mobility management applications. An operator may have different applications that use CAMEL mobility management
Mobility management triggers	This element contains a list of mobility events for which a notification shall be sent if the event occurs

registration. As is the case with other CAMEL subscription data, the sending of M-CSI to MSC is subject to the support of CAMEL phase 3 in the MSC. Table 5.32 lists the contents of M-CSI. In addition, M-CSI may contain the elements 'notificationToCSE' and 'csi-Active'. These elements are described in Section 5.8.

The following mobility events may be included in M-CSI:

- *Location update in VLR* – this trigger is used for the location update within VLR service area, from one location area to another.
- *Location update to other VLR* – this trigger is used when the subscriber moves to a new VLR service area, which may be located in the same PLMN or in another (e.g. roaming to another country). This mobility management event relates to MSC registration. The HLR is involved in this procedure; it may send M-CSI to the new MSC. The notification is sent from the new MSC, provided that that MSC supports CAMEL phase 3. If the new MSC does not support CAMEL phase 3, e.g. when the subscriber roams to another PLMN with no CAMEL support, then the gsmSCF does not receive a notification.
- *MS-initiated IMSI detach* – the IMSI detach event relates to the case where the subscriber switches off the MS. The subscriber remains registered with the MSC and the HLR is not informed. The MSC may notify the gsmSCF about the detach event.
- *IMSI attach* – the IMSI attach event relates to the case where the subscriber switches on the MS. The subscriber was already registered with the MSC, but had switched off the MS later. The MSC may notify the gsmSCF about the attach event.
- *Network-initiated IMSI detach* – the MSC may detach the subscriber from the network. The network-initiated detach, also referred to as *implicit detach*, may take place when the MS has not performed a period location update for a certain time.

The location update to another VLR does not result in detach and attach notifications.

5.5.3 Information Flows

The relevant MAP information flow is the *note MM event* MAP message. Table 5.33 lists the information elements that are reported to the gsmSCF in this MAP message. For encoding details of this message, refer to 3GPP TS 29.002 [103].

The information that is reported to the gsmSCF does not include call reference number (CRN) and MSC address. This is due to the fact that the mobility management event is not call-related. Hence, the gsmSCF cannot easily correlate a notification with an active call for the subscriber on whose behalf the notification is sent. The gsmSCF could, however, use MSISDN or IMSI for the correlation.

5.5.4 Service Examples

CAMEL mobility management may be used for services like Personal Number (PN) or Single Number Reach (SNR). A subscriber may be reachable on a variety of devices, under a single

Table 5.33 Information elements in MAP Note MM event

Information element	Description
Service key	The service key is obtained from M-CSI; it may be used to select the mobility management application
Mobility event	This element indicates which mobility event took place
IMSI	This element identifies the subscriber
MSISDN	This element identifies the subscriber. The mobility management application may use IMSI or MSISDN as index into a subscriber database
Location information	This element contains the location as applicable after the location update or IMSI attach. This element may be absent when (network-initiated) detach is reported. The reported location information may reflect 2G or 3G access; see Section 5.1.2.
Supported CAMEL phases	This element indicates which CAMEL phases are supported by the MSC that sends the mobility management notification. This element should indicate CAMEL phase 3 or CAMEL phase 4.

number. The PN or SNR service keeps track of the state and the location of subscriber. When a call arrives for the subscriber, the service uses the subscriber state and location information to determine to which device the call should be offered; see Figure 5.72 for a possible network implementation of PN.

The PN service as in Figure 5.72 may not require continuous mobility notifications for a subscriber to be received. The notifications may be required only when a particular PN feature is activated. The PN service may therefore use a MAP interface to the HLR to activate or de-activate the mobility management subscription for a subscriber, i.e. activate or de-activate M-CSI; see Figure 5.73. A requirement for activating M-CSI in HLR is that M-CSI is already provisioned for the subscriber. Activating M-CSI has the effect that the HLR sends M-CSI to the VLR where the subscriber is currently registered, provided that the VLR supports CAMEL phase 3. The PN service may use the same MAP interface to the HLR to verify the supported CAMEL phases in the VLR, in order to verify that M-CSI is actually sent to the VLR. De-activating M-CSI has the effect that the HLR removes M-CSI from the VLR. See Section 5.8 for this capability of the SCP (or service node) to activate and de-activate CAMEL subscription data in HLR.

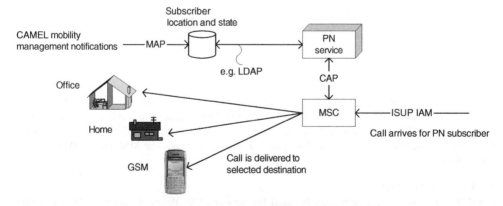

Figure 5.72 Example for CAMEL mobility management

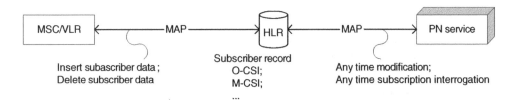

Figure 5.73 Activating and de-activating M-CSI

5.6 CAMEL Interaction with Location Services

5.6.1 Description

CAMEL interaction with location services is a means of obtaining location information of a subscriber. This feature may be compared with ATI; it uses a similar signalling method. ATI is generally used to determine the network in which a subscriber is located and the location area in which she currently resides. Location services are primarily meant for obtaining the geographic location of a subscriber; refer to Figure 5.74 for a comparison between ATI and location services.

The location information that may be obtained with ATI (MCC, MNC, LAC, CGI/SAI) may be used for service logic processing, e.g. when a subscriber is determined to be in the HPLMN (indicated by MCC + MNC), then call completion may be applied for that subscriber. The use of ATI to HLR may also result in geographical information, in the following ways:

- the MSC where the subscriber is located derives geographical location from the cell Id;
- the HLR that is processing the ATI derives geographical location from the cell Id;
- the gsmSCF that uses ATI and receives the location information, derives geographical location from the cell Id

Obtaining geographical information with ATI to HLR has the following shortcomings:

- The accuracy of the geographical information that may be obtained with ATI to HLR is determined by the size of the cell where the subscriber is located. If the cell has a radius of 5 km, then that is the uncertainty of the location information.
- When ATI is used, the most recently stored location information is received; the subscriber may since have moved within the location area of the VLR. This dilemma is resolved with active location retrieval.

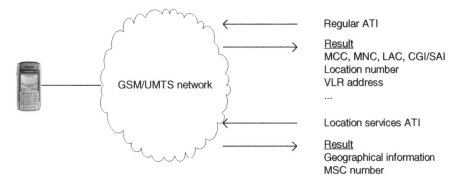

Figure 5.74 Comparison between regular ATI and location services

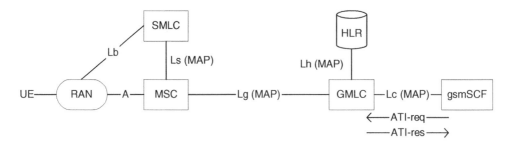

Figure 5.75   Location services architecture for GSM network

CAMEL interaction with LCS enables a service to obtain accurate location information; refer to Figure 5.75 for the LCS architecture in GSM network.

The gateway mobile location services Centre (GMLC) is normally located in the HPLMN. The serving mobile location services centre (SMLC) is located in the PLMN where the subscriber is located. For UMTS network, the SMLC functionality is located in the RNC. This implies that, if location service is used for roaming subscribers, then the VPLMN requires support for LCS. If the gsmSCF sends ATI to the GMLC whilst the subscriber is in a VPLMN that does not support LCS, then an ATI error component error will be returned to the gsmSCF.[41]

The interfaces and functionality of location services are specified in GSM TS 03.71 [36] (GSM-only), 3GPP TS 23.171 [88] (UMTS-only; 3GPP R99) and 3GPP TS 23.271 [92] (GSM and UMTS; 3GPP Rel-4 onwards). The Lc interface is used for CAMEL and consists of MAP ATI. When ATI is used between gsmSCF and GMLC, the gsmSCF may request only the subscriber's location, not the subscriber's state. For the sending of ATI to GMLC, the GMLC's address is used, not the subscriber's IMSI or MSISDN. The GMLC does not store subscriber data, so the gsmSCF may send ATI to a designated GMLC.

The information that may be returned to the gsmSCF may be obtained in various manners. For more information on positioning methods, refer to 3GPP TS 25.305 [98] (for WCDMA networks) and 3GPP TS 43.059 [126] (for GSM and EDGE networks). When the geographical information is returned to the gsmSCF, it contains the elements as listed in Table 5.34.

When the subscriber is registered with an MSC and attached to an SGSN, the HLR may provide the GMLC with both the MSC number and the SGSN number. Positioning of the subscriber may be done through the MSC or through the SGSN. When the GMLC sends the ATI result to the gsmSCF, it includes the number of the MSC or SGSN that was used for the subscriber positioning.

5.7 Active Location Retrieval

Active location retrieval (ALR)[42] is an enhancement to MAP ATI. With ATI, the gsmSCF may obtain the subscriber's location information as currently stored in the VLR. The information in the VLR is stored during the most recent location update procedure. The location update procedure may be due to the subscriber changing location to another location area, call establishment or periodic location update. Changes in location within the location area (LA) in which she currently resides are not reported to VLR; see Figure 5.76.

---

[41] The GMLC probably returns 'ati-NotAllowed'; the gsmSCF cannot deduce from that error code that ATI failed because the subscriber is currently registered in a VPLMN that does not support LCS. The GMLC could, as a vendor's option, use an extension containing the 'ATI-NotAllowedParam' parameter to indicate the reason why ATI is not allowed. See 3GPP TS 29.002 for ATI error details.

[42] Active location retrieval is also known as current location retrieval.

Table 5.34 Geographical Information

Information element	Description
Type of shape	The location of the subscriber is expressed by means of a *shape* that is located at a defined position. The subscriber is located within the perimeter of the shape. Details of the shapes that are used for location services are described in 3GPP TS 23.032 [76]. For CAMEL, only the shape 'ellipsoid point with uncertainty circle' may be used. The ellipsoid point is described by means of *latitude range* and *longitude range*
Latitude	The latitude is coded as 24-bit variable, with 1 sign-bit and 23 bits to reflect the angle (0–90°). The latitude is encoded as a *latitude range*
Longitude	The longitude is coded as 24-bit variable to reflect the angle ($-180 - 180°$). The longitude is encoded as a *longitude range*
Uncertainty code	This variable indicates a radius; the subscriber is located within the circle defined by the ellipsoid and the radius

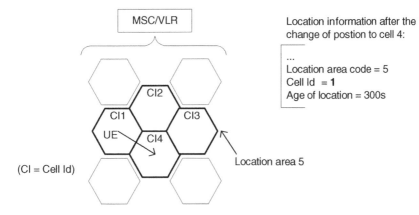

Figure 5.76 VLR location information after intra-LA change of position

In Figure 5.76, the subscriber (UE) moves from cell 1 to cell 4, both within location area 5. The cell Id in the location information in VLR no longer indicates the exact cell where the subscriber resides. With ALR, the gsmSCF may instruct the VLR to page the subscriber. The location information in the VLR will be refreshed and it will then include the current Cell id; see Figure 5.77.

The Age of Location parameter in Location Information indicates 0. That serves as another indication that the reported location is the current location. The parameters *current location* (in MAP ATI), *active location retrieval requested* (in MAP PSI) and *current location retrieved* (in location information) are all placed 'after the ellipsis'. This implies that this enhancement of the location retrieval process is syntactically backwards compatible. The MAP operation versions used for the active location retrieval procedure are the same as for the regular location retrieval procedure. If HLR or VLR does not support the transport of these parameters in the respective MAP operations, then the MAP respective operations are processed as for the regular location retrieval procedure. That implies that the location retrieval succeeds, but without paging; the gsmSCF obtains the information stored in the VLR. The gsmSCF deduces from the absence of *current location retrieved* that the subscriber was not paged.

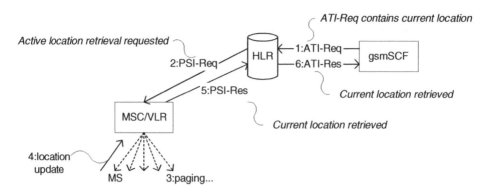

Figure 5.77 Information flows for active location retrieval. 1, gsmSCF sends an ATI request (Req) to HLR; ATI-Req includes the parameter *current location*. 2, HLR sends provide subscriber information (PSI) request to MSC; PSI-Req includes *active location retrieval requested*. 3, MSC pages the subscriber. For details of the location retrieval in MSC, refer to 3GPP TS 23.018 [75]. 4, The subscriber is successfully paged; the location register in VLR is updated. 5, MSC sends PSI-Result (Res) to HLR; PSI-Res includes *current location retrieved*. 6, HLR sends ATI-Res to gsmSCF; ATI-Res includes *current location retrieved*.

When ALR is requested whilst the subscriber is in radio contact with the MSC/VLR, e.g. during a call, the MSC will not perform paging; however, the stored location in VLR is in that case already the current location. If the subscriber is detached from the MSC, then no paging will take place either. The gsmSCF should therefore accompany the request for current location by a request for subscriber state; these requests may be included in the same ATI/ALR request. If paging is not performed due to the subscriber being busy, then the reported state ('CAMEL busy') may be used by the gsmSCF to deduce why paging did not take place. CAMEL phase 4 also includes active location retrieval for GPRS.

5.8 Subscription Data Control

Subscription data control is a means of controlling the interworking between CAMEL and some of the GSM supplementary services. Specific CAMEL services may contain functionality that has partial overlap with a corresponding GSM supplementary service. Examples include:

- *CAMEL call diversion* – this service controls call delivery to a called subscriber. When the establishment of a call to that subscriber fails due to the no reply, busy or not reachable condition, the CAMEL service forwards the call to an alternative destination. Such CAMEL service has overlap with GSM call forwarding supplementary service.
- *CAMEL access control* – this service applies rules that determine which calls a subscriber may establish. A subscriber may be barred from establishing international calls. Such a CAMEL service has overlap with GSM call barring supplementary service and operator determined barring.
- *Supplementary service provisioning* – an operator may offer a user-friendly interactive voice response service that enables subscribers to set their call forwarding or call barring data.

To control above-mentioned overlap of functionality, the following mechanisms were introduced in CAMEL phase 3: (1) any-time subscription interrogation (ATSI); (2) any-time modification (ATM); and (3) notify subscriber data change (NSDC). Although subscription data control is a CAMEL phase 3 feature, this does not imply that it may be used only in a CAMEL phase 3 call control service. ATSI, ATM and NSDC are MAP messages that may be used by a CAMEL service of

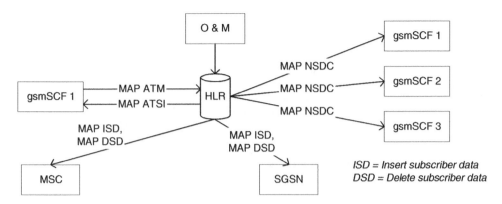

Figure 5.78 Subscription data control

arbitrary phase or call case. For example, ATSI may be used during a CAMEL phase 2 terminating call service. ATSI, ATM and NSDC may also be used outside the context of a call.

Subscription data control is specified in 3GPP TS 23.078 [83], section 'Control and interrogation of subscription data'; the corresponding MAP messages are defined in 3GPP TS 29.002 [103].

5.8.1 Network Architecture

Figure 5.78 depicts the architecture for the subscription data control functionality.

5.8.2 Any-time Subscription Interrogation

The gsmSCF may use ATSI to obtain subscriber data from the HLR. Whereas ATI is used to obtain subscriber data from VLR or from SGSN (see CAMEL phase 4), ATSI is used to obtain subscriber data from HLR. The gsmSCF uses IMSI or MSISDN to identify the subscriber. Addressing the HLR for ATSI is done in the same manner as for ATI. Table 5.35 lists the elements that may be requested with ATSI.

The HLR may restrict the use of ATSI to specific gsmSCFs, e.g. gsmSCFs in the same PLMN or gsmSCFs belonging to the same operator group. The U-CSI and UG-CSI cannot be requested with ATSI. The reason is that these subscription elements are defined for internal use in the HLR only; the exact structure of U-CSI and UG-CSI is not specified.

Support of ATSI is optional for an operator. An operator may decide to have partial implementation of ATSI, e.g. the implementation of ATSI could be limited to requesting the call forwarding and call barring data.

5.8.3 Any-time Modification

Any-time modification (ATM) may be used to modify subscription data in the HLR. Table 5.36 lists the elements that may be modified with ATM. Only one category of data may be changed in a single ATM request.

Each CSI in the HLR has a 'CSI-state' flag. The state flag, which may be set or reset with ATM, is used to activate or de-activate a CSI. This flag, which is for HLR-internal use only, indicates whether the CSI shall be sent to MSC or SGSN, e.g. when a subscriber has O-CSI with CSI-state 'inactive' and that subscriber performs a location update, then the HLR will not send the O-CSI to the MSC, even if other conditions for sending O-CSI to MSC are fulfilled. Further action to be taken by the HLR when the sending of O-CSI to MSC is prohibited by the CSI-state flag, such as denial of registration, is operator/vendor-specific.

Table 5.35 Information to be requested with ATSI

Element	Description
Call forwarding subscription	The gsmSCF may obtain the settings for one of the call forwarding categories (CF-U, CF-NRc, CF-B, CF-Nry)
Call barring subscription	The gsmSCF may obtain the settings for one of the call barring categories (BAOC, BOIC, BOIC-exHC, BAIC, BIC-Roam)
Operator-determined barring	The gsmSCF may obtain an indication of all the ODB categories for the indicated subscriber. Both the general ODB categories and the HPLMN-specific ODB categories may be obtained.
CAMEL subscription information	The gsmSCF may request the subscription data of any of the subscribed CAMEL services (O-CSI, T-CSI, etc.)
Supported CAMEL phases in MSC/VLR	This data element indicates which CAMEL phases are supported in the MSC/VLR for this subscriber. The supported CAMEL phases are reported to the HLR during the previous location updated procedure, data restoration procedure or stand-alone insert subscriber data procedure. If the subscriber has not yet registered in any MSC, then this data element is not available in HLR
Supported CAMEL phases in SGSN	This data element indicates which CAMEL phases are supported in the SGSN for this subscriber

Table 5.36 Data elements for ATM

Element	Description
Call forwarding subscription	The gsmSCF may change the settings for one of the call forwarding categories
Call barring subscription	The gsmSCF may change the settings for one of the call barring categories
CAMEL subscription information	The gsmSCF may activate or de-activate a specific CSI, except U-CSI and UG-CSI. The contents of the CSI, such as gsmSCF address or service key, cannot be modified with ATM. Neither can the gsmSCF provision or delete a CSI

Another parameter associated with each CSI, with call forwarding data and with call barring data is the 'notification-to-CSE flag'. This parameter is used for notify subscriber data change.

When ATM is used, the HLR performs the regular subscription data checks for the call forwarding and call forwarding data. Dependencies between call barring and call forwarding data are specified in 3GPP TS 23.082 [84]. The result of the ATM request is reported in the ATM response. If any request for subscriber data change was not performed, then the gsmSCF will be aware.

After the ATM request has been processed in HLR, the actions below may follow:

- The modified subscription data may trigger the HLR to update the subscriber data in MSC or SGSN. For example, if the gsmSCF activates O-CSI, then the HLR uses MAP insert subscriber data to send the O-CSI to VLR, provided that any other relevant conditions are fulfilled; if the gsmSCF de-activates O-CSI, then the HLR uses MAP DSD to remove O-CSI from VLR.[43]
- One or more gsmSCF entities are notified of the changed subscriber data. This is described in Section 5.8.4.

---

[43] Removing a single CSI from VLR for a subscriber may require a combination of MAP DSD and MAP ISD, depending on the CAMEL phase of the CSI.

Support of ATM is optional for an operator, as is the case for ATSI. An operator may have partial implementation of ATM, as is permissible for ATSI.

5.8.4 Notify Subscriber Data Change

The notify subscriber data change (NSDC) feature is used by a gsmSCF to receive a notification when specific subscriber data changes. NSDC uses the following mechanisms:

- A parameter 'notification-to-CSE flag' may be associated with each CSI, call forwarding data, call barring data and with ODB data. When this flag is present for a subscription element, then any change in that subscription element will be reported to one or more gsmSCF entities.
- The HLR has a list with gsmSCF addresses; when a subscriber data change notification needs to be sent, the HLR sends the notification to all gsmSCFs contained in the list. The change in subscriber data may result from operation and maintenance (O&M) action or from an ATM operation from a gsmSCF. In the latter case, the gsmSCF that used ATM to change the subscriber data will not receive a notification, even if that gsmSCF is included in the list with gsmSCF addresses; refer to Figure 5.79 for a graphical representation.

The notification-to-CSE flag for CSIs, call forwarding and call barring may be modified by a gsmSCF, with the ATM procedure. ODB also has a notification-to-CSE flag associated with it, but this flag cannot be modified by a gsmSCF.[44] When gsmSCF uses ATM to reset the notification-to-CSE flag for a particular CSI, then data change notifications for that CSI are suppressed for all gsmSCFs in the aforementioned list of gsmSCF addresses.

5.9 Enhancement to USSD

The USSD signalling between HLR and gsmSCF is specified in CAMEL phase 2. Subscriber identification in the USSD information flows is done with IMSI. That implies that the gsmSCF uses IMSI to identify a subscriber when it wants to initiate a USSD relationship with that subscriber. The IMSI of a subscriber is reported to the gsmSCF during CAMEL service invocation. Not all CAMEL services, however, use IMSI as subscriber identification. Therefore, a CAMEL service

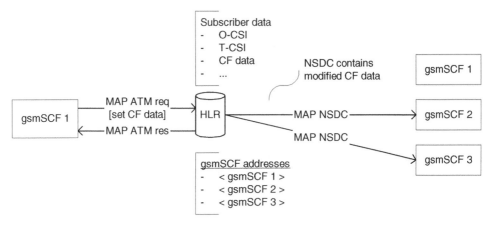

Figure 5.79 Data change notification resulting from ATM

---

[44] CAMEL phase 4 includes any-time modification for ODB.

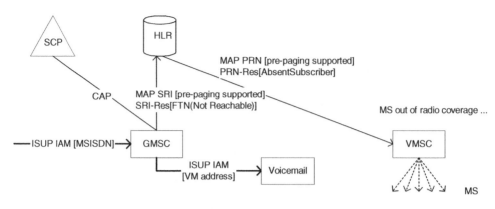

Figure 5.80   Pre-paging

that uses MSISDN for its subscriber registration would need to obtain the IMSI of the subscriber, before sending the USSD string to the HLR.

CAMEL phase 3 introduces the MSISDN in the USSD signalling between gsmSCF and HLR. This facilitates the use of USSD between gsmSCF and HLR in cases where the CAMEL service does not use IMSI for its service processing. The MSISDN may be used in the following USSD messages:

- *Unstructured SS request* (from gsmSCF to HLR) – either IMSI or MSISDN is used as subscriber identification.
- *Unstructured SS notify* (from gsmSCF to HLR) – either IMSI or MSISDN is used as subscriber identification.

The use of MSISDN for unstructured SS request and unstructured SS notify is for signalling between gsmSCF and HLR only. USSD signalling between HLR and MSC is always based on IMSI. The optional inclusion of MSISDN in *process unstructured SS request* (from HLR to gsmSCF) is specified in CAMEL phase 2.

5.10 Pre-paging

Pre-paging is introduced in 3GPP R99 as a means towards efficient call routing for forwarded calls; refer to Figure 5.80. The main principle of pre-paging is that, when the VMSC receives MAP PRN from HLR, it pages the subscriber. The VMSC provides the requested MSRN to HLR only when the subscriber is successfully paged. If paging fails, then the VMSC indicates to the HLR that the subscriber is not reachable. The HLR may now send a forwarded-to-number to the GMSC, resulting in the initiation of early call forwarding from GMSC.

The rationale of this mechanism is that a subscriber who is marked as 'idle' in VLR may in fact be out of radio coverage at that moment. If the periodic location update timer in VLR has not yet expired, then the VLR is not aware that the subscriber is out of radio coverage. The VLR would in that case provide the requested MSRN. The GMSC then routes the call to the VMSC, where paging will take place. The paging failure will now result in late call forwarding. With pre-paging, on the other hand, the call forwarding would have been early call forwarding.

If a CAMEL service was invoked from the GMSC before the HLR sends MAP PRN, then the early call forwarding, as a result of no paging response, is notified to the gsmSCF; see Chapter 4.

The execution of pre-paging in the VMSC has the effect that HLR and GMSC have to wait longer for MAP PRN-Result and MAP SRI-Result respectively. GMSC and HLR can indicate whether they support this longer response time, by including the parameter 'pre-paging supported' in MAP SRI and MAP PRN.

6

CAMEL Phase 4

CAMEL phase 4, the flagship of IN, is the fourth and last CAMEL phase developed by ETSI or 3GPP. CAMEL phase 4 was introduced in 3GPP Rel-5. Later releases of the 3GPP mobile network contain enhancements to CAMEL phase 4. These enhancements are described in Chapter 8. CAMEL phase 4 in 3GPP Rel-5 builds onto the service capability of CAMEL phase 3 in 3GPP R99. CAMEL phase 4 is both forward-compatible and backwards-compatible, compared with CAMEL phase 3. This implies:

- all service control capability that is included in earlier CAMEL phases is also available in CAMEL phase 4;
- CAMEL phase 4 services may be used in both GSM networks and UMTS networks.

A new network domain that is introduced in 3GPP Rel-5 is the IP Multimedia Subsystem (IMS). CAMEL phase 4 contains specific control capabilities for the IMS domain. CAMEL control of IMS is described in Section 6.9.

6.1 General

Comparable with CAMEL phase 3, CAMEL phase 4 is divided into a number of main functional areas that may be used independently of one another. Table 6.1 contains an overview of CAMEL phase 4 feature groups. Most of the feature groups are already specified as such in CAMEL phase 3 or earlier. However, the capabilities of these feature groups are enhanced in CAMEL phase 4.

For some features, there is no functional change in CAMEL Phase 4, compared with CAMEL phase 3. These features include: CAMEL interaction with USSD; supplementary service invocation notification; and CAMEL interaction with location services.

6.1.1 Specifications Used for CAMEL Phase 4

CAMEL phase 4 is specified in the following set of specifications:

- *3GPP TS 22.078* [66] – this specification contains the service definitions for CAMEL phase 4;
- *3GPP TS 23.078* [83] – this specification describes the information flows, subscription data, procedures, etc.;
- *3GPP TS 29.078* [106] – this specification specifies the CAP for call control, SMS control and GPRS control;

Table 6.1   CAMEL phase 4 features and capabilities

	Feature	Section
1	*Call control*	6.2
	Call party handling	6.2.2
	Network-initiated call establishment	6.2.3
	Interaction with basic optimal routing	6.2.4
	Alerting detection point	6.2.5
	Mid-call detection point	6.2.6
	Change of position detection point	6.2.7
	Flexible warning tone	6.2.8
	Tone injection	6.2.9
	Enhancement to call forwarding notification	6.2.10
	Control of video telephony calls	6.2.11
	Service change and UDI/RDI fallback	6.2.12
	Reporting of IMEI and MS Classmark	6.2.13
2	*GPRS control*	6.3
3	*SMS control*	6.4
	MO SMS control	6.4.1
	MT SMS control	6.4.2
4	*Mobility management*	6.5
	Mobility management for PS subscribers	6.5
5	*Any-time interrogation*	6.6
	Any-time interrogation for CS subscribers	6.6.1
	Any-time interrogation for PS subscribers	6.6.2
6	*Subscription data control*	6.7
7	*Mobile number portability*	6.8
8	*Control of IP multimedia calls*	6.9

- *3GPP TS 23.278* [93] – this specification describes the technical implementation of CAMEL control of IP multimedia calls;
- *3GPP TS 29.278* [111] – this specification specifies the CAP for CAMEL control of IP multi-media calls.

Besides the above CAMEL-specific 3GPP technical specifications, various other non-CAMEL specific 3GPP technical specifications contain CAMEL functionality, such as 3GPP TS 29.002 [103] (MAP), 3GPP TS 23.018 [75] (basic call handling), 3GPP TS 32.205 [117] (CDR formats), etc. Any reference to a 3GPP technical specification or technical report in the present section relates to 3GPP Rel-5, unless otherwise indicated.

6.1.2 Partial CAMEL Phase 4 Support

For CAMEL phases 1–3 the 'full CAMEL support' principle applies to core network nodes MSC, GMSC and SGSN. This principle implies the following:

- When an MSC or GMSC indicates to HLR, e.g. during location update or data restoration, that it supports a particular CAMEL phase, then that MSC or GMSC will support the full capability set that is specified for that CAMEL phase, in as far as applicable to the MSC or GMSC.
- When an MSC or GMSC initiates a CAMEL dialogue with the gsmSCF, that MSC or GMSC will support the full capability set that is specified for the application context used for that CAMEL dialogue.

- When an SGSN indicates to HLR, during location update or data restoration, that it supports CAMEL phase 3, then that SGSN will support the full capability set that is specified for CAMEL phase 3, as far as is applicable to the SGSN.
- When an SGSN initiates a CAMEL phase 3 dialogue with the gsmSCF, that SGSN will support the full capability set that is specified for the application context used for that CAMEL phase 3 dialogue.

Practically, this implies, for example, that, when an MSC indicates in MAP location update that it supports CAMEL phases 1–3, the HLR may send any combination of O-CSI, SS-CSI, TIF-CSI, D-CSI, VT-CSI, M-CSI and MO-SMS-CSI to that VLR. Furthermore, when the MSC sends CAP v3 InitialDP, the gsmSCF should be able to use any of the operations that are specified within CAP v3, provided that any applicable dialogue rule is fulfilled.

With the introduction of CAMEL phase 4, this principle could no longer be maintained. The complete capability set of CAMEL phase 4 is of such magnitude that MSC and SGSN vendors apply a phased implementation approach. Implementation of the complete CAMEL phase 4 functionality may be implemented during a few successive MSC or SGSN system releases. Hence, a 'CAMEL phase 4 MSC' may support the full CAMEL phase 4 capability or a defined subset thereof. Two mechanisms are used to control the CAMEL Phase 4 subsets.

Supported CAMEL 4 CSIs
When a subscriber registers in MSC or SGSN, the MSC or SGSN reports the following to the HLR:

- *Supported CAMEL phases* – the MSC or SGSN indicates which CAMEL phases it supports. This may be CAMEL phase 1 up to CAMEL phase 4. The industry convention is that, when an MSC or SGSN supports CAMEL phase n, that MSC or SGSN will also support the CAMEL phases prior to n. For the SGSN, there is only CAMEL phase 3 and CAMEL phase 4 defined.
- *Offered CAMEL4 CSIs* – when the MSC or SGSN indicates that it supports CAMEL phase 4, that MSC or SGSN will also indicate which parts of CAMEL phase 4 it supports. Table 6.2 lists the permissible CAMEL phase 4 subsets.

MT-SMS-CSI and MG-CSI are introduced in CAMEL phase 4; the other CSIs (O-CSI, T-CSI, D-CSI, VT-CSI) are specified in CAMEL phase 3, or earlier, but may also have a CAMEL phase 4 variant. MO-SMS-CSI, SS-CSI, TIF-CSI and M-CSI do not have a CAMEL phase 4 variant. That

Table 6.2 CAMEL phase 4 subsets

Reported subset	Description
O-CSI	The VMSC or GMSC is capable of initiating CAMEL phase 4 services for MO and MF calls
D-CSI	The VMSC or GMSC is capable of initiating CAMEL phase 4 subscribed dialled services
VT-CSI	The VMSC is capable of initiating CAMEL phase 4 services for MT calls
MT-SMS-CSI	The VMSC or SGSN has the capability to trigger a CAMEL phase 4 service for MT SMS
T-CSI	The GMSC is capable of initiating CAMEL phase 4 services for MT calls
MG-CSI	The SGSN support GPRS mobility management
PSI enhancements	The SGSN is capable of receiving MAP provide subscriber info (PSI). PSI is used for the ATI procedure

implies that an MSC that supports CAMEL phase 4 shall support MO-SMS-CSI, SS-CSI, TIF-CSI (if the MSC supports call deflection) and M-CSI. Likewise, an SGSN that supports CAMEL phase 4 shall support MO-SMS-CSI.

PSI enhancements relates to the SGSN's capability to receive and process MAP PSI, for the purpose of GPRS ATI. Although ATI for CS subscribers is also enhanced in CAMEL phase 4, there is no dedicated CAMEL phase 4 subset to signal that capability from MSC to HLR.

When the subscriber registers in the MSC or SGSN and the HLR has received the indication of supported CAMEL phases and supported CAMEL phase 4 subsets, the HLR shall take care not to send any non-supported CAMEL phase 4 (variant) CSI to that MSC or SGSN.

Offered CAMEL 4 Functionalities

When a CAMEL relationship is started between the gsmSSF (in VMSC or GMSC) and the gsmSCF, the gsmSSF reports to the gsmSCF the CAMEL phase 4 functionalities supported in that gsmSSF. 'Offered CAMEL 4 functionalities' indicates which CAP v4 call control capabilities are supported by the (G)MSC/gsmSSF. These capabilities are features that are introduced in CAMEL phase 4. The gsmSSF may report a combination of the CAMEL phase 4 functionalities in Table 6.3.

Each of the functionalities in Table 6.3 is described below. A CAMEL service logic will not use capability that is not explicitly indicated. Otherwise, service logic processing may fail. The offered CAMEL 4 functionalities indication does not replace other CAMEL dialogue rules. For example, when a dialled service is triggered with D-CSI and offered CAMEL 4 functionalities indicates, amongst others, 'alerting DP', then the CAMEL service is still not entitled to arm the alerting DP. The reason is that arming the alerting DP is not permissible for a dialled service.[1]

Table 6.3 Offered CAMEL 4 functionalities

Functionality	Description
Initiate call attempt	The gsmSCF is entitled to create additional call parties in an existing call
Split leg	The gsmSCF may place an individual call party on hold
Move leg	The gsmSCF may (re)connect an individual call party to the group
Disconnect leg	The gsmSCF may disconnect an individual call party from the call
Entity released	The erroneous release of an individual call party may be reported to the gsmSCF
DFC with argument	The gsmSCF may apply user interaction to individual call parties
Play tone	The gsmSCF may instruct the gsmSSF to play a tone to a call party or to the entire call group
DTMF mid call	The gsmSCF may instruct the gsmSSF to collect and report DTMF digits during the call
Charging indicator	The gsmSCF may receive the ISUP charging indicator when a call is answered
Alerting DP	The gsmSCF may instruct the gsmSSF to report the occurrence of the ISUP alerting event
Location at alerting	The gsmSCF may receive the served subscriber's location information when the alerting event is reported
Change of position DP	The gsmSCF may instruct the gsmSSF to notify the gsmSCF when the served subscriber changes location during the call
OR interactions	The gsmSCF may instruct the gsmSSF to apply basic optimal routing for the call
Warning tone enhancements	The gsmSCF may use the flexible warning tone in the charging operations
CF enhancements	The gsmSCF may receive the forwarded-to number in the call-forwarding notification

---

[1] When a gsmSSF supports enhanced dialled service (EDS), the dialled service may arm use the alerting DP, if supported by the gsmSSF. EDS is introduced in 3GPP Rel-6; see Chapter 8.

Table 6.4 Overview of CAMEL protocols for call control

CAMEL phase	CAMEL protocol
CAMEL phase 1	CAP v1
CAMEL phase 2	CAP v2
CAMEL phase 3	CAP v3
CAMEL phase 4, Rel-5	CAP v4
CAMEL phase 4, Rel-6	CAP v4
CAMEL phase 4, Rel-7	CAP v4

The introduction of the Partial CAMEL phase 4 support mechanism also allows the CAMEL phase 4 capability to be expanded without upgrading the CAMEL phase 4 protocol. This is reflected in Table 6.4.

Additional call control related functionalities is introduced in CAMEL phase 4 in 3GPP Rel-6 and Rel-7. An operator may introduce new CAMEL phase 4 functionalities in the network without upgrading the CAMEL subscription data in HLR. Chapter 8 describes the CAMEL phase 4 features in 3GPP Rel-6 and beyond.

6.2 Call Control

As is described in the chapters on CAMEL phases 1–3, each CAMEL phase supports a number of call cases. CAMEL phase 4 introduces two new call cases. Each call case has its own specific trigger mechanism and capability. Table 6.5 lists the call cases supported in CAMEL phase 4.[2]

Table 6.5 CAMEL phase 4 call cases

Call case	Description
MO call	CAMEL service triggered from VMSC as a result of an MO call from a subscriber with O-CSI or D-CSI or as a result of provisioning of N-CSI in VMSC
MT call	CAMEL service triggered from GMSC as a result of an MT call to a subscriber with T-CSI
MF call	CAMEL service triggered from VMSC or GMSC as a result of a MF call by a subscriber with O-CSI or D-CSI or as a result of provisioning of N-CSI in VMSC or GMSC
VT call	CAMEL service triggered from VMSC as a result of an MT call to a subscriber with VT-CSI. The VT call was introduced in CAMEL phase 3
NP call	The NP ('new party') call relates to an additional call party that is created within an existing call. The NP call was introduced in CAMEL phase 4
NC call	The NC ('new call') is a call that is initiated by a CAMEL service, by sending CAP operation 'initiate call attempt' to an MSC/gsmSSF. That MSC/gsmSSF will establish an outgoing call. This call case is also known as 'call out of the blue'. The NC call was introduced in CAMEL phase 4

---

[2] CAMEL phase 4 in 3GPP Rel-7 will introduce an additional call case.

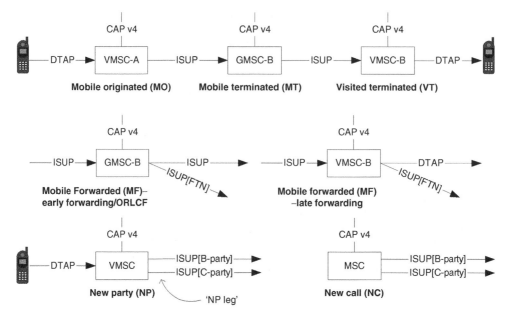

Figure 6.1   CAMEL phase 4 call cases

Figure 6.1 presents a graphical overview of these call cases.

6.2.1 Basic Call State Models

The BCSMs that are used in CAMEL phase 4 are further enhanced, compared with CAMEL phase 3. Figures 6.2 and 6.3 reflect the O-BCSM and the T-BCSM for CAMEL phase 4.

CAMEL phase 4 adds the DPs O_Term_Seized, O_Mid_Call and O_Change_of_Position to the O-BCSM. These DPs are part of specific features and are described in Sections 6.2.5–6.2.7. DPs O_Term_Seized and O_Mid_Call are also defined in the ITU-T and ETSI O-BCSM of Core INAP; DP O_Change_of_Position is specific for GSM and is not part of the O-BCSM of Core INAP.

CAMEL phase 4 adds the DPs Call_Accepted, T_Mid_Call and T_Change_of_Position to the T-BCSM. These DPs are described in Sections 6.2.5–6.2.7 respectively. DP T_Change_of_Position is specific for GSM and is not part of the T-BCSM of ETSI/ITU-T Core INAP.

The rules with respect to arming, disarming and reporting of DPs are the same as for previous CAMEL phases. Also, the rules for CAMEL relationships such as dialogue initiation, dialogue termination, monitor relationship vs control relationship, use of TCAP, etc., are not modified compared with previous CAMEL phases. One new arming related feature is the *implicit re-arming*.

6.2.2 Call Party Handling

Call Party Handling (CPH) is a toolkit that enables a CAMEL service to control individual parties in a call. In this manner, a CAMEL service may be used to build conference calls, transfer calls, play announcements during a call, etc. This is accomplished using the following set of CPH-specific capabilities: creating a new call or creating an additional party in an existing call; (re)connecting a party to the call group; placing a party on hold; and removing a party from a call. CPH may be used for speech calls only, i.e. calls with basic service = tele service 11. The basic service for a call is reported to the CAMEL service when that CAMEL service is initiated. CPH uses the concepts of call segment (CS) and call segment association (CSA) to control individual call parties, to control

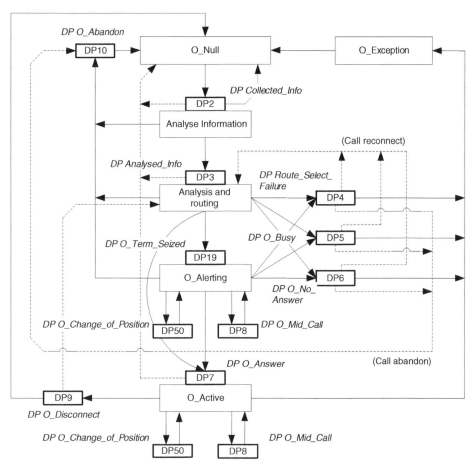

Figure 6.2 O-BCSM for CAMEL phase 4. Reproduced from 3GPP TS 23.078 v5.9.0, Figure 4.3, by permission of ETSI

a group of call parties and to control the entire call. When a call is established, e.g. an MO call, a call segment is created in the gsmSSF; see Figure 6.4.

A call segment represents a call connection for one or more call parties. For a two-party call, the MSC/gsmSSF typically has one call segment to connect calling party and called party. All parties that reside in one call segment have a speech connection with one another. The call segment that is created during call establishment is designated call segment 1, also known as the *primary call segment*. Call segment 1 may contain one, two or more call parties. All other call segments that exist in a call contain exactly one call party. Therefore, to place two or more parties in speech connection with one another, these parties are placed in call segment 1. Each call party is represented by a leg number. The following rules apply for the leg numbering.

- *Leg 1* – this is the calling party in the call; in an NC call, there is no calling party and hence no leg 1;
- *Leg 2* – this is the called party, resulting from the call establishment from the calling party; in an NC call, there is neither calling party nor called party and hence, no leg 1 or 2.
- *Leg > 2* – any leg that is created by the SCP has leg number 3 or higher.

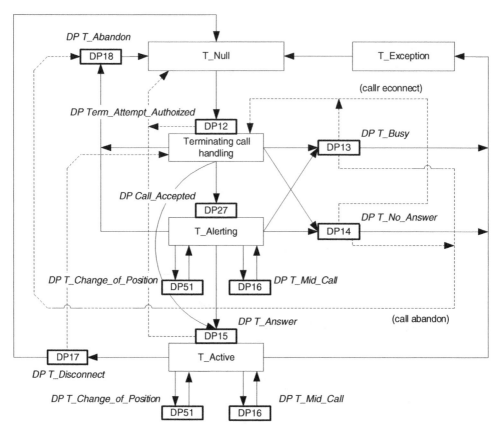

Figure 6.3   T-BCSM for CAMEL phase 4. Reproduced from 3GPP TS 23.078 v5.9.0, figure 4.4. by permission of ETSI

The CSA controls the various call segments that exist in the call. The call connection that is depicted in a call segment represents both a logical connection and a physical connection; refer to Figure 6.5.

CPH uses a number of CPH-specific CAP operations:

- *Split leg (SL)* – this CAP operation is used to split a leg off from call segment 1 and place it in a separate call segment. SL may also be used to move a leg to call segment 1 when call segment 1 does not exist. SL always involves the creation of a new call segment.
- *Move leg (ML)* – ML is used to move a leg from a separate call segment to call segment 1. The separate call segment is released after the execution of ML.
- *Disconnect leg (DL)* – DL is used to disconnect a leg from the call.
- *Initiate call attempt (ICA)* – ICA is used to create a new party in a call or to create a new call.

Some of the existing CAP operations are enhanced for CPH purposes, such as continue with argument, connect to resource, etc.

6.2.2.1 Creating an Additional Party in a Call
CPH allows the CAMEL service to create additional call parties in the call. New call parties are created in separate call segments, containing an O-BCSM instance for that new call party

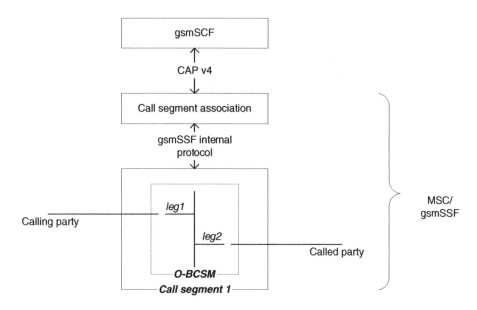

Figure 6.4   Call segment for an MO call

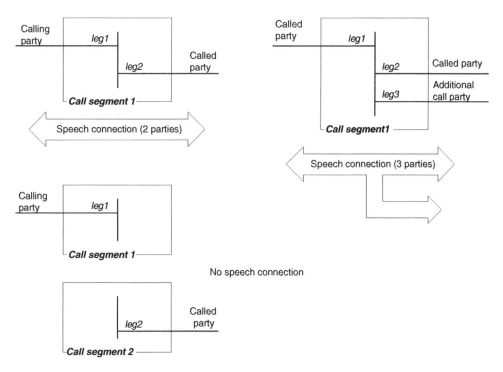

Figure 6.5   Relation between call segment configuration and speech connection

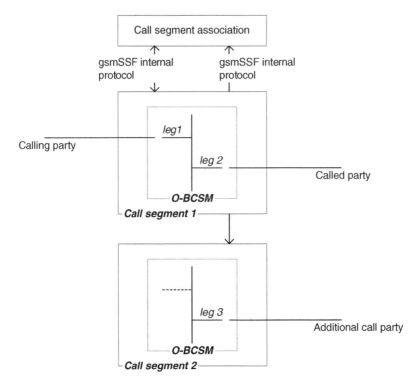

Figure 6.6   Call with multiple call segments

(Figure 6.6). When a new leg is created in a call, the call segment in which that leg is created may be call segment 2 or higher; the newly created legs are leg 3 or higher. The call segment for the new call party contains only an outgoing call leg. The call segment in fact also contains an incoming leg, but that leg is considered a *surrogate leg*; it has no practical implications. The call to the additional called party follows the normal call establishment processes, meaning that the MSC applies number analysis, route selection, etc. For this additional called party, the DPs defined in the O-BCSM may be armed and reported. However, since there is no incoming leg associated with this O-BCSM, DPs like O_Abandon cannot be armed for that O-BCSM. The call segment association (CSA) is now governing both call segments; the CSA has an interface with each individual call segment.

While the additional call leg is in a separate call segment, the CAMEL service may connect this call party to an announcement device. For example, when the additional call leg has reached the active state, i.e. the additional called party has answered, the CAMEL service may play an announcement to inform the person the she will be placed in conference (Figure 6.7).

Once the additional leg is created, it follows the regular call establishment and call clearing processes. This implies that the CAMEL service may create a follow-on call for a newly created call party, once that call party has disconnected. The creation of an additional party in the call follows a two-step approach: *step 1* – create new call segment containing the leg to be created; and *step 2* – route the call to the required destination. This is reflected in Figure 6.8.

In step 1 of the process of creating a new call party, the destination routing address is provided, along with the call leg Id and an indication of the call segment in which the new call party will be created. The CAMEL service will arm the DPs that are associated with establishment of an outgoing call: O_Answer, O_No_Answer, O_Busy and route select failure. This is required to ensure that

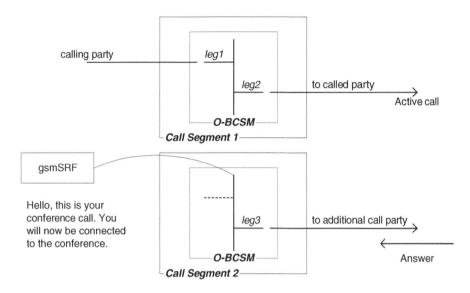

Figure 6.7   Announcement connection to additional call party

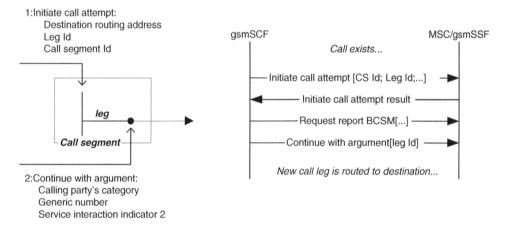

Figure 6.8   Two-step process for creation of new call party

the CAMEL service retains a relationship with this call leg (pre-arranged end rules apply). After arming these events, optionally followed or preceded by the sending of charging operations, the CAMEL service will send CWA. The sending of CWA constitutes step 2. That operation is used to start the routing of the call to the indicated destination. In addition, the CAP CWA may carry call-related parameters, such as calling party's category, additional calling party number (contained in generic number), etc.

6.2.2.2 Connecting a Party to the Call Group

The CAMEL service may place a newly created party into the main call group, i.e. connect leg 3 with legs 1 and 2. The action to place an individual call leg in a group may also be performed on

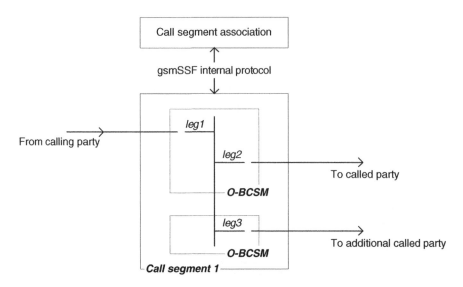

Figure 6.9 Call segment 1 with additional call leg

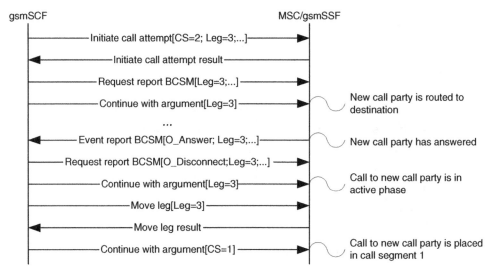

Figure 6.10 Connecting an additional call party to the group

a leg that was previously disconnected from the group, i.e. placed on hold. An example is given in Figure 6.9. After leg 3 is moved to call segment 1, call segment 2 is released by the CSA. Call segment 1 now contains three legs: one incoming leg and two outgoing legs. These three legs are in speech connection with one another. The signal sequence that is used to move the additional call party to the group is reflected in Figure 6.10.

In the example sequence flow in Figure 6.10, the CAMEL service creates an additional party, leg 3, in a call. When the newly created call party has answered the call, the CAMEL service uses the move leg operation to move leg 3 to call segment 1. Move leg is one of the 'CPH operations'.

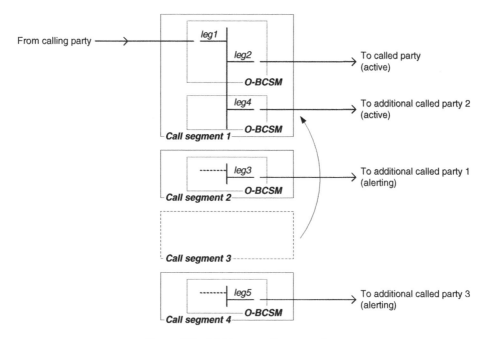

Figure 6.11 Multi-party call in setup phase

This has the effect in the above example, that call segment 1 is in 'suspended state' after the move leg operation. The CAMEL service must use CWA to get call segment 1 back into the call processing state.

Call segment 1 now has two O-BCSM instances: one for the connection from leg 1 to leg 2 and one for the additional called party. From this point onwards, the call may continue as a three-party call. Call-related events, such as disconnect leg, may be reported on the individual legs. The CAMEL service may move more legs into call segment 1. A CAMEL service may create a number of additional call legs and place each leg into call segment 1 as soon as the new call leg reaches the active state, i.e. the called party has answered. In Figure 6.11, the CAMEL service has created three additional parties in the call: call segment 2, containing call leg 3; call segment 3, containing call leg 4; and call segment 4, containing call leg 5. Call party 4 has answered the call and is placed into call segment 1; it has become part of the conference call. After call leg 4 is moved, call segment 3 no longer contains a call leg; call segment 3 is therefore released by the CSA. The other newly added call parties (legs 3 and 5) are still in call establishment phase and are kept in their respective call segments. A newly created call party may be moved to call segment 1 as soon as that call party has reached the alerting state.

An interesting case occurs when additional call parties are created in an MT call in a GMSC. This is depicted in Figure 6.12. The GMSC instantiates a T-BCSM that controls the incoming leg from the calling party and the roaming call forwarding to the called party (i.e. the leg to the VMSC of the called party). If the CAMEL service creates an additional party in the call, then an O-BCSM is instantiated for that party. When the additional call party is placed in call segment 1, the call segment 1 contains both a T-BCSM instance and an O-BCSM instance.

6.2.2.3 Placing a Party on Hold

A CAMEL service may place an individual call party on hold during a call. This is accomplished by creating an additional call segment and placing this call party in that call segment. Placing the

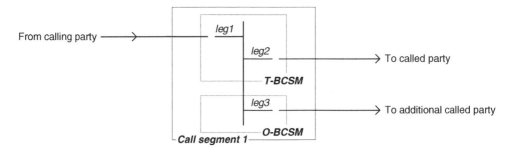

Figure 6.12   Combination of O-BCSM and T-BCSM in a single call segment

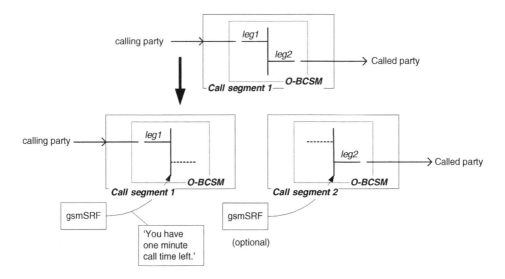

Figure 6.13   Placing one party on hold during a call

party in a separate call segment allows the CAMEL service to play an announcement to one party during the call. This is depicted in Figure 6.13.

In Figure 6.13 the CAMEL service places leg 2 in a separate call segment. The result is that legs 1 and 2 are no longer in speech connection. The CAMEL service may now connect call segment 1 to a gsmSRF in order to play an announcement to the calling party. The playing of the announcement to the calling party follows the user interaction mechanisms described in Chapter 4. That implies, amongst others, that the CAMEL services use CAP operations like connect to resource, play announcement, disconnect forward connection, etc.

Figure 6.14 contains the sequence flow for the example scenario in Figure 6.13. The separation of the calling party and the called party also has the effect that there are temporarily two O-BCSM instances, one O-BCSM instance containing leg 1 and one O-BCSM instance containing leg 2. When the playing of the announcement is complete, the CAMEL service may reconnect the calling and called party, by moving leg 2 back into call segment 1; the two call legs are then back into a single O-BCSM.

Placing a party on hold always involves the creation of a new call segment. It is not possible to create a 'side conference'. Only the primary call segment, i.e. call segment 1, may contain more than one call party.

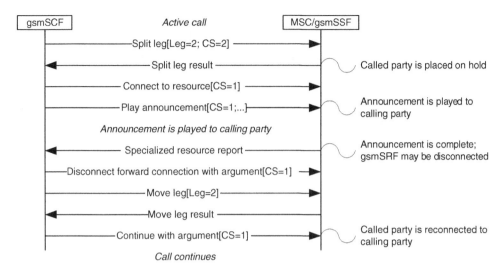

Figure 6.14   Sequence flow for mid-call announcement

The SRR depicted in Figure 6.14 does not contain a CS Identifier. SRR is a LINKED operation; it is *linked* to the previous PA operation. When CAMEL orders PA for multiple call segments (using multiple PA operations), then the linking of the CAP SRR to the CAP PA indicates to the CAMEL service for which Call Segment a CAP PA was completed.

6.2.2.4 Removing a Party from a Call

The CAMEL service may at any moment remove a party from the call, without affecting the other parties in the call. This mechanism may be used during a conference call; see Figure 6.15. In the example in Figure 6.15, the CAMEL service had already disconnected call party 3 from the conference call. The other parties in call segment 1 remain in speech connection with one another. Call party 5 is now placed in a separate call segment; this enables the CAMEL service to play an

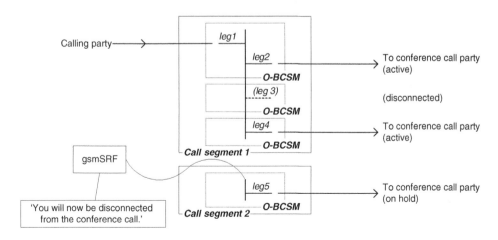

Figure 6.15   Removing parties from a call

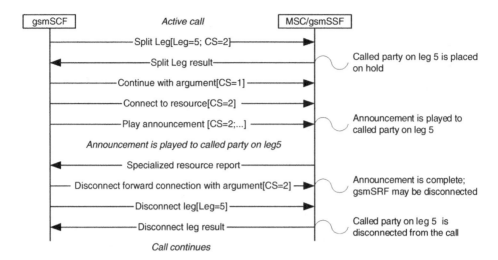

Figure 6.16   Disconnecting called party on leg 5 from the call

announcement to that call party prior to disconnecting that party from the call. Figure 6.16 presents an example sequence flow for removing a party from the call with pre-disconnection announcement.

By removing a party from the call, the CAMEL services also releases the 'logical view' of that call party, i.e. the CAMEL service cannot apply a reconnect on that leg. The CAMEL service may, however, create again a leg 3 in the call. That leg will be created in a separate call segment. When the leg has reached the alerting phase of the call, it may be moved to call segment 1.

Another practical case whereby a party is removed from a call is the ringback tone service. The ringback tone service entails a call being established to a subscriber and the normal ringback tone being replaced by a personalized ring tone. The ringback tone for an MT GSM call is normally generated in the VMSC of the called subscriber. The ringback tone service establishes the call with calling and called party in separate call segments. During call establishment, the ringback tone service injects the personalized ring tone into the speech connection to the calling party (Figure 6.17).

Such service may be triggered in the GMSC for the called party. The CAMEL service that is triggered with T-CSI for this call disconnects the original called party, i.e. call party 2, form the call. This called party is replaced by another call party, which is created in a separate call segment. Whilst call establishment is ongoing, the CAMEL service uses user interaction to play a personalized announcement towards the calling party. The regular ring tone that is generated from the VMSC of the called party and sent over ISUP in backwards direction, will not be heard by the calling party, since there is no speech connection between the called party leg and the calling party leg. When the called party answers the call, the CAMEL service moves the called party leg into call segment 1; the calling party and the called party are in speech connection from then onwards.

The reason for replacing the original called party leg with a newly created call leg is that the original called party leg may not be split off from the primary call segment (i.e. call segment 1) prior to reaching the active state. However, speech separation of calling and called party is already required during call establishment. The signalling sequence that may be used for this call case is shown in Figure 6.18.

In this service example, the CAMEL service is controlling both a T-BCSM instance and an O-BCSM instance in the GMSC. The T-BCSM instance contains leg 1; the O-BCSM instance contains leg 3. When the service creates the replacement call leg towards the called party, it may apply the following methods:

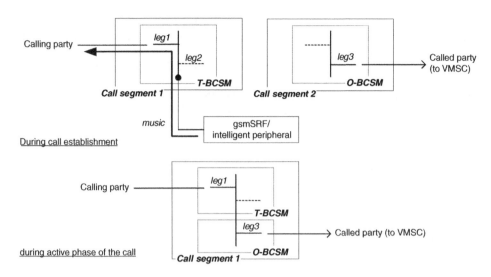

Figure 6.17   MT call establishment by called party substitution

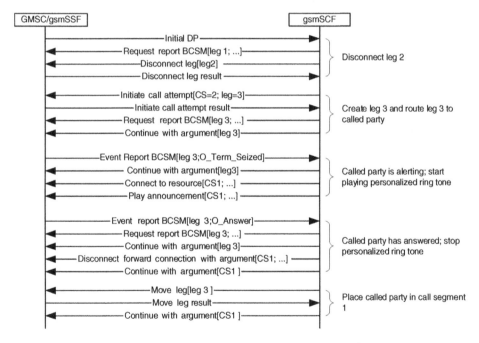

Figure 6.18   Signal sequence for personalized ring tone service from GMSC

- *MSISDN* – the service includes the called party's MSISDN in the CAP initiate call attempt operation. The GMSC uses the MSISDN to set up the call to the B-party. Here, the GMSC contacts the HLR again, which would result in the HLR sending T-CSI again to the GMSC. To prevent the HLR from sending T-CSI to the GMSC a second time, the service may include 'suppress T-CSI' in initiate call attempt. The GMSC will in that case include 'suppress T-CSI' in MAP SRI to the HLR. The HLR will, in response, obtain the MSRN from the VMSC and return the MSRN to the GMSC. The GMSC can now establish the call to the VMSC of the B-party.

- *MSRN* – the service sends MAP SRI, including 'suppress T-CSI', to the HLR, to obtain the MSRN. The service then copies MSRN to initiate call attempt. The call to the destination subscriber is now established from GMSC, using MSRN. The GMSC will not contact the HLR; it can directly route the call to the VMSC. The usage of MSRN when creating a call party is described further in Section 6.2.3.

In both methods, the call to the B-party is controlled through an O-BCSM instance. The call forwarding notification mechanism is defined for use within a T-BCSM instance.[3] This implies that, for leg 3, i.e. the leg to the called party, the GMSC will not send a notification to the CAMEL service in the case where call forwarding occurs in the GMSC for this call.

6.2.2.5 Parallel Alerting

Parallel alerting is one service example of CPH. When a subscriber is called, the CAMEL service may route the call to two destinations simultaneously; two phones may ring. When one phone answers the call, the CAMEL service will disconnect the call to the other destination. The person who answered is connected to the calling party. This is illustrated in Figure 6.19.

A CAMEL service may disconnect the original leg 2 that resulted from the call establishment in the GMSC, for the call to the destination subscriber. This leg 2 is replaced by two separate legs,

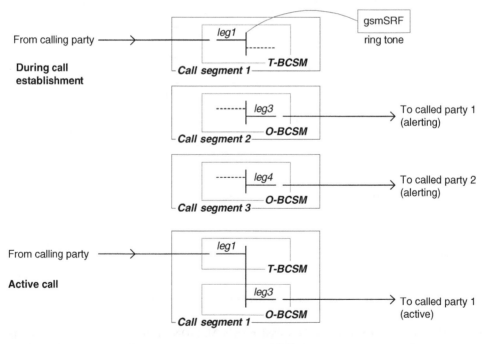

Figure 6.19 Using CPH for parallel alerting

---

[3] The parameters 'call forwarded' and 'forwarding destination number' are defined for T_No_Answer and T_Busy, but not for O_No_Answer and O_Busy.

each to one of the destinations. Whilst call establishment is in progress, the CAMEL service may connect a gsmSRF to call segment 1. The gsmSRF is used for ring tone generation. The reason is that the ring tone that is generated by either destination party is not transferred towards the calling party, since there is no backwards speech connection between calling and a called party. Therefore, the CAMEL service shall supply the ring tone. When called party 1 answers, that party is moved to call segment 1. The call to called party 2 is disconnected.

6.2.2.6 Signalling Transparency

When a CAMEL service creates a new party in a call, it uses ISUP parameters from the incoming leg to populate the corresponding parameters in the ISUP signalling for the new call party. If the call is created by the SCP (i.e. NC call), then there is no incoming leg and hence various ISUP parameters in the ISUP signalling for the new call party will be absent (Figure 6.20).

In the example in Figure 6.20, the ISUP IAM arriving at the GMSC contains the ISUP parameters UUS information; multi-level precedence and pre-emption (MLPP) information and calling geodetic location. UUS information[4] and calling geodetic location may be reported to the gsmSCF, but the gsmSCF cannot add, remove or modify these parameters in ISUP signalling. MLPP information[5] is not reported to gsmSCF. Both the outgoing call on leg 2 and the outgoing call on leg 3 will contain these parameters in ISUP IAM. Some other parameters for ISUP IAM, such as service interaction indicator 2 and additional calling party number, may be supplied by gsmSCF (using CAP CWA) for the outgoing legs and may differ between legs 2 and 3, as is the case in the example in Figure 6.20.

In addition to the transfer of parameters in ISUP IAM for a CPH call, ISUP messages during call establishment and during the active phase of the call need to be considered. Generally, ISUP messages may be transferred between the calling party and the called party. Some of the ISUP messages are transparent for the O-BCSM or T-BCSM. That is to say, these ISUP messages do

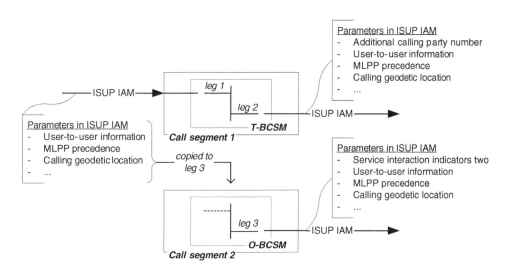

Figure 6.20   Re-use of ISUP parameters for new call parties

---

[4] See 3GPP TS 23.087 for user-to-user information. Reporting of UUS information to gsmSCF is introduced in CAMEL phase 4 in 3GPP Rel-6.

[5] See 3GPP TS 23.067 for enhanced multi-level precedence and pre-emption service (eMLPP).

Table 6.6   Transparently transferred ISUP messages (examples)

ISUP message	Description
Call progress (CPG)	An event has occurred during call setup which should be relayed to the calling party. CPG may contain charge indicator, called party's status indicator, etc.
Facility (FAC)	FAC may be used to inform the calling party about pending call transfer
Resume (RES)	RES indicates that the call is no longer on hold
Suspend (SUS)	SUS indicates that the call is temporarily placed on hold
User-to-user information (UUS)[6]	UUS information may be used during call set up and during active call, to transfer information between calling and called party

[6] CAMEL phase 4 in 3GPP Rel-6 specifies that user-to-user information may be reported to gsmSCF. That applies to the user-to-user information element in ISUP IAM, but not to separate user-to-user ISUP messages.

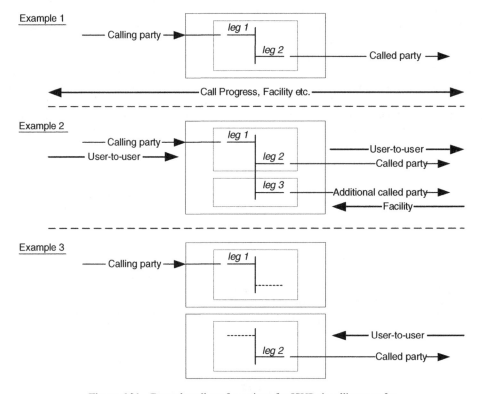

Figure 6.21   Example call configurations for ISUP signalling transfer

not have corresponding DPs in the BCSM. As a result, a CAMEL service may not be aware of the transfer of these messages through the ISUP link. Examples of messages are given in Table 6.6.

Transfer of these ISUP messages between calling and called party requires, however, that calling and called party are controlled by the same BCSM. Practically, this means that the transparent transfer of ISUP signalling occurs only between legs 1 and 2. Figure 6.21 shows a couple of examples of ISUP signalling transfer.

In example 1 in Figure 6.21, legs 1 and 2 reside in the same BCSM and are in speech connection; ISUP messages are transparently transferred between legs 1 and 2. In example 2, the ISUP message transfer between legs 1 and 2 still holds. There is, however, no ISUP message transfer, in either direction, between legs 1 and 3. Example 3 applies to a two-party call whereby the called party is temporarily placed on hold. Although the two legs belong to the same BCSM (not visible in the figure), ISUP messages are not transferred between them. The reason is that the two legs are not in speech connection at that moment.

6.2.3 Network-initiated Call Establishment

The initiate call attempt (ICA) operation that is used to create additional parties in a call may also be used to create a new call. Such a call is often referred to as a 'call out of the blue'. Use cases for this capability include, but are not limited to: person-to-person call establishment (e.g. internet 'click-to-talk' service); conference call establishment (e.g. corporate mail and agenda system, initiating a scheduled conference call); and content-push (e.g. news services; notification call). The gsmSCF initiates a call by sending CAP ICA to an MSC/gsmSSF. This MSC may be an arbitrary MSC. That is to say, the MSC does not need to be a VMSC, i.e. it has no VLR and is not connected to the RAN, and the MSC does not need to be a GMSC, i.e. it cannot handle MT call establishment. Figure 6.22 presents rudimentary network architecture for a network-initiated two-person call.

Figure 6.22 shows a case where the SCP initiates a call to a GSM subscriber in the HPLMN and a call to a PSTN subscriber, e.g. the owner of a web-based click-to-call service. The two calls are established in the MSC/gsmSSF and are placed in speech connection. Calls that are created by the SCP are always *outgoing calls*. That means that these calls are controlled through an O-BCSM instance. The dotted gsmSSF instances in Figure 6.22 indicate that the GMSC may invoke a CAMEL service as a result of receiving T-CSI from HLR; the VMSC may invoke a CAMEL service as a result of having VT-CSI in VLR. The CAMEL service may establish the call in an MSC/gsmSSF that is at the same time GMSC. That node can then do the HLR interrogation. Figure 6.23 shows the call segment configuration for this call example.

The new call parties have call leg 3 and higher; each leg is created in a separate call segment. A leg may be placed in call segment 1 (the primary call segment) as soon as the leg has reached the alerting state. Alternatively, one of the legs is created in call segment 1. The CAMEL service may connect individual call legs to an announcement device. The reason is that the respective call parties will not answer the call simultaneously; hence, an announcement is needed. A call party

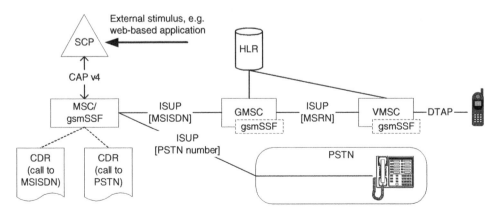

Figure 6.22 Network-initiated call establishment

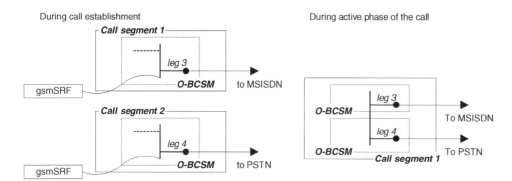

Figure 6.23 Call segment configuration for SCP-initiated two-party call

may be connected to the announcement device as soon as that call party has answered. The example presented in Figures 6.22 and 6.23 may also be used to set up a multi-party conference call. The CAMEL service creates separate outgoing call legs and places each call leg into the primary call segment as soon as that call party has answered the call.

When the gsmSCF initiates a call to a GSM subscriber, there is no connection with a VLR for that call, in the MSC/gsmSSF where the call is generated. The MSC that is handling the network-initiated call does not have to be the MSC where the subscriber is registered. And the gsmSSF does not have the IMSI of the served subscriber in any case. As a result, no GSM subscription check is performed for the network-initiated call, e.g. call barring supplementary service.

6.2.3.1 Using MSRN for Call Establishment

In the example in Figure 6.22, the CAMEL service uses MSISDN to establish a call to a mobile subscriber. The call to the mobile subscriber will be routed through a GMSC, for the purpose of HLR interrogation. The GMSC resides external to the MSC/gsmSSF that receives the initiate call attempt. Alternatively, the MSC/gsmSSF may act as a GMSC, i.e. contact the HLR. One advantage of that method is that the gsmSCF may include the suppress T-CSI parameter in initiate call attempt. The (G)MSC/gsmSSF will, as a result, include suppress T-CSI in MAP SRI, which has the effect that the HLR will not return T-CSI to the GMSC. In the end, if the network initiates a call to a subscriber, it may want to bypass a terminating CAMEL service.

Instead of using MSISDN to establish a call to the GSM subscriber, the gsmSCF may use MSRN (Figure 6.24). Using MSRN for network-initiated call establishment results in very efficient network usage. The gsmSCF uses MAP SRI to obtain the MSRN from the HLR, which obtains the MSRN from VLR with MAP provide roaming number (PRN). The use of MAP SRI and MAP PRN for this call case follows the rules for MT call establishment, for which MAP SRI is sent from GMSC. In fact, the HLR may not see a difference between MAP SRI from GMSC and MAP SRI from gsmSCF. MAP SRI may contain the parameter 'gsmSCF initiated call', indicating to the HLR that the MAP SRI originates from a gsmSCF. The gsmSCF may, in addition, include the parameter 'suppress incoming call barring', to stop the HLR preventing call establishment for this subscriber. This parameter is not recognized by a pre Rel-5 HLR; such an HLR may therefore apply the incoming call barring. The gsmSCF should include the parameter 'suppress T-CSI', to prevent the HLR from returning T-CSI to gsmSCF.

When the gsmSCF has obtained the MSRN, it may initiate the call directly in the MSC where the subscriber is located (see 1 in Figure 6.24). The MSC address may be derived from the MSRN. Alternatively, the gsmSCF may initiate the call in a designated MSC/gsmSSF (see 2 in Figure 6.24);

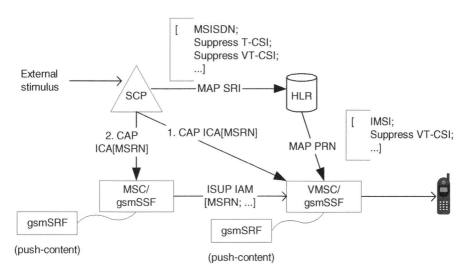

Figure 6.24   Using MSRN for call creation

that may be the case when the operator's network contains only a limited number of MSCs with the required CAMEL phase 4 capability. That MSC will then use the MSRN to route the call to the VMSC of the B-subscriber.

One further advantage of using MSRN for network-initiated call establishment is that the gsmSCF may suppress VT-CSI; the HLR includes 'suppress VT-CSI' in MAP PRN. For a network-initiated call, it may not be required or desirable to invoke the T-CSI or VT-CSI CAMEL service. When the SCP obtains the MSRN from HLR, it will include the call reference number (CRN), and gsmSCF address; the gsmSCF address takes the place of the GMSC address in MAP SRI. The SCP should also indicate in MAP SRI that ORLCF is not supported.

6.2.3.2 CAP Signalling

The network-initiated call establishment uses CAP v4 between gsmSCF and gsmSSF. This CAP v4 differs from the CAP v4 that is used for subscriber-initiated calls. Table 6.7 shows the ACs for CAP v4 call control between gsmSCF and gsmSSF.

A gsmSCF does not need to support both ACs. Support of capssf-scfGenericAC is needed in gsmSCF when the gsmSCF needs to be able to respond to CAP initial DP. Support of capscf-ssfGenericAC is needed in gsmSCF when the gsmSCF needs to be able to establish calls.

Network-initiated call establishment is not a subscribed service in the sense that the subscriber needs a CSI for it. The gsmSCF may at any time send CAP initiate call attempt to the gsmSSF to start a call. The gsmSCF may send initiate call attempt to any MSC/gsmSSF; even an MSC/gsmSSF in another network. Methods for selecting the MSC/gsmSSF to which initiate call attempt should be sent may include, but are not limited to:

- the gsmSCF uses designated MSC/gsmSSF;
- the gsmSCF uses ATI to determine the VMSC of the subscriber and bases choice of MSC/gsmSSF on VMSC address;
- the gsmSCF obtains an MSRN and uses MSRN for selection of MSC/gsmSSF, e.g. if MSRN is associated with an MSC in the HPLMN, then send initiate call attempt to that MSC; otherwise

Table 6.7   CAP v4 Application contexts for call control

```
capssf-scfGenericAC APPLICATION-CONTEXT ::= {
    CONTRACT                        capSsfToScfGeneric
    DIALOGUE MODE                   structured
    ABSTRACT SYNTAXES               {dialogue-abstract-syntax |
                                    gsmSSF-scfGenericAbstractSyntax}
    APPLICATION CONTEXT NAME        id-ac-CAP-gsmSSF-scfGenericAC}

capscf-ssfGenericAC APPLICATION-CONTEXT ::= {
    CONTRACT                        capScfToSsfGeneric
    DIALOGUE MODE                   structured
    ABSTRACT SYNTAXES               {dialogue-abstract-syntax |
                                    scf-gsmSSFGenericAbstractSyntax}
    APPLICATION CONTEXT NAME        id-ac-CAP-scf-gsmSSFGenericAC }
```

send initiate call attempt to a designated MSC in the HPLMN, from where the call will be routed to the destination VMSC.

The gsmSCF should take care that it sends initiate call attempt to an MSC/gsmSSF that supports the CAP v4 AC that is used for this service (capscf-ssfGenericAC). There is no defined mechanism for the gsmSSF to accept initiate call attempt from selected gsmSCFs, e.g. an operator may accept CAP initiate call attempt from its own SCP, but not from other operators' SCPs. When an MSC receives MAP PRN, it will not know whether the MAP PRN resulted from GMSC-initiated or SCP-initiated call establishment. When the gsmSSF receives CAP initiate call attempt, it may use the gsmSCF Address to determine whether the initiate call attempt should be accepted.

Compare this with subscriber registration in the MSC. The MSC may adapt the 'supported CAMEL phases' parameter in the MAP location update message to the IMSI of the subscriber who registers in that MSC. In that manner, the MSC may offer CAMEL capability to selected roaming partners.

6.2.3.3 Charging for Network-initiated Calls

The network-initiated call consists of one or more *outgoing* call legs. Each outgoing call leg may be subject to on-line charging control, as described in Chapter 4. In a two-person call, the gsmSCF may apply different charging for the two legs, depending on the destination numbers of the respective call legs. If one call leg is to an MSISDN, then the charge may also be made dependent on the location of the subscriber that is identified with MSISDN. Figure 6.25 reflects the various charging aspects of network-initiated calls.

The gsmSCF may use the CAP operations ACH, FCI and call information request (CIRq) for the individual call legs. The send charging information (SCI) operation is not used in a network-initiated call. The reason is that the MSC/gsmSSF has no direct link with the RAN to deliver the advice of charge information that is carried in CAP SCI. When the pre-paid warning tone in CAP ACH is used, the warning tone is played to the call leg that was created first for this call. If that call leg is no longer available at the moment that the warning tone timer expires, then the warning will not be played.

The MSC may create a mobile-originated call attempt (MOC) CDR for each outgoing call leg; refer to 3GPP TS 32.205 [117] for details.[7] In addition, the gsmSCF may create a service CDR. The CRN that is generally used in CAMEL control also applies for the network-initiated call. Since

---

[7] Implementations of the network-initiated call establishment may use different CDR type for this call case.

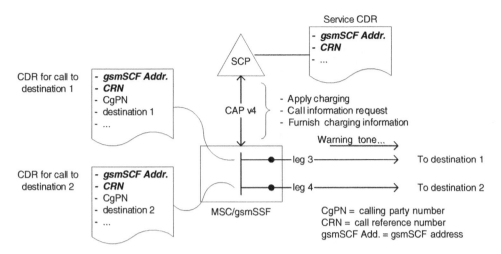

Figure 6.25   Charging for network-initiated calls

the gsmSCF is initiator of the call, it generates the CRN and sends it in CAP ICA to the gsmSSF. The CRN is accompanied by the gsmSCF Address. The combination of CRN and gsmSCF address forms a globally unique pair, comparable with the combination of MSC address and CRN for the subscriber-initiated call. The CRN and gsmSCF address are placed in the respective CDRs; this enables post-processing systems to correlate CDRs that are created for this call case. If the gsmSCF sends the CAP ICA with MSRN directly to the MSC/gsmSSF where the subscriber for whom the MSRN is obtained is registered, then that MSC may generate an MT call attempt CDR.

6.2.4 Optimal Routing of Basic Mobile-to-mobile Calls

The principle of BOR is described in Chapter 4. BOR entails, amongst others, GMSC and HLR applying appropriate logic to determine whether BOR may be applied for a call.

- *GMSC* – the GMSC will verify whether the called subscriber number (MSISDN) belongs to a PLMN for which it is entitled to interrogate the HLR.
- *HLR* – the HLR will verify whether the interrogating GMSC is entitled to contact this HLR. The HLR will also verify whether the current location of the called subscriber justifies the use of BOR for this call. Direct routing from the IPLMN of the calling party to the VPLMN of the called party may not necessarily be cost-effective, compared with routing the call to the HPLMN of the called subscriber first. In addition, the called party may subscribe to a CAMEL service for terminating calls. The HLR should therefore send the T-CSI to the GMSC in the IPLMN. Hence, CAMEL capability negotiation is required between HLR (in HPLMN) and GMSC (in IPLMN). If the GMSC does not support the required CAMEL capability for this subscriber, then the HLR may decide to instruct the GMSC to route the call via the HPLMN.

To alleviate GMSC and HLR of this decision-making process, the gsmSCF may instruct a gsmSSF to apply BOR. This instruction is provided to the gsmSSF in the 'Basic OR interrogation requested' parameter in CAP CON or in CAP CWA. (Figure 6.26).

In the example case in Figure 6.26, a subscriber from KPN Mobile Netherlands (KPN-NL) is calling a Dutch MSISDN. The SCP, which is controlling the MO call from the calling subscriber, ascertains that the called party belongs to KPN and currently resides in the same VPLMN as the

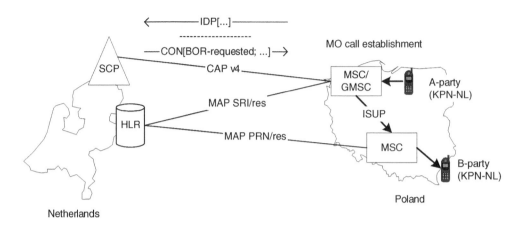

Figure 6.26   CAMEL-induced basic optimal routing

calling party. Therefore, the call may be established as a local call. The calling party therefore pays a local charge and the called party does not pay for receiving the call. Should the called party currently be roaming in Belgium, then direct routing from Poland to Belgium may not be cheaper than routing the call via The Netherlands. The SCP would in that case not instruct the gsmSSF to apply BOR.

For the MSC/gsmSSF in Poland to apply BOR, it also needs to be able to act as a GMSC. CAP IDP may indicate 'OR interactions' in offered CAMEL 4 functionalities. The presence of that parameter indicates to the gsmSCF that the MSC/gsmSSF may act as GMSC for the purpose of BOR. One other task that the SCP needs to carry out, before instructing the gsmSSF to apply BOR, is checking which network the called party belongs to. If a country uses mobile number portability (MNP), then the MSISDN of the called party does not indicate which network that subscriber belongs to.

When the calling party in the example in Figure 6.26 sets up the call to a Dutch mobile subscriber, she may not know whether that person currently resides in the same country. Hence, the caller may not know what the rate of the call will be. If advice of charge is used for the calling party, then the charge that is indicated on the display informs the subscriber that the call is a local call. Likewise, when the called party is alerted, she may not know that the call is set up with BOR and hence that no terminating call charge applies. Here as well, if advice of charge is used for the called party,[8] then the charge on the display indicates to the user that this call bears no cost for the called party.

Other relevant CAMEL service invocations remain valid for a call that is subject to CAMEL-induced BOR. This includes: the MSC/gsmSSF, when acting as GMSC, may receive T-CSI from the HLR in response to MAP SRI – the MSC will now invoke a terminating CAMEL service; the VMSC of the called party may have VT-CSI – the VMSC will in that case invoke a CAMEL service when receiving the call.

6.2.5 Alerting Detection Point

One of the enhancements to the O-BCSM and the T-BCSM, compared with CAMEL phase 3, is the introduction of the alerting detection points. The following DPs are used for the alerting event: O-BCSM: O_Term_Seized; and T-BCSM: T_Call_Accepted. The alerting DPs relate to

---

[8] CAMEL controlled advice of charge for MT calls requires VT-CSI in the VMSC. This is not reflected in figure 6.26. See Chapter 5 for advice of charge for MT calls.

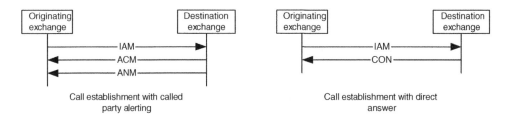

Figure 6.27   Example ISUP signal sequence flows

ISUP address complete message (ACM) and ISUP call progress (CPG). When the MSC has sent ISUP IAM, it waits for ISUP ACM. Reception of ISUP ACM serves as an indication that the call has reached its destination; it may, in addition, contain an indication that called subscriber is being alerted. Normal ISUP message sequence flows related to ISUP ACM are illustrated in Figure 6.27.

For a call to a GSM subscriber, the ISUP ACM is normally generated by the destination MSC. The ISUP ACM contains the indication 'subscriber free', which is carried in the 'backward call indicators' parameter. The sending of the ring back tone from VMSC towards calling party normally commences after ISUP ACM has been sent. When the called party answers, the ISUP ANM is sent in a backwards direction. When the call is answered by an electronic device, the ISUP ACM and ISUP ANM may be replaced by an ISUP CON. In that case, a CAMEL BCSM that was invoked for this call will transit directly from the point in call (PIC) analysis and routing (O-BCSM) or terminating call handling (T-BCSM) to the answer DP.

It may occur that an intermediate node generates an ISUP ACM as part of call processing. For example, when a call is set up to a GSM subscriber, the GMSC for the destination subscriber may invoke a terminating CAMEL service for pre-paid charging. If the CAMEL service processing includes user interaction, then the GMSC sends an ISUP ACM towards the calling party, at the time of starting the user interaction. This ISUP ACM will, however, not contain the parameter 'subscriber free'. Therefore, the O-BCSM in the originating MSC will not transit to the O_Term_Seized DP. When the call is eventually routed to the MSC of the destination subscriber, the VMSC will generate an ISUP ACM, including 'subscriber free', when the called part is alerted. The GMSC will convert the ISUP ACM to ISUP CPG, since the GMSC had already sent an ISUP ACM for this call. The ISUP CPG will, however, contain the 'subscriber free' indication. The O-BCSM now transits to the O_Term_Seized DP. Figure 6.28 reflects the above, plus additional call forwarding.

An exchange will generally send only one ISUP ACM in the backwards direction. If the exchange receives a subsequent ISUP ACM for this call, then the ISUP ACM is converted to an ISUP CPG. Information elements carried in the ISUP ACM, such as *backward call indicators, optional*

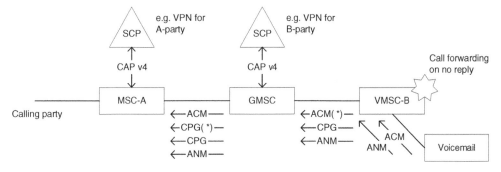

Figure 6.28   Generation of ISUP ACM and ISUP CPG

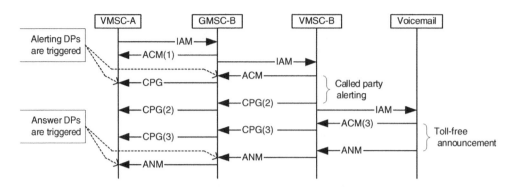

Figure 6.29   ISUP signal sequence diagram

backward call indicators and *user-to-user information*, are copied to corresponding information elements in CPG. In the example flow in Figure 6.28, the messages indicated with (asterisks) lead to the triggering of alerting DP in the MSC receiving that message. The respective CAMEL services may not be aware that the ISUP ANM following the alerting indication relates in fact to another destination than the original called party.[9] Figure 6.29 contains the ISUP signal sequence diagram associated with the example call case.

The ACM indicated with (1) relates to user interaction (e.g. announcement) for CAMEL service in the GMSC. The CPG indicated with (2) relates to the subscription option 'notification to the forwarding party', related to call forwarding. The forwarding MSC may, if the calling party subscribes to this feature, send a notification to the calling party that the call is forwarded. The ACM and CPG indicated by (3) relate to a toll-free voicemail announcement. The voicemail system may defer the sending of ANM for the duration of a welcome announcement. During this announcement, the calling party is not charged, since the originating exchange has not yet received the ANM. The VMSC converts the ACM from the voicemail box into CPG because the VMSC has already sent ACM for this call. The GMSC and the originating VMSC will not receive a second alerting indication.

When the gsmSSF sends an alerting indication to the gsmSCF, the gsmSSF may include the location information of the served subscriber in the notification. This applies only for a gsmSSF in an MO call or a gsmSSF in a VT call. This feature is particularly useful for the VT call case. When a CAMEL service is invoked as a result of VT-CSI, the location information that is included in CAP IDP is retrieved from the VLR. At that point of the call, the called subscriber has not yet been paged. The reported location information may therefore be out of date; the subscriber may have moved to another cell within the location area. The alerting event results from subscriber paging response. Paging response results in an update of the location register in VLR. The updated location information is included in the alerting notification.

6.2.6 Mid-call Detection Point

Mid-call detection is a means for the CAMEL service to obtain user input during a call without connecting the subscriber to a specialized resource function. A CAMEL service may instruct the gsmSSF to monitor for dual-tone multi-frequency (DTMF) digits in the speech path of the call. The detection and reporting of DTMF digits for a call is restricted to the MO call case and the VT call case. The detection and reporting of DTMF digits is transparent for the further transfer

---

[9] The 'destination address' in the O_Answer and T_Answer notification to the CAMEL service includes the destination address that was included in the ISUP IAM sent from the MSC. It does not include the connected number, as contained in ISUP ANM.

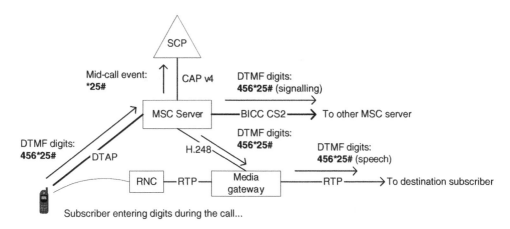

Figure 6.30 DTMF reporting in an MO call

of these digits in the speech path; the digits may be transferred to one or more call partners. See Figure 6.30 for a graphical representation.

In the example in Figure 6.30, a split-MSC architecture is used. The CAMEL service that is controlling the MO call has armed the 'mid-call event' in the gsmSSF. This implies that the gsmSSF will monitor the speech channel for the call for the presence of DTMF digits. The calling subscriber may use a digit combination to activate a particular feature, such as the creation of an additional call party during a conference call or the removal of a party from the conference call. The CAMEL service may define 'conditions' for the mid-call event to occur. For example, the condition may be that the digit string starts with a star and ends with a hash. Digits not matching this criterion are not reported to the CAMEL service, as shown in Figure 6.30. The monitoring and reporting of DTMF digits do not affect the transfer of these digits through the call. DTMF is transported 'out-band' over the radio network interface, i.e. they are conveyed in DTAP messages to the MSC. The MSC injects the DTMF as speech into the speech channel. In the case of split-MSC architecture, the MSC instructs the media gateway to inject the DTMF into the speech channel. If the MSC server uses bearer-independent call control (BICC) capability set 2 (CS2), then the DTMF may even be transported as an out-band signal over the BICC signalling to the destination exchange.

If a calling subscriber enters DTMF during a multi-party call, then the MSC injects the DTMF digits into the respective call legs for this call connection (Figure 6.31). In the example case in Figure 6.31, the digits that are generated by the calling party are distributed over the outgoing call parties in call segment 1, but not to the call party in call segment 2. The rationale is that DTMF digits are considered as in-band speech information and leg 5 is currently not in speech connection with the other call parties.

The CAMEL service may provide a set of criteria for the detection and reporting of DTMF digits. When the digits entered by the subscriber fulfill the criteria, the entered digits are reported to the CAMEL service. The criteria include:

- *minimum number of digits* – the subscriber shall enter a minimum number of digits before triggering can take place;
- *maximum number of digits* – when the subscriber has entered the maximum number of digits, triggering can take place;
- *start digit string* – digit collection by gsmSSF can start when this digit string (one or two digits) is entered;

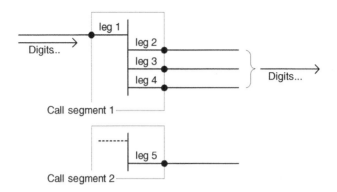

Figure 6.31   Distribution of DTMF digits over multiple call legs

- *end-of-reply digit string* – when this digit string (one or two digits) is detected, triggering can take place;
- *cancel digit string* – when this digit string (one or two digits) is detected, the gsmSSF clears the input buffer; this criterion may be used to enable the subscriber to cancel the input and start entering digits again;
- *interdigit timeout* – this is the maximum duration that the gsmSSF will wait for further input; when triggering occurs as a result of inter-digit expiry, the CAMEL service may regard the input as erroneous.

A service example of the mid-call event is presented in Table 6.8. When the gsmSCF arms the mid-call event, it may instruct the gsmSSF to automatically re-arm the mid-call event when the gsmSSF reports the detection of digits. In this manner, the gsmSSF is ready for the detection of the next DTMF string as soon as a DTMF string is reported.

6.2.7 Change of Position Detection Point

The location of a subscriber is reported to a CAMEL service at the start of a call. For an MO call, the calling subscriber's location is reported at initial DP; for an MT or VT call, the called

Table 6.8   Service example of DTMF detection during a call

Minimum number of digits = 3
Maximum number of digits = 8
Start digit string = ∗
End-of-reply digit string = #
Cancel digit string = 9
Interdigit timeout = 4s

Input	Result
5677	No triggering; start digit (∗) is not yet detected
5677∗25	No triggering; end of reply digit (#) is not yet detected, maximum number of digits (8) not yet reached
5677∗259	Input is cancelled due to cancel digit (9)
5677∗259∗25#	Triggering takes place as a result of end-of-reply digit (#); gsmSSF reports ∗25# to gsmSCF

subscriber's *approximate location* is reported at initial DP.[10] In addition, for a VT call, the called
subscriber's actual location may be reported in the alerting notification. In CAMEL phases 1–3,
any change of location during the call is not reported to the CAMEL service. In CAMEL phase 4,
the change of position detection point may be used by the CAMEL service to receive a notification
when the subscriber changes location during the call (Figure 6.32).

In the example in Figure 6.32, the subscriber initiates or answers a call when residing in cell 2.
During the call, the subscriber changes location to neighbouring cell 3, which is served by the same
MSC as cell 2. The CAMEL service had previously armed the change of position event; the gsmSSF
will therefore send a notification to the gsmSCF. The notification contains the location information
after the change of location. The subscriber subsequently performs a handover to another MSC
('MSC-B'). This will also lead to a notification from the gsmSSF. The location information is
reported from MSC-B to the anchor MSC, through MAP signalling. DTAP messages are relayed
through MSC-B. Should the subscriber perform further change of positions while in the service
area from MSC-B, these change of positions will also be notified to the gsmSCF. Communication
with the gsmSCF remains with the gsmSSF in the anchor MSC.

Use cases for this feature include:

- Home zone charging – when a subscriber moves in or out of her home zone during the call, then
  that may result in a change of call rate.
- When an operator applies national roaming, handover to another network in the same country
  may affect the call rate.
- Some operators in neighbouring countries have handover agreements; the subscriber may enter
  the PLMN in the neighbouring country during the call; this form of handover may affect the call
  rate and availability of services during the call.

To facilitate a change in tariff during the call, as a result of change of location or handover, an
apply charging report is sent to the gsmSCF when the change of location occurs, provided that a
charging report was pending at that moment. The gsmSCF may now apply a modified tariff for
the call.

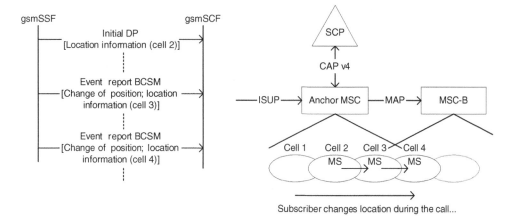

Figure 6.32 Change of position notifications during a call

---

[10] The reporting of the location information of the called subscriber for MT calls depends on the usage of
MAP PSI between HLR and VLR during call establishment.

When the change of position DP is armed, all changes of position of the subscriber lead to a notification to the gsmSCF. CAMEL phase 4 in 3GPP Rel-6 introduces 'change of position notification criteria', by means of which the operator can reduce the number of change of position notifications.

6.2.8 Flexible Warning Tone

CAMEL phase 4 specifies a flexible variant of the pre-paid warning tone that is introduced in CAMEL phase 2. When the operator uses that tone during charging control, the cadence of the warning tone is fixed. With the introduction of features such as call party handling, service logics benefit from differentiated warning tones. In this manner, different tone burst cadences may be used to denote different events.

The flexible warning tone has the format indicated in Figure 6.33. The usage of the flexible warning tone in CAP ACH is the same as the usage of the fixed pre-paid warning tone. The gsmSCF may instruct the gsmSSF to play the flexible warning tone before the expiry of a call period. In CAMEL phases 2 and 3, the warning tone is played 30 s before expiry of the call period for which the warning tone applies. In CAMEL phase 4, the start of the flexible warning tone may also be determined by the gsmSCF. The gsmSCF may supply the parameters in Table 6.9 to the gsmSSF.

Each parameter of the flexible warning tone has a default value; default values allow for efficient information transfer; omission of the parameter from the CAP operation implies the implicit transfer of the default value of this parameter. If all parameters of the flexible warning tone are

Table 6.9 Flexible warning tone parameters

Parameter	range	Default value
Warning period	1–1200 s	30 s
Tone duration	1–20 ds	2 ds
Tone interval	1–20 ds	2 ds
Tones per burst	1–3	3
Burst interval	1–1200 ds	2 ds
Number of bursts	1–3	1
ds = deci second.		

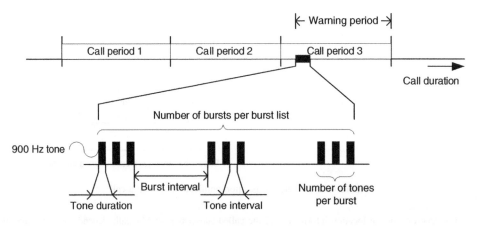

Figure 6.33 Flexible warning tone

omitted, then this flexible warning tone corresponds to the fixed warning tone from CAMEL phases 2 and 3.

The call duration control may be applied to individual call parties or to an entire call segment during user interaction. Hence, the flexible warning tone may also be played to individual call parties (Figure 6.34). Since the total duration of the flexible warning tone, including burst intervals, may extend to 270 s, the MSC ensures that the speech transmission is not blocked during the tone interval or during the burst interval. If the gsmSSF to which the flexible tone instruction is sent is located in an MSC Server, then the MSC server will instruct the media gateway to inject the warning tone. 3GPP TS 29.232 [110] contains the definition of the warning tone for the interface between MSC server and media gateway.

6.2.9 Tone Injection

Tone injection is a mechanism for a CAMEL service to play an *ad hoc* sequence of tones to an individual party or an entire call segment. The tone injection is not related to call duration control (CAP ACH). Tone injection may therefore be associated with a call party handling operation, such as adding a party to a call or removing a party from the call. Playing an audible tone when a call party is added to a call may be mandated by a national regulator. Tone injection is performed with the play tone (PT) operation. The tone injection to a call segment or an individual call party is not allowed when user interaction is ongoing for that call segment. The tone that may be played with the play tone operation has the same characteristics as the flexible warning tone that is related to CAP ACH. Figure 6.35 shows an example where the gsmSCF plays a tone to the group when a participant has entered the conference.

One tone sequence is played to the conference group; another tone sequence, with different characteristics, is played to the party that has just accepted the connection to the conference call, but is not yet connected to the group. If the CAMEL service wishes to play different tone sequences to different parties in call segment 1, then multiple play tone operations are required, each with a different tone definition.

3GPP TS 23.078 [83] specifies detailed rules for the gsmSSF to handle overlapping tone playing instructions, e.g. when tone playing for a leg is in progress and the CAMEL service orders the playing of another tone, then that other tone will replace the ongoing tone. If in such case the playing of the ongoing tone was the result of a previous play tone instruction, then the cessation of the playing of that tone will not lead to an operation error to the gsmSCF. Although the execution of the play tone operation may take up to 270 s (which corresponds to the duration of the longest possible flexible tone), an operation error may be reported by the gsmSSF only immediately after the reception of the play tone instruction. This is due to the fact that the play tone operation has a short operation timer; see 3GPP TS 29.078 [106] for operation timer values.

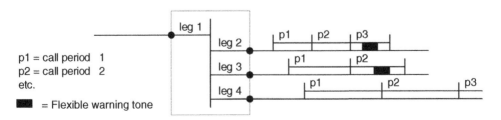

p1 = call period 1
p2 = call period 2
etc.

▬ = Flexible warning tone

Figure 6.34 Playing of flexible warning tone to individual call parties

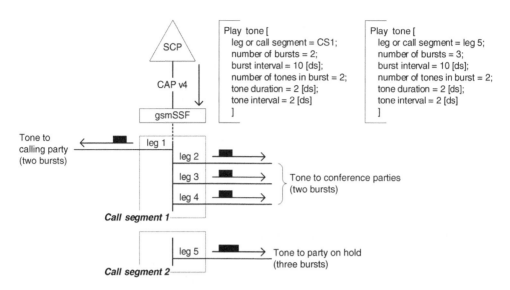

Figure 6.35   Tone injection to conference call participants

6.2.10 Enhancement to Call Forwarding Notification

Call forwarding notification was introduced in CAMEL phase 2. That feature means that the gsm-SCF may receive a notification when GSM call forwarding is taking place in the GMSC. In CAMEL phase 3, the call forwarding notification feature also applies to the VMSC, when a CAMEL service is invoked with VT-CSI. This notification, however, contains no indication about the forwarding destination. The notification may be used to stop roaming leg charging. In CAMEL phase 4, the forwarding notification also contains the forwarding destination. This is reflected in Figure 6.36.

The gsmSCF may use the forwarding destination number to determine whether the forwarding will be allowed and whether a CAMEL service will be invoked for the forwarding call leg. One practical example may be the case where the terminating leg service, triggered with T-CSI, handles the charging of the forwarding leg. The service uses the forwarding destination number to determine the charge for the forwarding leg and then instructs the gsmSSF to continue call handling.

When HLR sends an FTN to GMSC, it may accompany the FTN by an O-CSI; the O-CSI is used for invoking a CAMEL service for the forwarding leg. When the gsmSCF has received a

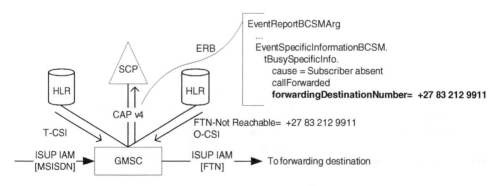

Figure 6.36   Notifying the forwarding destination number

call forwarding notification and instructs the gsmSSF to continue the call forwarding, the GMSC will use the O-CSI, if received from HLR, to invoke a CAMEL service for the forwarding leg. In CAMEL phases 2 and 3, the gsmSCF does not have the possibility to influence the trigger of this CAMEL service; the 'O-CSI applicable' parameter in CAP connect is applicable only for the CAMEL-induced call forwarding, but not for the GSM-based call forwarding. In order to enable the gsmSCF to allow GSM call forwarding to take place, but without CAMEL service for the forwarding leg, the parameter 'suppress O-CSI' is introduced in CWA. The gsmSCF may respond to the forwarding notification, containing the forwarding destination number, with CAP CWA, containing 'suppress O-CSI'. If O-CSI was received from the HLR together with the FTN, then the O-CSI will not be used for the forwarding leg. There is, however, no corresponding mechanism to suppress the invocation of a D-CSI service for the forwarding leg.

The forwarding notification with forwarding destination number also works in the following call cases:

- *Optimal routing* – when the GMSC receives MAP resume call handling (RCH) and notifies the gsmSCF, it includes the FTN that is included in MAP RCH.
- *IDP at T_Busy or T_No_Answer* – an MT call service may be triggered as a result of T-CSI containing TDP T_Busy or T_No_Answer. In that case, CAP IDP will contain the FTN that was received from HLR (early call forwarding), from VLR (late call forwarding), from VMSC (in the case of ORLCF) or from the called party (call deflection).[11]
- *IDP at terminating attempt authorized* – the HLR may accompany the T-CSI in the first MAP SRI result by an FTN. This may occur in the case of call forwarding unconditional or when the called subscriber's MS is not currently registered in any VLR. In that case, CAP IDP will contain the FTN that was received from HLR. In this case, however, there is no cause code present in CAP IDP and hence the gsmSCF may not know whether the pending GSM call forwarding is the result of unconditional call forwarding or the subscriber not being registered in a VMSC.
- *Short FTN* – the FTN that is registered in the HLR, or provided by the called party in the case of call deflection, is notified to the gsmSCF in unmodified form. Hence, if this FTN or deflected-to-number (DTN) is a short number, then the gsmSCF receives the short number in the notification. The rationale of the short forwarded-to number (short FTN) is that a CAMEL service in the forwarding leg translates the short FTN to the full number. The terminating CAMEL service, which had received the notification for the forwarding to the short FTN, will not be informed about the translated number.

6.2.11 Control of Video Telephony Calls

As described in Chapter 1, GSM may be used for establishment of calls of different basic service type. Examples of basic service types for GSM include: voice calls (TS11), automatic facsimile group 3 (TS 62), asynchronous general bearer service (BS20) and synchronous general bearer service (BS30). The capabilities of the GSM basic services are specified in 3GPP TS 22.002 [58] and 3GPP TS 22.003 [59]. One special basic service is BS30; BS30 allows for the establishment of data transfer channels with a transfer speed varying between 1.2 and 64 kb/s. The 64 kb/s synchronous data channel may carry a CS video telephony call. Video telephony calls are possible with the 2G RAN (with EDGE support; see Chapter 5) and with the 3G universal terrestrial radio access network (UTRAN). Table 6.10 describes the characteristics of a video telephony call. The values in brackets in the right-most column indicate the required value, expressed as binary number, of the corresponding variable in the low layer compatibility (LLC) parameter.

---

[11] The parameters 'eventTypeBCSM' and 'cause' in the initial DP operation will in this case indicate to the gsmSCF what forwarding event has occurred.

Table 6.10   Video telephony call basic service characteristics

GSM basic service characteristic	Setting	Corresponding variable in LLC
Fixed network user rate	64 kb/s	User rate (10000)
Access structure	Synchronous	Access (0)
Information transfer capability	Unrestricted digital information (UDI)	Information transfer capability (01000)
Rate adaptation	H.223 and H.245	User information layer 1 protocol (00110)

The bearer used for CS video telephony is in many aspects the same as a non-multimedia, 64 kb/s synchronous CS data bearer. The use of the H.223 and H.245 is transparent for most of the RAN and the core network. A UMTS mobile terminal that supports video telephony is referred to as a '3G-324M terminal'. ITU-T H.324 [131] defines protocols that may be used for transporting video telephony over a narrow-band, error-prone CS communication channel. 3GPP TS 26.111 [102] specifies the usage of H.324 for the UMTS network. Figure 6.37 represents how the video telephony is carried over a synchronous bearer.

The multiplexer in the 3G-324M terminal multiplexes the video signal, the audio signal and the control signals onto the 64 kb/s bearer. When the video call is established, the calling party and called party 'negotiate' between one another which video codecs shall be used for the call. This negotiation occurs 'in-band'; i.e. the negotiation forms part of the data transfer over the data bearer. Although a video telephony call is in many aspects identical to a 64 kb/s data bearer call, an operator may want to apply differentiated charging between these two types of calls. For that purpose, CAMEL phase 4 includes the LLC information element in CAP IDP. The CAMEL service may inspect the user information layer 1 protocol from the LLC and deduce the call type.

For MO video telephony calls, CAP IDP contains, amongst others, LLC and BC. The user information layer 1 protocol may be retrieved from LLC or from BC. For an MT call, the BC may be absent from ISUP IAM. In that case, the CAMEL service may retrieve the user information layer 1 protocol from LLC. Absence of the BC from ISUP IAM may occur when the video telephony call arrives from a fixed network. Fixed network operators as well as transit network operators should take care not to remove the BC from the ISUP signalling flow when video telephony may be transported through the network.

Video telephony calls may also be established over the TDMA access network (GSM) as opposed to the W-CDMA access network. H.223 and H.245 protocol for video telephony may be used with

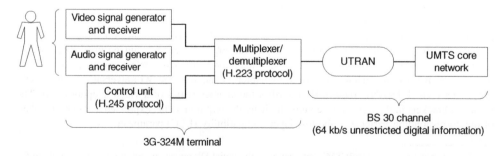

Figure 6.37   3G-324M terminal

32 or 64 kb/s data bearer. Video telephony over a 32 kb/s bearer could be offered over the GSM RAN, whereas an EDGE-capable RAN (GERAN[12]) may carry a 64 kb/s bearer for video telephony.

6.2.12 Control of SCUDIF Calls

3GPP Rel-5 contains a mechanism for switching between video telephony and speech telephony during an ongoing call or during call establishment. This feature is known as 'service change and UDI/RDI fallback' (SCUDIF). SCUDIF is specified in 3GPP TS 23.172 [89]. The rationale of SCUDIF is manifold; examples include:

- a subscriber with a 3G-324M terminal may want to establish a speech call and switch to video call during the conversation, after having agreed with her call partner that the call shall become a video call;
- a subscriber wants to establish a video telephony call, but network limitations or terminal limitations prevent the called party from receiving a video call; the call may automatically perform a fallback to speech call;
- a subscriber wants to establish a video telephony call, but the receiver wishes to answer the call as a speech call;
- the calling party wants to use video telephony capability for a limited portion of the call only, due to the higher cost of a video call compared with a speech call.

SCUDIF entails that a call is established with dual bearer capability. The following combination of bearer capability is available: (1) tele service 11 (speech); and (2) bearer service 30 (multimedia). Setting up a call with two bearer capabilities requires that the traffic channel support both the codecs for a speech call and the codecs for a video telephony call. At any moment, only one set of codecs is used in the traffic channel. When a service change takes place, e.g. from speech to video telephony, the traffic channel switches over to the other set of codecs.

Figure 6.38 depicts the establishment of the dual bearer capability call. When the subscriber establishes a dual bearer capability call, the UE provides two sets of bearer parameters (LLC1, HLC1, BC1 and LLC2, HLC2, BC2) to the MSC. The corresponding BS values (BS1, BS2) are not provided by the UE but are derived in the MSC, from the respective LLC, HLC and BC values.

SCUDIF uses the terms 'preferred service' and 'less preferred service' for the requested bearer capabilities. The call will be established with the preferred service (e.g. speech), with the possibility

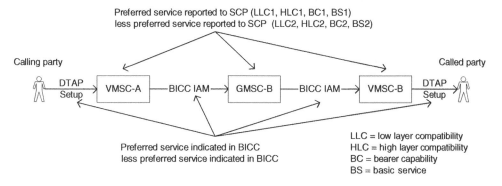

Figure 6.38 Establishment of dual bearer capability call

---

[12] 'GERAN' is often used as a generic term for a GSM (TDMA) RAN, which may or may not support EDGE.

of switching to the less preferred service at a later stage. If network capabilities or the called party's terminal capabilities prevent the establishment of a call with the preferred service (e.g. video), then the call will be established with the less preferred service. In that case, there is no possibility of switching to the preferred service at a later stage. Figure 6.38 depicts the use of BICC for call signalling. Video calls may also be established in a network that uses ISUP signalling. However, bearer service change during an active call is feasible with BICC only.

A SCUDIF call may be subject to CAMEL services; all CAMEL call scenarios are applicable to SCUDIF calls, except SCP-initiated calls. Figure 6.39 shows the invocation of an MO call CAMEL service, an MT call CAMEL service and a VT call CAMEL service. However, a SCUDIF call may also be subject to D-CSI and N-CSI service, depending on trigger conditions and network configuration.

HLR interrogation for a dual bearer capability call is described in detail in 3GPP TS 23.172 [89]. For CAMEL phase 4 in 3GPP Rel-5, there are two points in the call where the SCP is notified about the available bearer capabilities for the call: (1) call establishment; and (2) call answer.

6.2.12.1 CAMEL Interaction with SCUDIF at Call Establishment

When a SCUDIF call is established, CAP IDP contains the bearer related parameters for both requested bearers. That entails the initial DP containing:

$$\left.\begin{array}{l} \text{LLC1} \\ \text{HLC1} \\ \text{BC1} \\ \text{BS1} \end{array}\right\} \text{Related to preferred service}$$

$$\left.\begin{array}{l} \text{LLC2} \\ \text{HLC2} \\ \text{BC2} \\ \text{BS2} \end{array}\right\} \text{Related to less preferred service}$$

At this moment, the SCP does not know which of the two services the call will be established with. The SCP may verify that the subscriber is entitled to use these services. The SCP does not have the capability to force the call to continue as a single bearer service call, e.g. in the case where a speech call is allowed, but not a video call.[13]

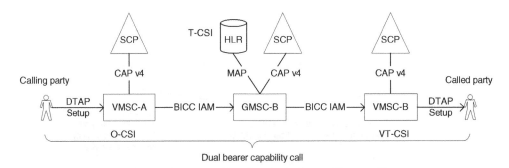

Figure 6.39   CAMEL service interaction with dual bearer capability call

---

[13] The SCP-controlled service change is planned for introduction in CAMEL phase 4 in 3GPP Rel-7.

6.2.12.2 CAMEL Interaction with SCUDIF at Call Answer

A call that starts as a SCUDIF call, i.e. with dual bearer capability, may continue as a SCUDIF call or may fall back to single service. Fallback to single service may occur during call establishment, prior to call answer. At call answer, the gsmSSF sends an answer notification to the gsmSCF, provided that the answer event (DP O_Answer or DP T_Answer) is armed. The answer notification, conveyed in the ERB CAP operation, may contain an indication of the available bearer capability at the moment of answer. The following information may be reported to the gsmSCF:

- *Basic service* – the call has made a fallback to single bearer capability. The basic service included in the answer notification indicates which service is available for the remainder of the call; either speech (BS = TS11) or video (BS = BS30).
- *Basic service and basic service 2* – both requested bearer capabilities are still available. The reported *basic service* indicates the bearer capability that is now active for the call; the *Basic Service 2* indicates the bearer capability that is not active, but is available for the call.

The gsmSCF may use the reported active service to determine the charge for the call. Should the service, however, change during the call, then the CAMEL service will not be notified.[14] If the call started off as a single bearer capability call, then the answer notification does not need to contain either of these information elements.

6.2.12.3 Conditional Triggering

For some call cases, the gsmSSF may apply conditional triggering based on the basic service for the call. These cases include:

- *MO call in VMSC* – basic service criteria from O-CSI may apply at DP collected info;
- *MT call in HLR* – basic service criteria from T-CSI may apply at reception of MAP send routing information;
- *MF call in VMSC or GMSC* – basic service criteria from O-CSI may apply at DP collected info.

If a SCUDIF call is established, then the gsmSSF or HLR will consider both requested basic services for the conditional triggering check. The basic service condition is fulfilled if the basic service list contains TS11, BS30 or both. TS11 and BS30 may be included as specific basic service or as part of a basic service group. 3GPP TS 22.078 [66] contains further specific rules for SCUDIF calls, e.g. related to user interaction, tone injection.

6.2.13 Reporting IMEI and MS Classmark

A CAMEL phase 4-compliant MSC may report equipment-related information to the SCP at service initiation. These elements may be included in CAP IDP. The MSC may report the following information.

IMEI

The international mobile equipment identifier (IMEI) contains the serial number of the MS or UE in use by the served subscriber. The IMEI may also include a software version (SV); the SV relates to the version of the terminal software.[15] The structure of the IMEI(SV) is given in Section 1.4.

---

[14] Notification of change of bearer capability during a call was introduced in CAMEL phase 4 in 3GPP Rel-6.

[15] A GSM/UMTS terminal contains operating system (OS) and terminal software. The SV relates to the version of the terminal software, not the OS.

The IMEI and IMEISV are both encoded as an octet string with length 8. When the IMEI does not contain the SV, the fifteenth digit in the IMEI (the spare digit) shall have the value 0. The value of the spare bit may therefore be used by the SCP to determine whether the reported IMEI contains an SV. Further details of the IMEI(SV) may be found in 3GPP TS 23.003 [73] (structure), 3GPP TS 22.016 [61] (allocation) and 3GPP TS 29.002 [103] (encoding).

A CAMEL service may use the IMEI for various purposes, such as:

- IMEI verification, e.g. by contacting an EIR;[16]
- maintaining subscriber capability database;
- adapting service logic processing; the CAMEL service may offer certain functionalities only for certain terminal types;
- help desk; a help desk agent may refine their advice to the type of terminal of the calling subscriber.

MS Classmark 2
The MS Classmark 2 is a descriptor of radio network-related capabilities of the MS or UE in use by the served subscriber. GSM specifies three types of classmark parameters:

- *MS classmark 1* – this element contains characteristics of the MS or UE that are generally independent of the frequency band used for the GSM or UMTS network. Examples include power transmission capabilities.
- *MS classmark 2* – this element contains the same parameters as MS classmark 1. MS classmark 2 contains additional parameters related to the capability of the MS/UE for, amongst others, supplementary services. Examples include support for MT-SMS, support for voice broadcast service, support for voice group call service and support for location request.
- *MS classmark 3* – this element contains various parameters that indicate the capability of the MS related to GSM and UMTS radio signalling.

A CAMEL service may use the MS classmark 2 to decide which services it may use.

The inclusion of IMEI and MS classmark 2 in CAP IDP applies only for a CAMEL service for an MO call. If IMEI and MS classmark 2 of a subscriber are needed in other call cases or outside the context of a call, then a CAMEL service may use the CAMEL phase 4 version of ATI (see Section 6.6).

The MS sends the MS classmark 2 to the MSC at, amongst others, call establishment. Hence, the MSC can include this parameter in IDP. Should the subscriber detach from the MSC and re-attach with a different MS, then the MS classmark 2 of the current MS will be reported. The MSC will ensure that it obtains the IMEI when the subscriber attaches to the network, since a subscriber may re-attach with a different IMEI.

6.3 GPRS Control

A CAMEL phase 4 compliant SGSN may report the served subscriber's IMEI or IMEISV to the SCP at service initiation. The SGSN shall ensure that it obtains the IMEI from the MS when the subscriber attaches to the network. The inclusion of the GPRS MS class in the initial DP GPRS (IDP-GPRS) operation is introduced in CAMEL phase 3 already. The GPRS MS Class consists of the following two elements:

- *MS network capability* – this parameter contains various indicators related to MS capability, such as support for SMS or support of encryption-particular algorithms.

---

[16] The interface between SCP and IMEI is not standardized, but an enhanced version of ATI could be used for that purpose.

• *MS radio access capability* – this parameter contains various indicators related to radio signalling capabilities of the MS. Examples include supported access technologies and transmission power capabilities.

The inclusion of the IMEI in CAP IDP-GPRS is done in a backwards-compatible manner. The IMEI is placed after the ellipsis in the argument of CAP IDP-GPRS. As a result, the same CAP application context may be used for GPRS control in CAMEL phase 4. A further result is that no new version of GPRS-CSI is defined in CAMEL phase 4; the GPRS-CSI in CAMEL phase 4 still indicates CAMEL capability handling = CAP v3. If an SCP does not support the reported IMEI in IDP-GPRS, then that SCP will ignore that parameter, if received.

6.4 SMS Control

Whereas CAMEL phase 3 includes control of MO-SMS, CAMEL phase 4 includes both control of MO-SMS and control of MT-SMS. For CAMEL control of MO-SMS, only a very small enhancement is introduced.

6.4.1 Mobile-originated SMS Control

CAMEL phase 4 specifies the inclusion of the IMEI and classmark parameters in CAP IDP-SMS. The exact parameters to be reported differ slightly between MSC-based SMS and SGSN-based SMS (Table 6.11).

Similar to GPRS control, the inclusion of the IMEI and MS class information in CAP IDP-SMS is done in a backwards-compatible manner. IMEI and MS class are placed after the ellipsis in the argument of CAP IDP-SMS (Figure 6.40).

The same CAP application context is used for MO-SMS control in CAMEL phase 4 as for MO-SMS control in CAMEL phase 3. Consequently, no new version of MO-SMS-CSI is defined for CAMEL phase 4. Even in CAMEL phase 4, MO-SMS-CSI contains CAMEL capability handling = CAP v3. If an SCP does not support the new parameters in IDP-SMS, then that SCP ignores those

Table 6.11 Reporting of IMEI and MS class information for SMS

Network type	Reported parameters
MSC-based SMS	IMEI(SV), MS classmark 2
SGSN-based SMS	IMEI(SV), GPRS MS class

```
   extensions            [13] Extensions {bound}          OPTIONAL,
   ...,
   smsReferenceNumber    [14] CallReferenceNumber         OPTIONAL,
   mscAddress            [15] ISDN-AddressString          OPTIONAL,
   sgsn-Number           [16] ISDN-AddressString          OPTIONAL,
   ms-Classmark2         [17] MS-Classmark2               OPTIONAL,
   gPRSMSClass           [18] GPRSMSClass                 OPTIONAL,
   iMEI                  [19] IMEI                         OPTIONAL,
   calledPartyNumber     [20] ISDN-AddressString          OPTIONAL
   }
```

Figure 6.40   Inclusion of IMEI and MS class information in IDP-SMS

parameters. The handling of IMEI and MS class information for SMS is equivalent to the handling of these parameters in CS call control.

6.4.2 Mobile-terminated SMS Control

For CAMEL phase 3, the support of MO-SMS control is a mandatory feature for the MSC or SGSN. That means that, if an MSC indicates in the location update procedure to the HLR that CAMEL phase 3 is supported, then that MSC will support CAMEL control of MO-SMS. For CAMEL phase 4, support of control of MT-SMS is optional. The MSC will report in the location update procedure to the HLR whether CAMEL phase 4 is supported. If the MSC supports CAMEL phase 4, then the MSC will also report which CAMEL features it supports. One of theses features may be control of MT-SMS. Figure 6.41 reflects the architecture for CAMEL control of MT-SMS.

Short messages may be delivered to a subscriber via an MSC or an SGSN, depending on the registration state of the subscriber, i.e. whether the subscriber is registered with an MSC, attached to an SGSN or both. The choice of SMS delivery via MSC or SGSN also depends on the capability of the network. The MSC or SGSN through which the SMS is delivered to the subscriber may have an smsSSF. The smsSSF is the interface between gsmSCF and MSC or SGSN. The smsSSF in the MSC is identical to the smsSSF in the SGSN.

SMS delivery applies the *optimal routing* principle. This entails that the SMS is sent from the SMSC of the originator of the SMS, directly to the MSC or SGSN of the receiver of the SMS, without passing through the HPLMN of the receiver of the SMS (Figure 6.42).

In the example in Figure 6.42, a Canadian subscriber is roaming in Australia. Subscribers of various other networks may send an SMS to that roaming subscriber. These SMSs are sent directly from the SMSC of the initiator of the SMS to the MSC in Australia where the Canadian subscriber is currently registered. The various SMSCs contact the HLR in Canada to obtain the MSC address of the destination subscriber. The SMS itself, however, does not pass through an SMSC of the HPLMN in Canada.

6.4.2.1 Normal Sequence Flow for SMS Delivery

SMS delivery originates at the SMSC. For point-to-point SMS, the sender of the SMS normally submits the SMS to the SMSC of her HPLMN. However, when the sender of the SMS is outside

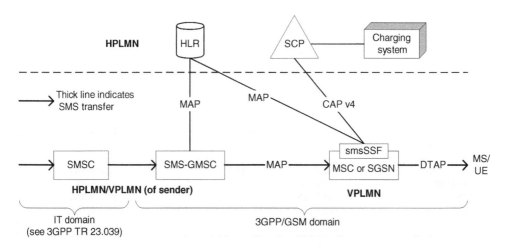

Figure 6.41   Architecture for MT-SMS control

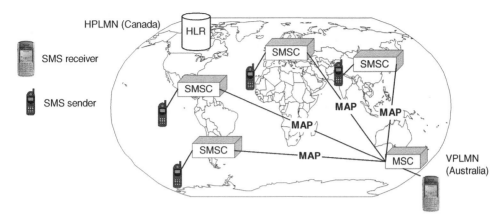

Figure 6.42   SMS delivery to roaming subscriber

her HPLMN, then she may also submit the SMS to an SMSC in that VPLMN. Most operators do not allow submission of an SMS to an SMSC other than their own SMSC.

The SMS-GMSC forms the interface between the SMSC and the 3GPP/GSM core network. The interface between SMSC and the SMS-GMSC is not specified in 3GPP. Commonly used protocols for interfacing to the SMSC are defined in 3GPP TR 23.039 [78]. The SMS-GMSC is often integrated with the SMSC. For delivering an SMS to the intended recipient of that SMS, the SMS-GMSC uses MAP signalling to contact the HLR of that recipient. The sending of the MAP message from SMS-GMSC to HLR is based on the MSISDN of the destination subscriber for the SMS. The MSISDN of the destination subscriber is used as GT to address the HLR. A signalling transfer point (STP) in the HPLMN of the destination subscriber takes care of translating the GT into the signalling point code (SPC) of the HLR for this subscriber.[17]

When the HLR has returned the IMSI of the subscriber and the MSC address or SGSN address, the SMS-GMSC may forward the SMS to the MSC or SGSN of the destination subscriber. The MSC or SGSN delivers the SMS to the subscriber, provided that the MS of that subscriber is switched on and there are no other conditions that prevent delivery of the SMS to that subscriber. This sequence flow is presented in Figure 6.43.

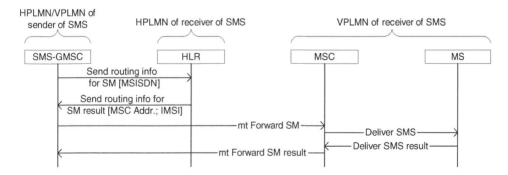

Figure 6.43   MAP signalling for successful SMS delivery via MSC

---

[17] Mobile number portability may have the result that an additional entity, the signalling relay function, is used to ensure that the MAP message arrives at the correct HLR.

The MAP messages send routing info for SM and MT forward SM are specified in 3GPP TS 29.002 [103]. Details of the delivery of the SMS over RAN, between MSC and MS or between SGSN and MS, are specified in 3GPP TS 24.008 [94] and 3GPP TS 24.011 [95].

6.4.2.2 Subscription Data for CAMEL Control of SMS Delivery

The HLR may send MT-SMS-CSI to the MSC during registration of the subscriber in the MSC or SGSN. Normal CAMEL negotiation rules apply; this implies that the MSC or SGSN reports the supported CAMEL phases and supported CAMEL phase 4 CSIs to the HLR. If the MSC or SGSN indicates to the HLR that it supports CAMEL phase 4, including support for MT-SMS, then the HLR may send MT-SMS-CSI to that MSC or SGSN. Table 6.12 shows the contents of MT-SMS-CSI.

6.4.2.3 State Model

The invocation of a CAMEL service for SMS delivery includes the instantiation of a state model in the smsSSF. Delivery of the SMS, including CAMEL control, occurs in accordance with the state transitions that are defined in this state model. Figure 6.44 shows the state model. The meaning of the DPs is described in Table 6.13.

When an SMS arrives in the MSC or SGSN, MT-SMS-CSI is used to invoke the CAMEL service. The address of the gsmSCF is retrieved from MT-SMS-CSI. The CAMEL service starts in DP SMS_Delivery_Request. At this point in the state model, the subscriber has not yet been paged. The location information that is reported in IDP-SMS is therefore not the *current location*. Rather, it is the location information that is currently stored in the VLR or SGSN. As a further

Table 6.12 Contents of MT-SMS-CSI

Element	Description
gsmSCF address	Global title of the gsmSCF to which CAP IDP-SMS will be sent
Service key	The service key (SK) identifies the CAMEL service to be invoked. The SK will be included in CAP IDP-SMS
Trigger detection point	Indication of the DP in the SMS state model where triggering will take place. This parameter will be set to SMS_Delivery_Request
CAMEL capability handling	Protocol to be used for CAMEL control of MT-SMS. This parameter will be set to CAP v4
Default SMS handling	The default SMS handling indicates what action the smsSSF will take in the case of communication failure between smsSSF and gsmSCF
Trigger conditions	Set of conditions that must be fulfilled before triggering takes place
CSI State	This parameter indicates whether the CSI is active or not. See Section 5.8 for explanation of this parameter
Notification flag	This parameter indicates whether a notification will be sent to gsmSCF when the CSI changes. See Section 5.8 for explanation of this parameter

---

[18] Call barring for MT-SMS is checked in the HLR of the destination subscriber, not in the MSC or SGSN.

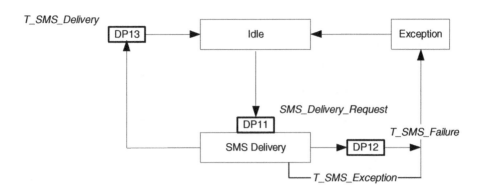

Figure 6.44 State model for CAMEL control of MT-SMS. Reproduced from 3GPP TS 23.078 v5.9.0, Figure 7.3, by permission of ETSI

Table 6.13 Detection points for the MT-SMS state model

Detection Point	Description
SMS_Delivery_Request	The MSC or SGSN has received an SMS from the SMSC. The CAMEL service is started[18]
T_SMS_Failure	A failure has occurred in the SMS delivery. The failure may have occurred internally in the MSC or SGSN or externally, e.g. in the MS
T_SMS_Delivery	The SMS is successfully delivered to the MS

consequence, the MSC or SGSN does not report the IMEI or MS class information to the SCP for MT SMS control.[19]

When the CAMEL service for MT-SMS control is invoked, its behaviour is similar to that of the CAMEL service for MO-SMS control. The only control that the SCP may assert on the SMS delivery is modifying the CgPN. The SCP does not have the capability to forward the SMS to another destination. The SCP may instruct the smsSSF to send a notification when the SMS is delivered or when delivery has failed. Figure 6.45 shows a typical signal sequence between smsSSF and gsmSCF for successful SMS delivery through MSC.

In Figure 6.45, the SMS delivery event is armed as EDP-R. When that event has occurred and been reported, the smsSSF FSM transits to the state 'waiting for instructions'. The gsmSCF may now use a furnish charging information SMS to place free-format data in the CDR. The gsmSCF will finally respond with CAP continue SMS in order to close the CAMEL dialogue.

If the SMS could not be delivered to the subscriber due to reasons such as no paging response, SIM card full, etc., then a T_SMS_Delivery event is reported to the gsmSCF. After the gsmSCF has responded with CAP Continue SMS, the MSC or SGSN returns *mt Forward SM error* to the SMS-GMSC, indicating the reason for failure. If the SMS that could not be delivered has a *validity*

---

[19] IMEI and MS classmark may, however, be present in the CDR related to MT-SMS. These elements may be obtained after paging the subscriber.

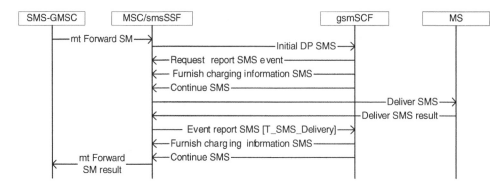

Figure 6.45 CAP signalling for successful SMS delivery via MSC

period larger than 0, then the SMSC places the SMS in a 'failed SMS queue'. In addition, the SMS-GMSC will request that the HLR set a monitor in the MSC or SGSN. If, for example, the SMS delivery failed due to SIM card full and the MSC detects that this condition is no longer prevalent, then the SMSC-GMSC is notified of the removal of the delivery failure condition. The SMSC-GMSC will now re-attempt to deliver the pending SMS. The re-attempt to deliver the SMS will result in re-invocation of a CAMEL service. Figure 6.46 shows an example whereby the gsmSCF bars delivery of an SMS due to insufficient credit.[20]

The cause code that is included in CAP Release SMS is used to set the value of the MAP error code that is returned to the SMS-GMSC. Table 6.14 lists the error values that may be returned to the SMS-GMSC, as a result of a release SMS from the gsmSCF.

6.4.2.4 Conditional Triggering

CAMEL control of MT SMS relates to two types of SMS (see also the section on MO-SMS in Chapter 5):

- *SMS-Deliver* – this type of SMS is a regular MT SMS. The SMS may be point-to-point or may be network generated, e.g. an information SMS such as stock exchange information or sports result.

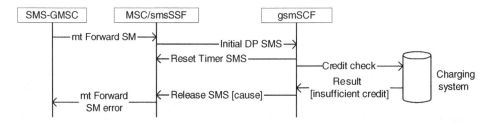

Figure 6.46 SMS delivery failure due to insufficient credit

---

[20] The sender of an SMS may set the validity period when creating the SMS on her mobile phone, depending on the capabilities of the mobile phone. Machine-generated 'content SMS' with information that has a short life span, should have an equally short validity period. This prevents content SMS being delivered to a subscriber when the content of that SMS is no longer valuable.

Table 6.14 MAP error values for CAMEL SMS barring

Value	Identifier	Comment
0	Memory capacity exceeded	The CAMEL service should not use this error, as it may have the effect that the SMSC places the SMS in a queue for deferred delivery attempt
1	Equipment protocol error	This error may be used. This is probably the most adequate error value to be returned
2	Equipment not SM equipped	This error may be used
3	Unknown service centre	This error is used for MO-SMS; the MT-SMS CAMEL service should not use this error cause
4	Service centre congestion	This error is used for MO-SMS; the MT-SMS CAMEL service should not use this error cause
5	Invalid SME address	This error is used for MO-SMS; the MT-SMS CAMEL service should not use this error cause
6	Subscriber not service centre subscriber	This error is used for MO-SMS; the MT-SMS CAMEL service should not use this error cause.

- *SMS-Status-Report* – this type of SMS contains an indication that a previously submitted SMS has arrived at its destination, e.g. a mobile subscriber. A status report may be requested by the initiator of an SMS.

Both SMS types are sent from SMSC to MS and may result in the triggering of a CAMEL service. An operator may, however, limit the triggering of a CAMEL service to one SMS type only, e.g. SMS-Submit. This is accomplished by including trigger criteria in MT-SMS-CSI. If MT-SMS-CSI does not contain trigger criteria, then triggering of a CAMEL service for MT-SMS is unconditional. The smsSSF includes the SMS type in CAP IDP-SMS. Hence, the CAMEL service may adapt its behaviour to the SMS type, e.g. the CAMEL service may not charge for SMS status report.

6.4.2.5 Charging

SMS reception is normally free of charge. The MSC or the SGSN where the subscriber is registered when receiving an SMS may generate a CDR for the SMS delivery. The operator will, however, not charge the subscriber's HPLMN operator for the SMS delivery. SMS may, however, be used for the transfer of chargeable information. A CAMEL service may use the SMSC address and the CgPN from initial DP SMS to determine that an SMS will be charged.[21]

The invocation of a CAMEL service for SMS delivery may result in the generation of an SMS CDR. Refer to 3GPP TS 32.205 [117] (SMS-MT record, for MSC-based SMS) and 3GPP TS 32.215 [118] (S-SMT-CDR, for SGSN-based SMS). These CDRs may include CAMEL information. The CDR that is generated in the MSC or SGSN may be correlated with a service CDR that is created by the SCP. This correlation may be done with the SMS reference number (SRN); the SRN is generated by the MSC or SGSN and included in the CDR and included in CAP IDP-SMS.

6.5 Mobility Management

CAMEL phase 4 introduces mobility management for GPRS, also known as PS mobility management. Figure 6.47 presents the architecture for PS mobility management.

---

[21] It may occur that the calling line identity is removed in ISUP signalling transit networks. It is, however, unlikely that the CgPN is removed in MAP signalling for SMS transfer.

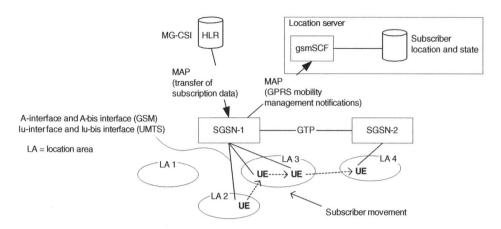

Figure 6.47 Architecture for GPRS mobility management

Analogous to CS mobility management, the PS mobility management is used to receive notifications from the PS core network, related to the mobility state of a subscriber. PS mobility management is subscription-based, using the MG-CSI subscription element in the HLR. MG-CSI may be sent to the SGSN when a subscriber attaches to the SGSN, i.e. during GPRS location update. The conditions for sending MG-CSI from HLR to SGSN are as follows:[22]

- the SGSN reports to the HLR that it (the SGSN) supports CAMEL phase 4;
- the SGSN reports to the HLR that it supports MG-CSI;
- the subscriber has MG-CSI in HLR.

6.5.1 Subscription Data

MG-CSI has the same structure as M-CSI,[23] which is described in Chapter 5. When a subscriber has both M-CSI and MG-CSI, then it is likely that the M-CSI and the MG-CSI contain the same gsmSCF address. This has the result that the CS mobility management event notifications and the PS mobility management event notifications are directed to the same SCP. Alternatively, M-CSI and MG-CSI may have different gsmSCF addresses, that are translated into the same signalling point code of the location server.

MG-CSI has a different set of mobility management events that may be reported to the gsmSCF than M-CSI. Table 6.15 lists the PS mobility events that may be included in MG-CSI. The event value is the enumerated value that is used in the event notification, to denote the event that has occurred.

The RAU events may occur in idle mode or when the subscriber has one or more PDP contexts active. The sending of a mobility management event notification is subject to the presence of MG-CSI in the SGSN and the marking in MG-CSI of the related mobility management event. Should a subscriber perform an inter-SGSN RAU to an SGSN that does not support CAMEL phase 4, then the HLR will not send MG-CSI to that SGSN and consequently, there will not be a 'routing

---

[22] There may be additional, operator-specific conditions in HLR that prevent the sending of MG-CSI to the SGSN.

[23] The ASN.1 definition of M-CSI is identical to the ASN.1 definition of MG-CSI.

[24] The 'SGSN service area' is the part of the RAN served by an SGSN; the expression is not related to the 'service area' being a group of cells in a 3G RAN.

Table 6.15 GPRS mobility management events

Event value	Mobility event identifier	Description
128	Routing area update in same SGSN	The subscriber performs a RAU within the service area of an SGSN[24]
129	Routing area update to other SGSN – update from new SGSN	The subscriber performs a RAU to another SGSN. The subscriber has moved from the service area of one SGSN to the service area of another SGSN. The old SGSN will not send a mobility management event notification
130	Routing area update to other SGSN – disconnect by detach	The subscriber detaches from the SGSN and re-attaches to another SGSN. The SGSN that the subscriber was previously attached to may have sent a mobility management event notification at the time of detach
131	GPRS attach	The subscriber attaches to an SGSN or the subscriber performs an RAU after the SGSN had marked this subscriber as 'network initiated transfer to MS not reachable for paging'. The SGSN may have sent a mobility management event at the time of detach or when the SGSN marked the subscriber as 'MS not reachable for paging'
132	MS-initiated GPRS detach	The subscriber initiates a detach from the SGSN, i.e. the subscriber switches her MS off
133	Network-initiated GPRS detach	The SGSN detaches the subscriber. This may occur when the subscriber is marked 'network initiated transfer to MS not reachable for paging' for a pre-defined period of time, or the HLR instructs the SGSN to detach the subscriber
134	Network-initiated transfer to MS not reachable for paging	The subscriber has not performed a RAU for a pre-defined period of time.

area update to other SGSN – update from new SGSN' event (129) notification. However, it may be assumed that, within one PLMN, all SGSNs have the same level of CAMEL support.

Not all intra-SGSN location updates may be reported to the gsmSCF. This depends on the access type, 2G vs 3G.

6.5.1.1 2G Radio Access

The location update between cells belonging to the same location area is not reported to SGSN. Hence, such location updates do not result in notification to gsmSCF. The location update between cells belonging to different location areas is reported to SGSN. Such location updates may result in notification to gsmSCF.

6.5.1.2 3G Radio Access

The location update between cells belonging to the same service area is not reported to the SGSN either. Such a location update does not result in notification to gsmSCF. The location update between cells belonging to different service areas belonging to the same location area is not reported to the SGSN either. Such a location update does not result in notification to gsmSCF. The location update between cells belonging to different service areas belonging to different location areas is reported to the SGSN. Such a location update may result in notification to gsmSCF.

6.6 Any-time Interrogation

In CAMEL phases 1–3, ATI is restricted to retrieval of subscriber information from the CS domain. In CAMEL phase 4, ATI may also be used for retrieval of subscriber information from the PS domain. At the same time, ATI for the CS domain is enhanced. When gsmSCF sends ATI to HLR, it may include the domain, i.e. CS or PS. If no domain indication is provided, then CS domain is assumed. The gsmSCF cannot request information from both domains in one ATI request.

6.6.1 ATI for CS Domain

ATI for the CS domain is enhanced with the capability to obtain additional information. The gsmSCF may request the HLR to obtain IMEI and MS classmark from the VLR (Figure 6.48). The gsmSCF will indicate in the ATI request which information elements are required:

- location information – this is part of CAMEL phase 1; location information in ATI request may be accompanied by current location for the purpose of ALR; ALR is part of CAMEL phase 3;
- subscriber state; this is part of CAMEL phase 1;
- IMEI;
- MS classmark.

6.6.1.1 IMEI

The MSC may read the IMEI from internal subscriber information store. If the MSC does not have the IMEI of this subscriber available, then it uses the DTAP message *identity request* to obtain the IMEI or IMEISV. The *identity request* DTAP message may be used to request various identifiers, amongst which are the IMEI or IMEISV. The MSC should in this case request the IMEISV. The MS may, in response, return IMEI or IMEISV.

6.6.1.2 MS Classmark

The MS reports its MS classmark 2 to the MSC every time a call is established, an SMS is transferred, etc. The MS classmark 2 may also be sent to the MSC at attach or at period location update.[25] Hence, the MSC may have the requested MS classmark 2 available in the internal subscriber information store. Otherwise the MSC will request the UE to report its MS classmark 2.

If the subscriber's MS is switched off and the MSC does not have the IMEI of this subscriber available, then the MSC does not report the requested IMEI. When the gsmSCF requests IMEI or MS classmark, it may be advisable to also request the subscriber state. If IMEI or MS classmark is not provided and the reported subscriber state indicates *not reachable*, then the gsmSCF may deduce that the IMEI or MS classmark could not be provided due to the not reachable condition.

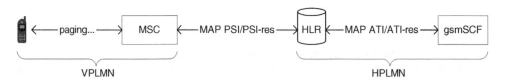

Figure 6.48   Retrieval of IMEI and MS classmark of subscriber

---

[25] This applies for UMTS network, but not for GSM network.

Although the capability to provide IMEI and MS Classmark in MAP PSI-res is an optional
CAMEL phase 4 feature, this capability of the MSC is not explicitly indicated in the *offered
CAMEL4 CSIs* parameter in MAP update location message to HLR. If the MSC does not support
this capability, then it will ignore a request for IMEI or MS classmark. The gsmSCF may for this
reason accompany a request for IMEI or MS classmark with a request for location information. If
IMEI or MS classmark is not provided and location information indicates that subscriber is in a
VPLMN, then the gsmSCF may deduce from the VLR address that the MSC was not capable of
providing IMEI or MS classmark.

6.6.2 ATI for PS Domain

ATI for the PS domain has a similar information flow to ATI for CS domain (Figure 6.49). The
support of MAP PSI by the SGSN is explicitly indicated to the HLR during GPRS location update.
In this manner, the HLR knows whether it can send PSI to the SGSN. If sending PSI to the SGSN
is not possible, then the HLR will return an error when it receives ATI for the PS domain for
this subscriber. In addition, the gsmSCF may use ATSI to obtain the supported CAMEL phases
and CAMEL phase 4 capabilities of the SGSN. The ATSI result may include 'PSI enhancements',
which indicates that the SGSN supports PSI.

When the gsmSCF indicates the PS domain in ATI, the HLR sends MAP PSI to the SGSN of
the subscriber, instead of to the MSC. The following information may be requested from a PS
subscriber:

Location Information
The SGSN shall supply the stored location information of the subscriber. The location information
for GPRS access is described in Chapter 5. If the request for location information is accompanied
by current location, then the SGSN will attempt to page the subscriber, in order to refresh the
location register and return the current location.

Subscriber State
The SGSN will report the state of the PS subscriber. The following states may be reported:

- detached – the subscriber is currently not attached to the SGSN;
- CAMEL attached, MS not reachable for paging – the subscriber is attached, without active PDP
  context, but the SGSN has not received a periodic location update for a pre-defined duration;
- CAMEL attached, MS may be reachable for paging – the subscriber is attached, without active
  PDP context; the subscriber is assumed (by SGSN) to be reachable;
- CAMEL PDP active, MS not reachable for paging – the subscriber is attached with one or more
  active PDP contexts, but there has not been radio contact for a pre-defined duration;
- CAMEL PDP active, MS may be reachable for paging – the subscriber is attached with one or
  more active PDP contexts; the subscriber is assumed (by SGSN) to be reachable;
- Network_Determined_Not_Reachable – the HLR has indicated that the network can determine
  from its internal data that the MS is not reachable.

Figure 6.49 Any-time interrogation for PS domain

If the subscriber is not registered with an SGSN, then the HLR returns the value *network determined not reachable*. If the subscriber has one or more PDP contexts active, then the SGSN will also supply a set of descriptors per active PDP context. The elements that are reported per PDP context include, amongst others, PDP type, PDP address, APN, GGSN address and QoS. When an internet-based application receives a request for information from a user who identifies herself with MSISDN, then ATI may be used to retrieve PDP context information. The application can then verify that the IP address of this user is currently in use by the MSISDN that was provided to the application.

IMEI
The request of IMEI(SV) from SGSN follows the rules that apply for the request of IMEI(SV) from MSC.

GPRS MS Class
The request of GPRS MS class from SGSN follows the rules that apply for the request of MS classmark 2 from MSC.

6.7 Subscription Data Control

Minor enhancements are made to subscription data control in CAMEL phase 4. For a primer on subscription data control, refer to Section 5.8. The following capability is added in CAMEL phase 4.

- *ATSI* – the gsmSCF may request the HLR to return the MT-SMS-CSI and MG-CSI of a subscriber.
- *ATM* – the gsmSCF may request the HLR to modify the subscriber's ODB settings. The gsmSCF may set both the generic ODB categories and the operator-specific ODB categories. ODB categories are not specific per basic service. Hence, the gsmSCF can provide just one set of generic ODB categories and one set of operator-specific ODB categories per subscriber. ODB categories are defined in 3GPP TS 22.041 [63]. The MAP protocol definition of any-time modification is forwards-compatible for new ODB categories that may be introduced in later 3GPP releases.
- *Notify subscriber data change (NSDC)* – no change, compared with CAMEL phase 3.

CAMEL does not specify a mechanism to inform the gsmSCF about the supported ATSI, ATM and NSDC capability in HLR. However, gsmSCF and HLR are normally owned by the same operator. It is an operator's option to use the CAMEL subscription data control MAP messages, in accordance with the capability of the operator's HLR.

Three more CSIs were introduced in CAMEL phase 4, O-IM-CSI, D-IM-CSI and VT-IM-CSI; see section 6.9. These CAMEL subscription elements reside in the home subscriber server (HSS). Refer to 3GPP TS 23.278 [93] for the use of ATSI, ATM and NSDC to/from HSS.

6.8 Mobile Number Portability

Number portability (NP) is a national requirement in many countries; it enables a subscriber of a telecommunication provider to port her service to another telecommunication provider in that country, but retain her telephone number. Mobile number portability (MNP) is the implementation in the GSM network of the NP concept. MNP is introduced in GSM Release R98. NP or MNP mean that a subscriber has a telephone number belonging to one operator (*number range holder network*), while that subscriber is a subscriber of another operator (*subscription network*).[26] At least the following NP cases may be defined:

---

[26] The terms '*number range holder network*' and '*subscription network*' are defined for MNP, but also apply functionally to NP.

- number portability between two GSM networks;
- number portability between two PSTNs;
- number portability between GSM and PSTN.

NP between GSM and PSTN is less common than NP between homogeneous networks. Denmark is one example of a European country that uses mobile-fixed NP.

One aspect of many telecommunications networks is that a subscriber is identified with a *number*, whereby the number is part of a *number range*. The number by means of which a subscriber is identified has various purposes, including:

- *routing* – when a call is established, the routing of that call through the network is based on the number that is used to identify that subscriber, e.g. a call to +31 65 ... is routed to the network from KPN Mobile ('65') in The Netherlands ('+31').[27]
- *network identification* – the number indicates the network that a subscriber belongs to, e.g. a subscriber with number +31 13 ... is associated with a local exchange in a particular region of the PSTN from KPN Netherlands.
- *tariff* – a subscriber from T-Mobile Netherlands may be charged a lower rate for calls to other T-Mobile Netherlands subscribers than for calls to Orange Netherlands subscribers.

The introduction of NP affects the above associations of a telephone number.

6.8.1 Call Routing

MNP is specified in 3GPP TS 22.066 [65] and 3GPP TS 23.066 [82]. When a call is established to a GSM subscriber, the MSISDN that is used to contact that subscriber is no longer an indication of the network that that subscriber belongs to. Hence, the network will take provision that the correct HLR is interrogated for this call. Owing to the nature of MNP, whereby correct handling of calls to ported subscriber requires cooperation between operators, one country normally has one MNP method. The method of MNP in a country may be determined by the national regulator. 3GPP TS 23.066 [82] specifies various methods for routing calls to subscribers who are ported to another GSM network. Two of these methods are described underneath.

6.8.1.1 All Call Query

Refer to Figure 6.50 for a graphical representation of two possible all-call-query implementations in the VMSC. The all call query method means that a VMSC queries a number portability database (NPDB) for all calls that may be subject to MNP. 3GPP TS 23.066 [82] specifies that CS1 may be used for this purpose (see description on IN-based MNP solutions in 3GPP TS 23.066). The VMSC will determine whether the call may be subject to MNP. That implies that the VMSC ascertains whether the called number is part of the 'portability domain'. The portability domain comprises the telephone numbers within a country that may be ported between operators. Normally, the NDC part of a number indicates the operator the number belongs to. 3GPP specifies that the portability domain comprises GSM networks only. However, an operator could equally perform the NP check for numbers that belong to fixed networks. Figure 6.51 contains an example signal flow for an NP

---

[27] This example does not consider BOR.

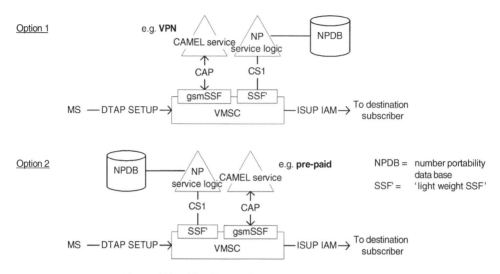

Figure 6.50   All call query for mobile number portability

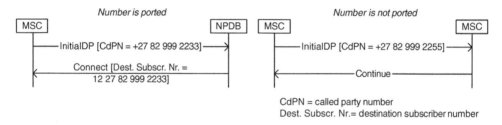

Figure 6.51   NP check for ported and for non-ported number

check for a call to a ported number, as well as an example signal flow for an NP check for a call
to a non-ported number.

The '12' that is returned for the NP check for the query for the call to +27 82 999 2233 is
a routing number (RN). This RN is used to route the call to the subscription network of the
called subscriber. Routing numbers for MNP are nationally standardized; hence the MSC will be
configured to use the RN to select a route to the subscription network.

The MNP check may be performed before CAMEL service triggering or after CAMEL service
triggering; the latter method is more common. Either method has its advantage.

MNP Check Prior to CAMEL Invocation
The RN is obtained before CAMEL service invocation and may be included in the initial DP for
the CAMEL service. This may be useful for CAMEL pre-paid service. The pre-paid service may
use RN to determine the destination network for the call and apply the corresponding tariff.

MNP Check after CAMEL Invocation
If the CAMEL service may modify the destination of the call, then it is advantageous to perform
the NP check after the CAMEL service. If the CAMEL-provided destination of the call is ported
to another network, then the NP check ensures that the RN is obtained to route the call to the
subscription network of the called subscriber.

The effect of performing the NP check at the originating MSC is that, when the call is routed out of the MSC, the call is routed to the correct network; no further NP check is required. For calls that are established in the PSTN, the NP check may be performed by a local exchange or by a transit exchange. This method of performing NP check in the originating network is also known as '*NP check at the source*'. If a call originates from another country, then the international border exchange may perform the NP check.

6.8.1.2 Signalling Relay Function

In countries that do not perform NP checks at the source, a call may arrive at a GMSC of a network that is not the subscription network of the called subscriber. The reason is that an MSC uses the called party number in ISUP IAM to determine whether it will act as GMSC for that call, i.e. contact the HLR for routing information. If the called party number falls within the MSISDN number range of the operator in which network the MSC resides, then the MSC will act as a GMSC. This may have the effect that the GMSC contacts the HLR for a call to a subscriber whose *number* belongs to that operator, but whose *subscription* resides with another operator. Figure 6.52 shows two examples of the related network diagram.

GMSC-A denotes a GMSC of the PLMN that the MSISDN of the called subscriber belongs to. That GMSC decides to send MAP SRI to the HLR. The SS7 network of the operator contains an MNP signalling relay function (SRF). The SRF has the function to intercept the MAP SRI and to verify whether the called subscriber belongs to this network.

If the called subscriber does not belong to this network, then the MNP SRF may return a routing number (RN) to the GMSC. The GMSC uses the RN to route the call to the subscription network of MSISDN-B. A GMSC in MSISDN-B's network performs the same action, i.e. sends the MAP SRI towards the HLR. In this case, the MNP SRF in that network determines that the called subscriber belongs to that network and forwards the MAP SRI to the HLR for that subscriber.

Alternatively, the MNP SRF relays the MAP SRI to the HLR of MSISDN-B in PLMN-B. In that case, GMSC-A may receive an MSRN for the called subscriber. GMSC-A could in this case also receive CAMEL trigger information, such as T-CSI.

The method of relaying ISUP IAM to the subscription network of the called subscriber has the advantage that terminating call handling is performed in the called subscriber's HPLMN. Hence, the operator may apply proprietary call handling mechanisms between GMSC and gsmSCF. However, the method of relaying MAP SRI to the subscription network of the called subscriber results in more efficient call routing through ISUP. The latter method may be compared with BOR.

More call routing mechanisms at the GMSC for number portability are described in 3GPP TS 23.066 [82]. 3GPP TS 23.066 [82] also specifies non-call-related routing for ported subscribers,

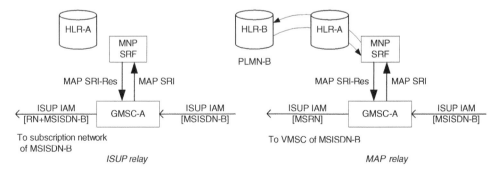

Figure 6.52   HLR interrogation through MNP SRF

such as HLR query by SMS-GMSC, gsmSCF-initiated USSD signalling based on MSISDN, any-time interrogation, etc.

ITU-T also specifies number portability methods; refer to ITU-T Q.769.1 [139], ITU-T Q-series supplement 3 [146], ITU-T Q-series supplement 4 [147] and ITU-T Q-series supplement 5 [148].

6.8.1.3 Flexible HLR Allocation

The MNP SRF in some operators' networks is also used for non-MNP purposes. In those networks, the SRF intercepts all MAP signalling directed to an HLR in that network. Examples of such signalling include:

- *Location update* – the IMSI of the served subscriber is used for HLR addressing.
- *MT call handling* – the MSISDN of the served subscriber is used for HLR addressing.
- *Any-time interrogation* – the IMSI or MSISDN of the served subscriber is used for HLR addressing.

The SRF contains tables that associate an HLR with each IMSI and MSISDN. The operator may move user subscription information from one HLR to another without impacting the MSISDN or IMSI. When moving subscription information to another HLR, the SRF is reconfigured for the new HLR address. This flexible allocation HLR may be useful for: HLR upgrade or replacement; and placing specific subscriber groups, e.g. UMTS subscribers, in a designated HLR.

6.8.2 MNP SRF Query by gsmSCF

For various call cases, an on-line charging system based on CAMEL has to rate a call to a number within the home country's number portability domain. If a call is established to a number that is recognized as an MSISDN, then the CAMEL service may not know whether the call will be routed to a subscriber in the HPLMN or to a subscriber in another PLMN in that country. For these cases, the gsmSCF may query the MNP SRF (Figure 6.53).

The gsmSCF queries the MNP SRF to obtain the subscriber's portability status. If the subscriber is ported, then the MNP SRF returns the RN related to the subscriber's current subscription network. The gsmSCF then uses the RN to determine the rate for a call to that subscriber. The MNP SRF query may also be used for other services, such as SMS on-line charging or VPN. 3GPP TS 23.078 [83] lists the exact information that may be returned by the MNP SRF to the gsmSCF.

The MNP SRF query is done with MAP ATI. ATI is enhanced; the ATI request may contain an indication that MNP information is requested; the ATI result may contain the requested MNP information. The MNP SRF that is queried with MAP ATI may in practice be an NPDB without SRF capability. Although the gsmSCF may obtain the MNP status of the called subscriber, the core

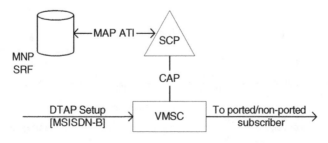

Figure 6.53   MNP SRF query by gsmSCF

network must still apply MNP handling to route the call to the subscription network of the called subscriber.

When the MNP SRF receives MAP ATI, this MAP message may or may not be destined for the MNP SRF. The following cases may be distinguished:

- *MAP ATI is sent to MNP SRF for MNP query* – the gsmSCF uses a translation-type (TT) value in the SCCP signalling towards the MNP SRF, indicating that the MNP SRF will act as the signalling end point for this message. Refer to ITU-T Q.713 [135] for a description of TT. For the MNP SRF query, a national TT value should normally be used. The MNP SRF will unpack the SCCP message and process the MAP message (MAP ATI).
- *MAP ATI is sent to HLR* – the gsmSCF uses a TT value in the SCCP signalling towards the HLR, indicating that the MNP SRF will act as a signalling relay point for this message. The MNP SRF will use its MNP database to forward the SCCP message to the HLR for this subscriber. A similar TT value is used when the gsmSCF sends other MAP messages to the HLR, using MSISDN for addressing the HLR.

The TT value to be used for addressing the MNP SRF as signalling end point is not standardized by ITU-T, but is allocated nationally. In general, when an entity needs to send a message to the HLR and uses IMSI to address the HLR, then there is no MNP issue. When a subscription is ported to another operator, the MSISDN may be ported to the other operator, but not the IMSI. Hence, the IMSI always reliably indicates the operator a subscriber belongs to.

6.8.3 Non-standard MNP Solutions

Some networks use non-standard solutions for MNP. One example involves the enhancement to CAMEL service invocation (Figure 6.54). In Figure 6.54, the STP intercepts the CAP IDP directed to the SCP. The CAP IDP may in fact travel through an STP in the HPLMN of the served subscriber in any case, for the translation of the GT of the gsmSCF to the SPC of the SCP. Before forwarding the CAP message to the SCP, the STP unpacks the SCCP message and, if needed, queries the MNP SRF. If the called party BCD number contains a number of a ported subscriber, then the

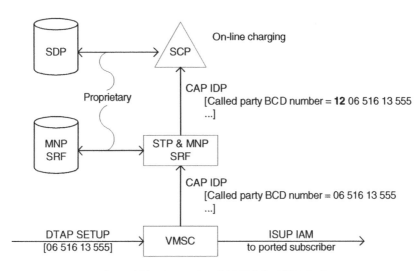

Figure 6.54   Intercepting CAP IDP for MNP check

MNP returns the RN for that subscriber. The STP places the RN in front of the called party BCD number and passes the CAP IDP on to the SCP. The SCP now receives

CAP IDP [Called Party BCD Number = **12** 06 516 13 555; . . .]

instead of

CAP IDP [Called Party BCD Number = 06 516 13 555; . . .].

The SCP may parse this destination number through the rating table in the SDP, which determines the cost of the call. The RN prefix (in this example RN = 12) for this number is taken into account by the SDP. Since the destination number is modified in CAP IDP, but not in ISUP signalling, the network must still take care of applying MNP for routing the call to the ported subscriber.

6.9 Control of IP Multimedia Calls

CAMEL control of IP multimedia calls forms a bridge between the CS mobile network and the IP multimedia network. The IMS is the IP-based communication system for mobile networks. Although IMS, which was introduced in 3GPP release Rel-5, is specified for the mobile network, it may also be used for wireline networks. In fact, the nature of IMS facilitates arbitrary access methods to be used. Table 6.16 specifies the main 3GPP technical specifications for IMS.

The highlighted technical specifications are the CAMEL specifications for IMS control. The reader may consult the specifications listed in Table 6.16 for a description of IMS. Many aspects of IMS are defined by the IETF. The standards that are published by the IETF are contained in *request for comments* (RFC). 3GPP TS's refer to RFCs for many of the GPRS- and IMS-related functionality. RFCs may be obtained from www.ietf.org. A good tutorial on IMS is *The*

Table 6.16 3GPP specifications for IMS

3GPP TS	Title
22.228 [70]	Service requirements for the IMS
22.250 [71]	IP multimedia subsystem group management
22.340 [72]	IP multimedia subsystem messaging
23.228 [91]	IP multimedia subsystem
23.278 [93]	**CAMEL – IP multimedia system interworking**
24.228 [96]	Signalling flows for the IP multimedia call control based on session initiation protocol and session description protocol
24.229 [97]	Internet protocol multimedia call control protocol based on session initiation protocol and session description protocol
29.228 [108]	IP multimedia subsystem Cx and Dx interfaces; signalling flows and message contents
29.229 [109]	Cx and Dx interfaces based on the diameter protocol; protocol details
29.278 [111]	**CAMEL application part specification for IP multimedia subsystems**

3GPP TS 24.228 is discontinued after Rel-5.

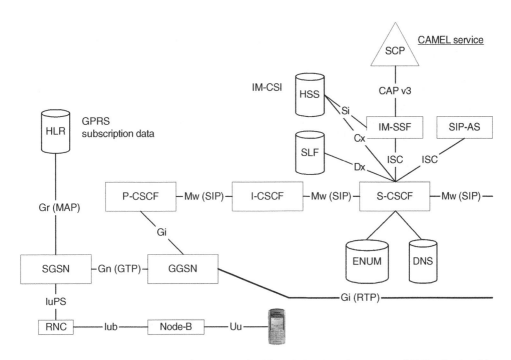

Figure 6.55   Architectural overview of IMS network with CAMEL control. *Note*: not all IMS-related entities and interfaces are reflected

3G IP multimedia subsystem (IMS): merging the internet and the cellular worlds [169]. Figure 6.55 contains a network overview of IMS, including the CAMEL-specific entities.

The radio network and core network entities (Node-B, RNC, SGSN, HLR, GGSN) are described in previous chapters. The other IMS-related entities are described underneath. IMS subscribers may use GPRS to establish a PDP context. The PDP context may be used to carry the session initiation protocol (SIP) messages to and from the IMS subscriber. SIP is used to establish a call between the calling party and the called party. The calling and called party may exchange their respective IP addresses and codec information; the call parties may then exchange payload through the internet, using the exchanged IP addresses.

HSS
The home subscriber system (HSS) is the subscription database for IMS. IMS subscribers' subscription data is permanently stored in the HSS. The HSS is often considered the successor of the HLR. That implies that HLR is contained in HSS. However, the IMS operator does not need to be the same as the GPRS/UMTS operator. In that case, HLR and HSS are owned and operated by different operators.

SLF
The subscriber locator function (SLF) is a node that maintains the link between subscriber and HSS. SLF(s) may be used in a network that has multiple HSSs; when an entity such as I-CSCF or SIP-AS needs to contact the HSS, that entity needs to contact the SLF first, to find out in which HSS the subscriber's data is stored.

S-CSCF
The serving call session control function (S-CFCF) is the main traffic control node for the IMS network. A subscriber that is 'IMS-registered' has registered herself with an S-CSCF in the home

network of the IMS operator. The S-CSCF receives subscription data from HSS at the time of IMS registration. When the subscriber initiates an IMS session or when an IMS session is initiated towards the subscriber, the SIP signalling passes through the S-CSCF.

P-CSCF
The Proxy-CSCF (P-CSCF) serves as the entry point into the IMS network. In the case of IMS via GPRS, the P-CSCF forms the link between the GPRS network and the IMS domain. The P-CSCF is always located in the same network as the GGSN. As described in Section 5.3, when a GPRS subscriber establishes a PDP context (PDPc), the GGSN for that PDPc may be located in the VPLMN or in the HPLMN of that subscriber. As a result, the P-CSCF may also be located in the VPLMN or in the HPLMN.

I-CSCF
An I-CSCF may be used during IMS registration, to ensure that the subscriber's subscription data from her designated HSS is transferred to the S-CSCF. An operator may also use the I-CSCF for the purpose of *topology hiding*. Topology hiding means that addresses of entities of that operator's network are shielded from other networks.

DNS
The domain name server (DNS) is the IP entity that may be queried for translating a SIP URI to an IP address. The IP address may be needed to forward an IMS call to the destination subscriber's network.

ENUM Server
The ENUM Server enables an operator to establish calls in the IMS network, whereby the destination subscriber is identified with a telephony universal resource locator (TEL URL), rather than with a SIP URI. This is an important capability for an IMS network, since many users of an IMS network will be addressable with a TEL URL, in conjunction with a SIP URI. One subscriber may be addressed as follows:

sip: wendy.smith@steelworks.com; or
tel: +44 20 356 7856.

In the case that the subscriber is addressed with her E.164 number, the ENUM Server converts the E.164 number into URI. The URI may subsequently be provided to the DNS in order to obtain an IP address for routing a call to that subscriber. The URI associated with the above example number would be:

6.5.8.7.6.5.3.0.2.4.4.e164.arpa[28]

SIP-AS
A SIP-application server (SIP-AS) is an application server for the IMS network. IMS calls that are handled by a S-CSCF may be routed through a SIP-AS. The SIP-AS may assert control over an IMS call, comparable to the way in which an SCP may assert control over a CS call. The interface between the S-CSCF and the SIP-AS is the IMS service control (ISC) protocol. The criteria for routing an IMS call through a SIP-AS are received from the HSS, during IMS registration. These criteria, the initial filter criteria (IFC), indicate which IMS calls shall be routed through one or more SIP-AS's.

---

[28] ARPA = Advanced Research Projects Agency, the organization that developed the ARPA network (ARPANET), the predecessor of the Internet.

IM-SSF

The IM-SSF forms the link between the IMS network and the CAMEL service environment; it is the entity that allows for running CAMEL services in an IMS network. The functioning of the IM-SSF will be described in more detail in Section 6.9.2.

6.9.1 Rationale of CAMEL Control of IMS

CAMEL call control is deployed in many GSM networks worldwide. When a GSM network operator deploys an IMS network, that operator may want to utilize the investment in CAMEL-based service network also for the IMS network. In this manner, one service platform may control at the same time a GSM network and an IMS network. Depending on network configuration, a subscriber may have both GSM access and IMS access. When the subscriber is in the office, she is registered as an IMS user via the company's LAN, through which IP connectivity is obtained. The subscriber may then use her IMS registration for establishing outgoing calls and receiving incoming calls. When that same subscriber is not in the office, she may use GSM access for outgoing and incoming calls. In both cases, a CAMEL service may control the call for this subscriber. This is reflected in Figure 6.56. Not all entities or subsystems such as RAN, HLR, HSS, P-CSCF, I-CSCF, etc., are included in the figure.

For CAMEL control of IMS calls, the IM-SSF functions as an interface (protocol converter) between IMS and GSM. The control capability that is available for IMS calls is not the same as the control capability that is available for GSM calls. The CAP protocol that is used by the IM-SSF has a reduced capability set compared with the CAP protocol that is used in GSM.

In the example of Figure 6.56, when the subscriber is attached to the GSM network, she may establish calls by using an E.164 number (MSISDN number, PSTN number, etc.) as destination address. Likewise, the subscriber is reachable with her MSISDN. When the subscriber is registered as IMS user on her office/home computer, she may establish calls by using a URI, such as john.smith@steelworks.com. She may also use a Tel URL for setting up a call, e.g. +27 83 212 65498. The subscriber is reachable on her IMS client by means of the URI under which she is registered, e.g. wendy.smith@steelworks.com or through an E.164 number.

When a subscriber has both GSM access and IMS access, the operator can determine whether the subscriber is reachable with MSISDN, URI or both. The service network, containing the SCP, can determine whether a call for that subscriber will be delivered to her GSM phone or to her IMS

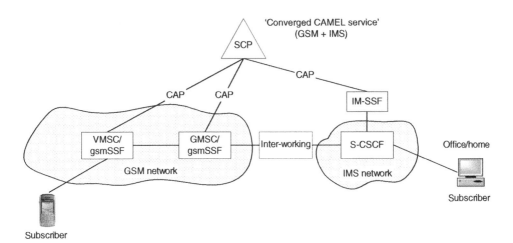

Figure 6.56 Converged CAMEL service (GSM + IMS)

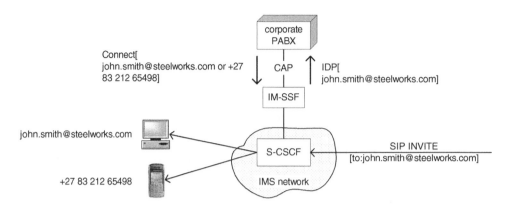

Figure 6.57 Corporate PABX for IMS access and GSM access

phone. This may depend on the status of the subscriber (e.g. IMS registered, GSM attached etc.), on her location (in the office, at home, etc.) and subscription settings (Figure 6.57).

6.9.2 The IM-SSF

The IM-SSF is the interface between the IMS domain and the CAMEL service environment. It converts between ISC and CAP. ISC is the protocol that is used in the IMS network, between the S-CSCF and a SIP-AS. The SIP-AS is the entity in the service layer of the IMS network; IMS calls may be routed through a SIP-AS. The SIP-AS may apply operator-specific control to a call. The SIP-AS may therefore be compared with the SCP in the GSM network.

The IM-SSF acts as SIP-AS towards the IMS core network. At the same time, the IM-SSF acts as gsmSSF towards the gsmSCF. The gsmSSF in the IM-SSF is referred to as 'imcnSSF'. The imcnSSF may invoke an O-IM-BCSM instance or a T-IM-BCSM instance for an IMS call. O-IM-BCSM and T-IM-BCSM are IM-SSF equivalent of O-BCSM and T-BCSM respectively. Figure 6.58 depicts the interfaces of the IM-SSF.

Si

The IM-SSF uses the Si interface to obtain subscription data of a subscriber. The Si interface is specified in 3GPP TS 23.278 [93]. It consists of the MAP messages ATSI and NSDC. The Si

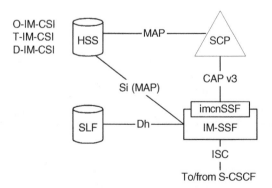

Figure 6.58 Interfaces to/from IM-SSF

interface is used when an IMS subscriber registers with the IM-SSF and when CAMEL subscription data in the HSS changes.

Dh

This is an interface that may be used by IM-SSF to find out which HSS to contact for subscription information of a particular subscriber. The Dh interface is specified in 3GPP TS 23.228 [91].

ISC

The ISC interface between the S-CSCF and the IM-SSF is used during IMS registration and during IMS call handling. ISC is specified in 3GPP TS 23.228 [91].

CAP v3

The IM-SSF may invoke a CAMEL service in the gsmSCF, using CAP v3. CAP v3 is the only version of CAP that may be used by the IM-SSF. The CAP v3 that is used for IMS control is derived from the CAP v3 that is used for call control, but is not identical. It will be described in detail in Section 6.9.5.

MAP

As an operator option, MAP may be used between the gsmSCF and the HSS e.g. to obtain subscription data (O-IM-CSI, VT-IM-CSI, D-IM-CSI). MAP messages for gsmSCF are described in 3GPP TS 23.078 [83] and 3GPP TS 23.278 [93].

6.9.3 Registration

When a subscriber registers as IMS user, the S-CSCF receives *initial filter criteria* (IFC) from the HSS. IFC consists of criteria that define when the S-CSCF will route an IMS session through a SIP-AS. IFC are subscriber-specific. Refer to 3GPP TS 23.228 [91] for the procedures that are used to send the IFC from HSS to S-CSCF. The IFC have the following purposes for CAMEL control of IMS.

6.9.3.1 Registration

When the S-CSCF receives IFC from HSS at IMS registration, it analyses the IFC and determines whether it will send a SIP Register method[29] to a SIP-AS. For CAMEL subscribers, the IFC will contain an indication that the S-CSCF will send a SIP Register method to the IM-SSF. The IP address of the IM-SSF is included in the IFC. From the S-CSCF point of view, the IM-SSF is a SIP-AS.

When the IM-SSF receives the SIP Register method, it will retrieve the subscriber's CAMEL data from the HSS. The IM-SSF sends MAP ATSI to the HSS and receives one or more of O-IM-CSI, VT-IM-CSI and D-IM-CSI. The IM-SSF stores the received CAMEL subscription data internally. The IM-SSF may need to contact the SLF to obtain the address of the HSS for this subscriber (Figure 6.59).

The IM-SSF uses IMSI to identify the subscriber to the HSS. This requires that the IMSI of the subscriber is carried in the SIP Register method to the IM-SSF. The IMSI is associated with *mobile* subscribers. If the IMS subscriber registers through a PC, then no IMSI will be available. In that case, an adaptation to the IM-SSF registration is needed to identify the subscriber.

When a subscriber has registered with an IM-SSF, the HSS stores the address of that IM-SSF. That enables the HSS to send CAMEL data to that IM-SSF, in the case that the CAMEL data for that subscriber has changed. The HSS uses the notify subscriber data change MAP message for that purpose.

---

[29] SIP messages are referred to as 'methods'.

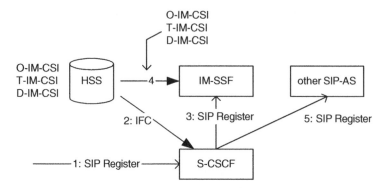

Figure 6.59   Registration at the IM-SSF

6.9.3.2 Call Control

When the IMS subscriber establishes or receives a call, the S-CSCF uses the IFC, which it received during registration, to determine whether call control will be given to a SIP-AS. For CAMEL subscribers, the IFC will contain the address of the IM-SSF where the subscriber was registered during IMS registration. The IFC and the CAMEL IMS data will be further aligned as follows:

- If a subscriber has O-IM-CSI and/or D-IM-CSI, then the IFC for that subscriber should contain settings that indicate that the S-CSCF shall route the SIP Invite for an originating call, to the IM-SSF.
- If a subscriber has VT-IM-CSI, then the IFC for that subscriber should contain settings that indicate that the S-CSCF will route the SIP Invite for a terminating call, to the IM-SSF.

6.9.4 IMS Call Control

Figure 6.60 reflects the signalling that takes place when an IMS subscriber who is also registered with the IM-SSF, originates an IMS call. The subscriber is already registered with IM-SSF, so IM-SSF has stored the CAMEL data for the subscriber; therefore, the IM-SSF does not need to contact the HLR during call handling.

When the IM-SSF receives the SIP Invite from the S-CSCF, it invokes an instance of the O-IM-BCSM or an instance of the T-IM-BCSM. If the SIP Invite relates to an *originating* IMS call, then an O-IM-BCSM instance is created; otherwise, if the SIP Invite relates to a *terminating* IMS call, then a T-IM-BCSM instance is created. The SIP Invite from S-CSCF may include the IMSI of the served IMS subscriber; the IM-SSF uses this IMSI to select the corresponding O-IM-CSI or VT-IM-CSI. The parameters that are needed to invoke the CAMEL service, such as gsmSCF Address, Service Key etc., are obtained from the O-IM-CSI, D-IM-CSI or VT-IM-CSI.

The IM-SSF converts the SIP Invite it receives from S-CSCF, into a CAP IDP. From this point onwards, the IM-SSF serves as a relay between S-CSCF and SCP: SIP methods (and responses) from S-CSCF are converted to CAP operations to SCP and CAP operations from SCP are translated to SIP methods to S-CSCF. Analogous to a CS call, an IMS call may be an *originating* IMS call or a *terminating* IMS call.

- *Originating IMS call* – when an IMS subscriber establishes a call, the S-CSCF where that subscriber is registered and that is processing the IMS call, may use the IFC of that subscriber to route the SIP Invite to the IM-SSF. The IM-SSF may then invoke a CAMEL service with O-IM-CSI, D-IM-CSI or both.

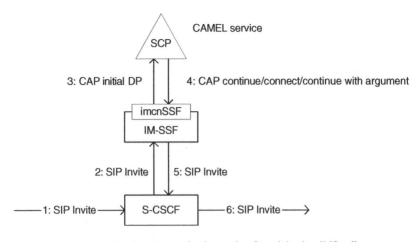

Figure 6.60 CAMEL service invocation for originating IMS call

- *Terminating IMS call* – when an IMS subscriber receives a call, the S-CSCF where that subscriber is registered and that is processing the terminating IMS call, may use the IFC of that subscriber to route the SIP Invite to the IM-SSF. The IM-SSF may then invoke a CAMEL service with the VT-IM-CSI. IMS does not have a concept similar to the distinction between GMSC and VMSC for terminating CS calls. Terminating IMS call handling takes place in the S-CSCF of the called IMS subscriber. For that reason, there is only VT-IM-CSI and not T-IM-CSI.

Figure 6.61 shows an example of a call case where originating and terminating subscriber both have a CAMEL service for originating call and terminating call.

The DPs in the BCSMs in the imcnSSF relate to SIP methods that traverse through the IM-SSF during an IMS call. The thick line in Figure 6.61 indicates the path of the SIP methods during an IMS call. The IM-SSF acts as a back-to-back user agent (B2BUA); refer to 3GPP TS 23.228 [91] for a description of B2BUA. This means that all SIP methods that are transported between calling and called subscriber for a particular IMS call are sent through the IM-SSF. This applies even when

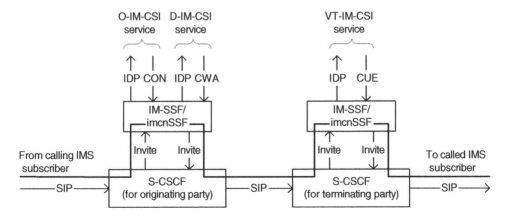

Figure 6.61 Originating and terminating CAMEL IMS services

Table 6.17   SIP methods and responses and related detection point in BCSM

SIP method/response	Detection point
O-IM-BCSM	
Invite	DP Collected_Info
	DP Analysed_Information
4xx (except 401, 407, 408, 480, 486),	DP Route_Select_Failure
5xx,	
6xx (except 600, 603)	
486 busy here	DP O_Busy
600 busy everywhere	
408 request timeout,	DP O_No_Answer
480 temp unavailable,	
603 decline	
200 OK	DP O_Answer
Bye	DP O_Disconnect
Cancel	DP O_Abandon
T-IM-BCSM	
Invite	DP Terminating_Attempt_Authorized
4xx (except 401, 407, 408, 480),	DP T_Busy
5xx,	
6xx (except 603)	
408 request timeout,	DP T_No_Answer
480 temp unavailable,	
603 decline	
200 OK	DP T_Answer
Bye	DP T_Disconnect
Cancel	DP T_Abandon

the CAMEL service has already terminated for this call. Table 6.17 contains the relation between SIP methods and responses and the related DP in the BCSM.

When the passing of a SIP method through the IM-SSF results in the occurrence of a DP, this may result in the sending of a CAP operation to the gsmSCF, such as initial DP or event report BCSM. The O-IM-BCSM and T-IM-BCSM apply the same CAP dialogue rules as the O-BCSM and T-BCSM for CS call control. For example, the gsmSCF may arm detection points or request reports. Pre-arranged and basic-end rules apply.

6.9.5 CAMEL Application Part for IMS Control

CAP v3 is the only version of CAP that is allowed for the IM-SSF. The IM-SSF was introduced in 3GPP Rel-5. Since the purpose of the IM-SSF is not the introduction of *new CAMEL services*, but the usage of *existing CAMEL services*, 3GPP decided that CAP v4 should not be specified for IM-SSF. The CAMEL subscription elements for IMS (O-IM-CSI, VT-IM-CSI, D-M-CSI) contain a CAMEL Capability Handling indicator, but that parameter will always have the value 'CAP v3'. Owing to the conversion between SIP and CAP, not all the parameters specified for CAP are available in an IMS environment (Figure 6.62).

Table 6.18 lists the CAP v3 parameters that are available in IDP, CON and CWA. The reader will notice that this is a small subset of CAP v3 that is used for CS call control. The main reason for the limited capability set is that many parameters that are defined for CAP relate to corresponding parameters in ISUP/BICC. Most of these parameters do not have an equivalent in SIP and hence,

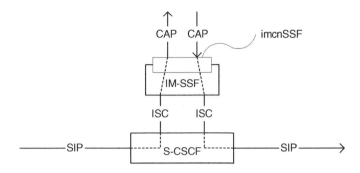

Figure 6.62   Signalling data loss between SIP/ISC and CAP

Table 6.18   Information elements in CAP v3 for IMS

Initial DP	Connect	Continue with argument
Media type info list		
Called party number		
Called party URL		
Calling party number		
Calling party URL		
Calling party category	Calling party category	Calling party category
Call gap encountered		
SIP call Id		
Cause		
Event type BCSM		
IMSI		
IP SSP capabilities		
IM-SSF address		
Original called party Id	Original called party Id	
Original called party URL	*Original called party URL*	
Redirecting party Id	Redirecting party Id	
Redirecting party URL	*Redirecting party URL*	
Redirection information		
Service key		
Subscriber state		
Time and timezone		
	Destination routing Address	
	Destination routing address URL	

cannot be carried in IDP, CON or CWA. For the available parameters in the other CAP operations for IMS control, refer to 3GPP TS 29.278 [111].

The parameter names in *italics* in Table 6.18 are specific to the CAMEL control of IMS. These parameters relate to information that may be copied from or to corresponding parameters in SIP messages on the ISC interface. Table 6.19 contains clarification for these specific parameters.

The application context used for CAMEL control of IMS is derived from CAP v3 for call control. Some IMS-specific parameters are added to CAP. This has resulted in that CAP v3 for IMS is not compatible with CAP v3 for call control. Hence, a CAMEL phase 3 SCP has to be adapted to the CAP v3 that is used for IMS control.

Table 6.19 IMS-specific parameters on CAP

Parameter	Description
Media type info list	This parameter is retrieved from the session description protocol (SDP) field in the SIP Invite. It contains the media type data, IP port number, transport protocol and media format sub fields. The media type info list may be compared to the bearer capability that is used for call control
Called party URL	This parameter contains a SIP URL and identifies the called subscriber. The format is defined in IETF RFC 2806 [167]. It is copied from the <to:> field or Request URI in the SIP Invite
Calling party URL	This parameter contains a SIP URL and identifies the calling subscriber. The format is defined in IETF RFC 2806 [167]. It is copied from the <from:> field or P-Asserted-Identity in the SIP Invite
SIP call Id	This parameter is a unique identifier that is generated by the S-CSCF and included in the SIP Invite to the IM-SSF. It may be compared with the CRN for CS call control. The gsmSCF may place the SIP call Id in a service CDR
Original called party URL	Call forwarding may occur in the IMS domain. The SIP Invite message may in that case contain the original called URL
Redirecting Party URL	In the case of call forwarding in IMS, this parameter contains the URL of the party on whose behalf the forwarding is performed
Destination routing address URL	The gsmSCF may provide an alternative destination to the IM-SSF in the form of a destination URL

Table 6.20 Supported IMS call cases

Call case	Subscription data	Triggering method
Originating IMS call	O-IM-CSI	The S-CSCF for the originating IMS subscriber sends a SIP Invite to the IM-SSF. The IM-SSF checks whether the subscriber has O-IM-CSI with DP collected info. If DP collected info is present, then the corresponding CAMEL service is triggered. If O-IM-CSI contains trigger criteria, then these will be fulfilled, otherwise no triggering will take place at this DP. O-IM-CSI may also contain DP route select failure. If the establishment of an IMS call fails, then a CAMEL service may be invoked, provided that no other CAMEL service is active at that moment. Table 6.17 specifies for which SIP message DP route select failure may occur. DP route select failure may have trigger conditions associated with it
	D-IM-CSI	After a CAMEL service is triggered as a result of O-IM-CSI, a *dialled* service may be triggered, as a result of D-IM-CSI. D-IM-CSI contains a set of numbers. Triggering takes place when the destination number of the IMS call corresponds with a number in D-IM-CSI
Terminating IMS call	VT-IM-CSI	A S-CSCF that is handling a terminating IMS call may route the call through IM-SSF. IM-SSF uses VT-IM-CSI to trigger a CAMEL service. VT-IM-CSI may also contain trigger information related to the failure of IMS call establishment. Table 6.17 specifies for which SIP methods DP T_Busy or T_No_Answer may occur

IM-SSF may be used for both terminating registered calls and terminating unregistered calls.

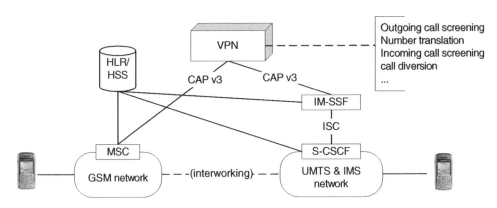

Figure 6.63 Combined GSM and IMS service control

6.9.6 Supported Call Cases for IMS Control

The supported call cases for CAMEL control of IMS are defined by the applicable subscription data. The call cases in Table 6.20 are supported.

The ASN.1 definitions of O-IM-CSI, D-IM-CSI and VT-IM-CSI are identical to O-CSI, D-CSI and VT-CSI, respectively. An operator may therefore re-use existing CAMEL phase 3 O-CSI, D-CSI and VT-CSI data for IMS control. O-CSI, D-CSI and VT-CSI need to be ported or copied to the HSS for that purpose. CAMEL control of IMS does not include serving network dialled services; compare N-CSI for call control. CAMEL services like freephone or premium rate, which may be offered with N-CSI in CS network, cannot be ported to the IMS network, unless using proprietary service triggering methodology.

6.9.7 Service Example

Figure 6.63 shows an example of a VPN service that controls both CS subscribers and IMS subscribers. Subscribers of a VPN group may be connected to the GSM network or may be registered with the IMS network. When the VPN subscriber is calling through GSM, a CAMEL phase 3 service is invoked; this service may apply services such as outgoing call screening and number translation. For incoming calls, VPN may apply incoming call screening and call diversion, etc. When the same or another subscriber registers with IMS, then she may enjoy the same services. When TEL URLs are used in IMS for calls from and to the VPN subscriber, VPN may use the same number translation and number screening tables as for GSM control. A VPN subscriber could be addressable both with an E.164 number (e.g. +27 83 212 9911) and with a URL (john.smith@abc.com). CAP v3 for IMS has the capability to receive a URL from the IMS network and to send a URL to the IMS network. Should the operator wish to utilize that enhancement to CAP v3 for IMS, an update of the VPN service is required.

7

Charging and Accounting

It requires no explanation that charging is one of the most important aspects in the GSM and UMTS network. If services cannot be charged, they cannot be offered! The architecture of the GSM and UMTS network provides mechanisms to accurately record the usage of services by users. These users may be own subscribers or inbound roaming subscribers. CAMEL interacts in a number of ways with the charging in the mobile network.

7.1 Architecture

One pivotal aspect of charging in GSM is the generation of CDRs. Figure 7.1 reflects which logical entities may generate CDRs. This figure considers CDRs related to CS domain, PS domain and SMS. Entities for the generation of CDRs for multimedia messaging service (MMS) or IMS are not included.

The CDRs that are generated by the CS or PS core network entities are sent to the billing system for further processing. The CDRs from the PS domain entities, i.e. SGSN and GGSN, are generated by the charging gateway function (CGF). The CGF collects charging information from SGSN and GGSN via the Ga interface and forwards the data, after conversion to CDR, to the billing system. (Refer to 3GPP TS 32.240 [119] for a description of the Ga interface.) The CGF may be a separate entity in the network or may be integrated in one or more SGSNs/GGSNs.[1] The generation of CDRs by the SCP is not specified in 3GPP. However, CAMEL contains specific functionality that facilitates the generation of CDRs by the SCP.

7.2 Call Detail Records

CDRs contain information that accurately reflects the usage of the network by a served subscriber. This served subscriber may belong to the operator's own network or another network, e.g. the MSC may generate CDRs for home subscribers and for inbound roaming subscribers. CDRs are generated by an MSC during call handling (e.g. MO call) and during non-call-related activities (e.g. call forwarding registration). CDRs may be used by post-processing systems for tasks like: subscriber charging; accounting (settlement between operators); statistics; and system monitoring.

---

[1] The definition of the CGF for PS domain results in uniform PS-related CDRs for the billing system. CDRs from the CS core network may differ per node vendor.

CAMEL: Intelligent Networks for the GSM, GPRS and UMTS Network   Rogier Noldus
© 2006 John Wiley & Sons, Ltd

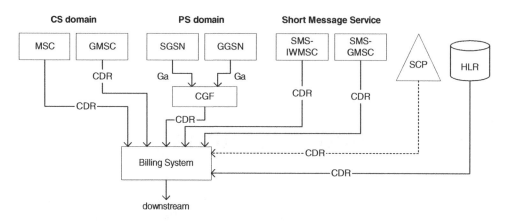

Figure 7.1   Logical entities for CDR generation

Table 7.1   Technical specifications for CDR generation

GSM/3GPP release (CAMEL phase)	Technical specification
ETSI R96 (phase 1) ETSI R97 (phase 2) ETSI R98 (phase 2)	GSM TS 12.05 [57]; event and call data
3GPP R99 (phase 3)	3GPP TS 32.005 [114]; 3G call and event data for the CS domain 3GPP TS 32.015 [115]; call and event data for the PS domain
3GPP Rel-4 (phase 3)	3GPP TS 32.200 [116]; charging principles
3GPP Rel-5 (phase 4)	3GPP TS 32.205 [117]; charging data description for the CS domain 3GPP TS 32.215 [118]; charging data description for the PS domain
3GPP Rel-6 (phase 4)	3GPP TS 32.240 [119]; charging architecture and principles 3GPP TS 32.250 [120]; CS domain charging 3GPP TS 32.251 [121]; PS domain charging 3GPP TS 32.298 [122]; CDR parameter description

Table 7.1 shows for each GSM and 3GPP release the technical specifications in which the structure and contents of CDR are specified. These specifications are available from www.3gpp.org.

GSM R97 and R98 contain, besides GSM TS 12.05 [57], GSM TS 12.15 (GPRS charging). CAMEL however, has no impact on that specification, since CAMEL control of GPRS is introduced in CAMEL phase 3 (3GPP R99).

7.2.1 Overview of Call Detail Records

Not all CDRs that are specified in the various GSM TSs and 3GPP TSs are impacted by CAMEL. Table 7.2 contains an overview of which CDR may be impacted by CAMEL services. For details of the CDRs, refer to 3GPP TS 32.250 [120] (for CDRs generated in MSC) or 3GPP TS 32.251 [121] (for CDRs generated in SGSN and GGSN).

For CDRs related to CPH, see Section 7.7.

Below is a short description of the above-listed CDRs.

Table 7.2 Overview of CDRs with CAMEL impact

Activity	Entity	CDR type
MO call	VMSC	Mobile-originated call attempt ('MOC')
MF call	VMSC, GMSC	Mobile-originated call forwarding attempt ('MOC forward')
MT call	GMSC	Roaming call attempt[2]
		Terminating CAMEL call attempt
	VMSC	Mobile-terminated call attempt ('MTC')
		Terminating CAMEL call attempt
Transit call	MSC	Transit call attempt
MO-SMS	MSC	Short message service, mobile-originated ('SMS-MO')
	SGSN	SGSN short message, mobile-originated ('S-SMO-CDR')
MT-SMS	MSC	Short message service, mobile-terminated ('SMS-MT')
	SGSN	SGSN short message mobile-terminated ('S-SMT-CDR')
GPRS session	SGSN	Mobile station mobility management data in SGSN ('M-CDR')
PDP Context	SGSN	SGSN PDP context data ('S-CDR')
	GGSN	PDP context charging data in GGSN ('G-CDR')

[2] This CDR is in some specifications referred to as 'roaming call forwarding' (RCF) CDR.

Mobile-originated Call Attempt
A MOC CDR is generated for an MO call established in the MSC. If the call is subject to CAMEL control, then the CDR contains data related to the CAMEL service. The MOC CDR may contain CAMEL data for O-CSI, D-CSI and N-CSI service.

Mobile-originated Call Forwarding Attempt
A MOC forward CDR is generated in VMSC or GMSC for a forwarded call. The CDR may contain data related to the CAMEL services for this call.

Roaming Call Attempt
The roaming call CDR is generated in the GMSC for the roaming leg of an MT call. The operator shall decide whether the roaming call CDR will be created for all MT calls or only for MT calls when the subscriber is roaming outside the HPLMN. If an MT call in the HPLMN is free of charge, then generation of this CDR may be needed only in the roaming call case.

Mobile-terminated Call Attempt
The MTC CDR is generated in the VMSC for the roaming leg to a mobile station. An operator shall decide whether the VMSC shall generate an MTC CDR for inbound roaming subscribers only or for all terminated calls.

Terminating CAMEL Call Attempt
CAMEL data for the MT call is contained in the terminating CAMEL call attempt CDR. This CDR may be associated with the roaming call CDR in GMSC (for T-CSI CAMEL service) or with the MTC CDR in VMSC (for VT-CSI service).

Transit Call Attempt
A transit call CDR may be generated by an MSC that is handling a call that is not an MO, MF or MT call. CAMEL interaction with transit calls is introduced in CAMEL phase 4 in 3GPP Rel-7: trunk-originated calls. An arbitrary MSC in the network may trigger a CAMEL service when certain conditions are fulfilled. One example of such a call is PABX connection to the MSC. Calls originating from the PABX may trigger a CAMEL VPN service.

MO Short Message Service
The submission of an SMS to the SMSC results in the generation of an SMS-MO CDR (MSC-based SMS submission) or S-SMO-CDR CDR (SGSN-based SMS submission). These MSC- and SGSN

CDRs are defined in different specifications and differ slightly. The CDR may contain CAMEL data related to this SMS submission.

MT Short Message Service

The delivery of an SMS to the MS or UE may result in the generation of an SMS-MT CDR (MSC) or S-SMT-CDR CDR (SGSN). Since SMS delivery is normally free of charge, an MSC or SGSN may be configured not to generate an MT-SMS CDR. If the CDR is generated, it may contain data related to a CAMEL service for this SMS delivery.

Mobile Station Mobility Management Data in SGSN (M-CDR)

The M-CDR reflects the GPRS attached and detached state of a subscriber. It does not reflect data transfer. This CDR may contain data related to a 'scenario 1' CAMEL GPRS service; see Section 5.3.

SGSN PDP Context Data (S-CDR)

The S-CDR reflects the usage of a PDP context, including a CAMEL service that may be invoked for this PDP Context. This CDR may contain data related to a 'scenario 2' CAMEL GPRS service. It is possible that two CAMEL services are invoked for a PDP context: one service at PDP context establishment and another service at PDP context establishment acknowledgement. This depends on trigger data (GPRS-CSI) and on service behaviour. However, the CDR definition for the PDP context allows for a single CAMEL service invocation only.

GGSN PDP Context Data (G-CDR)

Although CAMEL control of PDP context resides purely in the SGSN, 3GPP Rel-6 specifies the capability of transferring CAMEL data from SGSN to GGSN, for inclusion in the G-CDR. Refer to description 'PS furnish charging information' in 3GPP TS 32.251 [121] and 3GPP TS 32.299 [123].

7.2.2 CAMEL-related Parameters in CDRs

For each CDR, precise definitions indicate which elements will be placed in each CDR, as a result of CAMEL service invocation. Examples of parameters that may be present in call-related CDR are:

- *gsmSCF address* – this parameter contains the GT that is used to invoke the CAMEL service.
- *service key (SK)* – this parameter contains the SK that is used for the CAMEL service.
- *CRN* – the CRN is allocated by the MSC and serves as an identification of a call. It may be used for off-line CDR correlation.
- *MSC address* – this parameter is used together with the CRN; it is the address of the MSC that allocated the CRN.
- *Default call handling* – this parameter is present when DCH occurred for the call. A billing system may filter out CDRs that contain this parameter and may forward these CDRs to a post-processing system, e.g. for charging reconciliation.
- *Level of CAMEL service* – this parameter contains an indication of the complexity of a CAMEL service. The parameter reflects whether on-line charging operations are used.
- *Number of DP encountered* – this parameter has the same rationale as level of CAMEL service. The higher the value of this parameter, the more processing the MSC had to do when serving the subscriber. A VPLMN operator may charge the HPLMN operator of the served subscriber depending on the complexity of the CAMEL service invoked by that subscriber.
- *Free-format data* – this parameter contains the free-format data that was supplied by the CAMEL service in the FCI procedure.
- *CAMEL destination number* – this parameter contains the destination number that was received from the SCP (in CAP connect) for connecting the call.

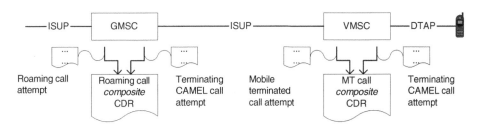

Figure 7.2   Generation of composite CDRs

- *Changed parameters* – this parameter indicates which parameters were changed by the CAMEL service, with CAP connect or CAP continue with argument. The CDR definition allows indicating that the CgPN has been modified, but the CAMEL protocol does not allow for modifying the CgPN. Furthermore, multiple Generic Number (GN) modifications may be indicated in the CDR, while CAMEL allows only for modifying GN 6 (additional CgPN). These capabilities may be considered as forward-compatible functionality for possible extension of the CAMEL protocol. When a CAMEL service changes the redirection information, the CDR reflects this by indicating only the call forwarding counter that was contained in the redirection information. This is unfortunate. Other parameters in redirection information that were (necessarily) provided by CAMEL are not reflected in the CDR. Examples include redirecting indicator and redirecting reason.

7.2.3 Composite CDRs

For certain call cases, multiple CDRs may be produced by one network entity. One example is the MT call in GMSC, for which roaming call attempt CDR and terminating CAMEL call attempt CDR may be produced. In addition, in some vendors' MSCs, the charging data for an MO CAMEL call is split into two or more CDRs: one CDR for call-related data, and another CDR for CAMEL service related data. The serving MSC may combine these CDRs into a 'composite CDR'. See Figure 7.2 for the creation of composite CDRs for a terminating call.

The generation of a composite CDR, as opposed to the generation of two single CDRs, alleviates the billing system from having to correlate the two CDRs. The generation of composite CDRs is not specified in the 3GPP charging specifications (e.g. 3GPP TS 32.250 [120]), but is a commonly applied method.

7.3 Transfer Account Procedure Files

The TAP file is a file format that is used for the transfer of call data between operators. The purpose of the transfer of the TAP files is, as the name implies, accounting between operators. The TAP file format is defined by the GSM Association (www.gsm.org). TAP is defined in the following *permanent reference documents* (PRDs):[3]

- PRD TD.32 [161]; 'Rejects and returns process';[4]
- PRD TD.57 [162]; 'Transferred account procedure data record format specification, version 3;
- PRD TD.58 [163]; 'TAP 3 implementation handbook';
- PRD TD.60 [164]; 'General scenarios for the transferred account procedure for data record format specification, version 3'.

---

[3] PRD is the document naming convention of the GSM Association.

[4] The acronym 'TD' in the file name refers to 'TADIG', which is the GSM Association's *transferred account data interchange group*.

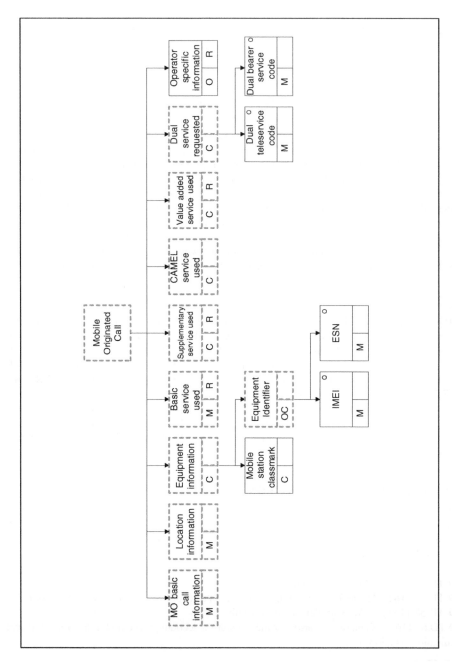

Figure 7.3 TAP file structure for MO call. Reproduced from PRD TD.57 v3.10.1, Figure 3.6 by permission of GSM Association

TAP file format is mainly specified in PRD TD.57 [162]. The TAP file format resembles the CDR format. That is to say, many of the data fields from the CDR have an equivalent data field in the TAP file. There is, however, no formal mapping between CDR and TAP. Whereas CDRs that are generated in a PLMN may differ per vendor or operator, the TAP files that are exchanged between operators will conform to the format specified in PRD TD.57 [162]. Figures 7.3 and 7.4 show the structure of a TAP file for an MO call. The CAMEL information for the call is included in the element 'CAMEL service used'.

Most of the elements in *CAMEL service used* are copied from the MOC CDR, such as gsmSCF Address (CAMEL server address), service key, reference number, etc. Figure 7.5 reflects how CDRs that are generated by an MSC in a VPLMN are transferred to a Billing System in the HPLMN of the served subscriber.

The 'billing system' may consist of various functional entities and nodes and varies per vendor and operator. The billing system may not forward all CDRs to the CDR formatter, for conversion to TAP file. Conversion to TAP may be needed only for call data that needs to be sent to another operator. Within an operator's own network, other call detail formats than TAP may be used.

Support for CAMEL is introduced in TAP version 2; support for CAMEL phase 2 is available in TAP version 3. As from TAP version 3 onwards, ASN.1 is used to define the TAP file format. The ASN.1 definition of the various data fields includes extension markers [the three dots (...) in the data element definition; also referred to as 'ellipsis']. The extension marker allows the data element definition to be extended when new data fields are added to the CDR that need to be copied to the TAP file. Figure 7.6 shows an example data structure for a TAP file that is used to convey data related to an MO call.

7.4 Inter-operator Accounting of CAMEL Calls

The CDR that is converted into TAP format and sent to another operator's network normally includes the price of that call. The price information may include currency information, tax information and discount information. This price does not have to be the price that is levied against the served subscriber. Rather, it is the price that the serving PLMN operator charges the HPLMN operator of the served subscriber. The HPLMN operator uses the TAP file to charge the subscriber. The method of determining the price that is charged towards the subscriber, based on the price indicated in the TAP file, is operator-specific.

When a call from a roaming subscriber is subject to CAMEL control, that does not have to affect the price indicated in the TAP file, unless special agreements are in place. The TAP file will contain additional CAMEL data, but the VPLMN operator charges the same amount to the HPLMN operator for the actual call. If special agreements are in place between the operators, then data elements such as level of CAMEL service or free-format data may be considered for the call charging (accounting) between the operators. If the roaming subscriber is a pre-paid subscriber, then she has paid for the call already during the call. In that case, the HPLMN will not forward the TAP file to the accounting system for generating a bill for the subscriber (Figure 7.7).

One example where CAMEL may be used to affect the price for roaming calls is the following. A roaming subscriber may have a CAMEL VPN service. When the subscriber establishes a call, the VPN service uses FCI to place free-format data in the CDR. When the billing system of the VPLMN operator calculates the price of the call, it may take the free format data into account and apply a discount for that call. This method requires mutual agreement between the involved operators.

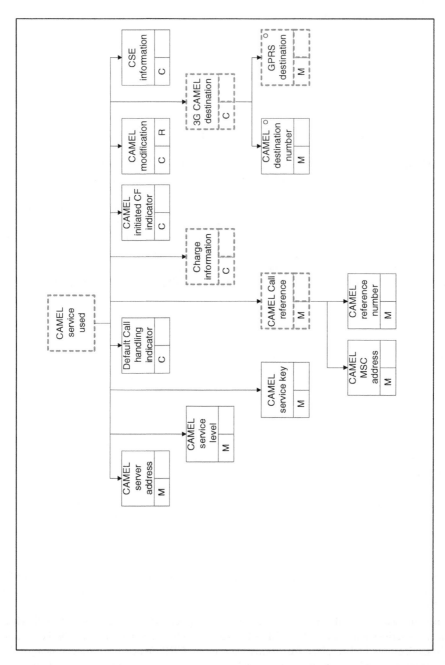

Figure 7.4 CAMEL information in TAP file for MO call. Reproduced from PRD TD.57 v3.10.1, Figure 3.12, by permission of GSM Association

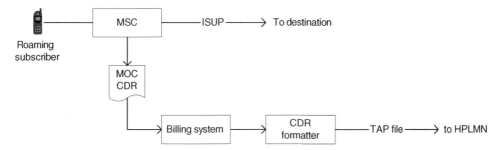

Figure 7.5   CDR to TAP conversion

```
DataInterChange ::= CHOICE
{
    transferBatch TransferBatch,
    notification  Notification,
...
}
TransferBatch ::= [APPLICATION 1] SEQUENCE
{
    batchControlInfo        BatchControlInfo                OPTIONAL,

    accountingInfo          AccountingInfo                  OPTIONAL,
    networkInfo             NetworkInfo                     OPTIONAL,
    messageDescriptionInfo  MessageDescriptionInfoList      OPTIONAL,
    callEventDetails        CallEventDetailList             OPTIONAL,
    auditControlInfo        AuditControlInfo                OPTIONAL,
...
}
CallEventDetailList ::=  [APPLICATION 3] SEQUENCE OF CallEventDetail

CallEventDetail ::= CHOICE
{
    mobileOriginatedCall    MobileOriginatedCall,
    mobileTerminatedCall    MobileTerminatedCall,
    supplServiceEvent       SupplServiceEvent,
    serviceCentreUsage      ServiceCentreUsage,
    gprsCall                GprsCall,
    contentTransaction      ContentTransaction,
    locationService         LocationService,
...
}
```

Figure 7.6   Example TAP file structure. Reproduced from PRD TD.57 v3.10.1, Section 6.1 by permission of GSM Association

```
MobileOriginatedCall ::= [APPLICATION 9] SEQUENCE
{
    basicCallInformation      MoBasicCallInformation      OPTIONAL,
    locationInformation       LocationInformation         OPTIONAL,
    equipmentIdentifier       ImeiOrEsn                   OPTIONAL,
    basicServiceUsedList      BasicServiceUsedList        OPTIONAL,
    supplServiceCode          SupplServiceCode            OPTIONAL,
    thirdPartyInformation     ThirdPartyInformation       OPTIONAL,
    camelServiceUsed          CamelServiceUsed            OPTIONAL,
    operatorSpecInformation   OperatorSpecInfoList        OPTIONAL,
...
}
CamelServiceUsed ::= [APPLICATION 57] SEQUENCE
{
 camelServiceLevel          CamelServiceLevel               OPTIONAL,
    camelServiceKey         CamelServiceKey                 OPTIONAL,
    defaultCallHandling     DefaultCallHandlingIndicator    OPTIONAL,
    exchangeRateCode        ExchangeRateCode                OPTIONAL,
    taxInformation          TaxInformationList              OPTIONAL,
    discountInformation     DiscountInformation             OPTIONAL,
    camelInvocationFee      CamelInvocationFee              OPTIONAL,
    threeGcamelDestination  ThreeGcamelDestination          OPTIONAL,
    cseInformation          CseInformation                  OPTIONAL,
...
}
```

Figure 7.6 *Continued*

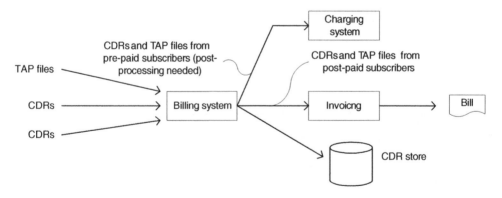

Figure 7.7 Separation of CDRs and TAP files

The billing system determines from the contents of the CDR or TAP file whether the CDR or TAP file relates to a pre-paid subscriber. Examples include:

- *CAMEL information* – the presence of a particular service key and gsmSCF address indicates that CAMEL pre-paid service was invoked for the call.
- *Free-format data* – the CAMEL pre-paid service may include free-format data in the CDR, indicating that the call was charged already.
- *IMSI* – some operators use a dedicated IMSI range for pre-paid subscribers.

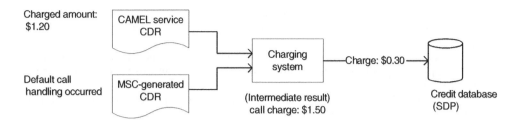

Figure 7.8 Processing CDRs with default call handling indication

Some TAP files from subscribers who are identified as pre-paid subscriber may still require post-processing. This may be the case when DCH occurs during service invocation or during the call. If DCH occurs, then the MSC indicates this in the CDR. The billing system of the HPLMN operator may filter out these CDRs and forward them to the charging system. Details relating to post-processing of these CDRs are not defined by CAMEL; a possible method is presented in Figure 7.8.

The MSC-generated CDR is used by the charging system to calculate the price to be levied to the subscriber. The price of the call may depend on the subscription type of the subscriber. The calculated price is compared with the amount already paid by the subscriber for this call; that amount is obtained from the CAMEL service CDR. This would require that the CAMEL service generate a service CDR indicating the charged amount for the call. The above-described method of post-processing of CDRs requires correlation of CDRs; see Section 7.5.

7.4.1 Clearing House

Figure 7.9 shows transfer of TAP files between operators, through a clearing house.

The clearing house acts as broker between operators. Not all operators make use of the services from a clearing house, but it is a quite common method. Normally, an operator is associated with one clearing house only. The TAP files that are generated in a PLMN and that need to be sent to another operator are forwarded to the operator's associated clearing house. The tasks of the clearing house include:

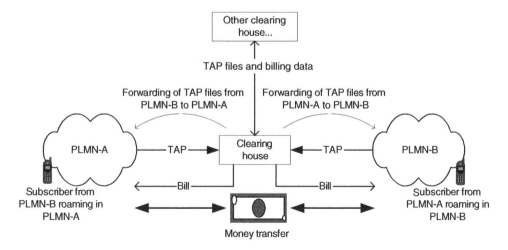

Figure 7.9 Transfer of TAP files via a clearing house

- forwarding TAP files to the HPLMN operator; assuming that each operator is associated with one clearing house only, the transfer of TAP files to and from an operator may run through two clearing houses;
- applying cross-charging between operators; the clearing house determines how much each operator should pay to any other operator for their respective roaming subscribers; the cross-charging includes currency conversion;
- data format conversion; where needed, the clearing house may apply reformatting of TAP files to suite individual operators' needs.

7.4.2 CAMEL Invocation Fee

A VPLMN operator may charge the HPLMN operator for the use of CAMEL in its network. That is to say, the VPLMN operator not only charges the HPLMN operator for the regular call charges, but also a fee for using CAMEL capability. The rationale is that the VPLMN operator has to have some additional capability available in its network (i.e. CAMEL capability) and wants to charge for the use of that capability. Although CAMEL does not specify the usage of the CAMEL invocation fee, the following mechanisms are available through the CDRs and the TAP file.

- *CAMEL invocation fee* – this data field contains an absolute amount. The operator may charge a fixed fee for invoking a CAMEL service, or charge a fee that depends on the complexity of the CAMEL service. This field is not present in the CDR, but is determined outside the MSC and is placed in the TAP file.
- *CAMEL service level* – this data field, which is obtained from the CDR, may be used to determine the CAMEL invocation fee.
- *Number of detection points* – this data field, which is obtained from the CDR but which is not present in the TAP file, may also be used to determine the CAMEL invocation fee.

7.5 Correlation of Call Detail Records

The chapters on CAMEL phases 1–4 include regular references to the CRN and MSC address. The present section explains how CRN and MSC address may be used to correlate CDRs that are generated by different nodes or functional entities. Figure 7.10 shows CDR generation and correlation for an MO call.

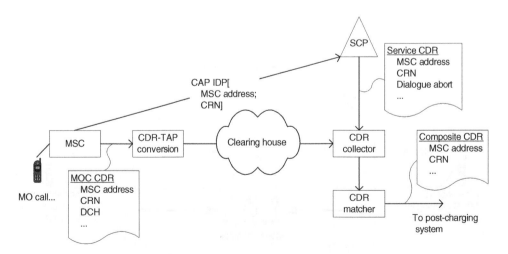

Figure 7.10   CDR correlation for mobile originated call

In the example in Figure 7.10, the CDR collector in the HPLMN has determined that CDR correlation is needed, because: the MOC CDR contains DCH; or the CAMEL service CDR indicates that dialogue abortion occurred and that CDR post-processing is needed, e.g. for charging.

7.5.1 Call Reference Number

The CRN is an octet string with variable length between one and eight octets. An MSC uses an internal counter to generate the CRN. Each CRN that is generated in an MSC is unique within that MSC. For every call that is established in an MSC, a CRN is generated by that MSC. The CRN is strictly associated with the MSC that allocates the CRN. Each call is therefore uniquely identified by a combination of CRN and MSC address. The CRN + MSC address is written to the CDR for the call and is reported to the CAMEL service, by means of inclusion in the CAP IDP operation. Hence, the SCP may write CRN and MSC address to a service CDR.

MSCs do not have to use the full eight octets when generating the CRN. An example of CRN format is:

CRN = <1 octet MSC identifier> + <4 octets variable part>[5]

Such a CRN will start repeating after 4 294 967 296 calls. If the CRN generator in an MSC is used for other purposes as well, e.g. an ISUP network reference number or SMS reference number, then the CRN will start repeating earlier. Hence, the MSC should take care to use an appropriate amount of octets for the (variable part of the) CRN. The usage of the CRN in the various call cases is as follows.

- *MO calls* – see Figure 7.10 and previous descriptions.
- *MT calls* – see Figures 7.11 and 7.12.

The GMSC allocates the CRN during MT call handling and includes CRN and GMSC address in MAP SRI to HLR. At this point, the GMSC does not know whether the HLR will return T-CSI. Hence, CRN and GMSC address are also included in MAP SRI for non-CAMEL calls. The GMSC also includes CRN and GMSC address in MAP SRI for the following purposes:

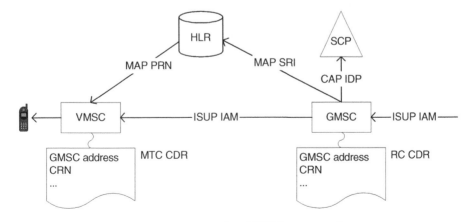

Figure 7.11 Transfer of CRN for an MT call

---

[5] Even though this sample CRN contains a one-octet MSC identifier, the CRN should still be associated with an MSC address.

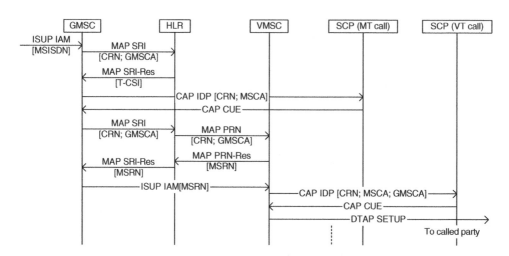

Figure 7.12   Signal sequence diagram for CRN + MSC address for MT/VT call

- inclusion of CRN and GMSC address in MTC CDR in VMSC;
- VT call CAMEL service: inclusion of CRN & GMSC address in CAP IDP from VMSC to SCP;
- optimal routing at late call forwarding (ORLCF).

CAP IDP resulting from T-CSI includes CRN and MSC address for inclusion in a service CDR from the SCP. Since HLR does not store data from the first MAP SRI, the second MAP SRI again includes CRN and GMSC address. The HLR sends MAP PRN to VMSC and includes CRN and GMSC address in MAP PRN. Now the VMSC has CRN and GMSC address available for terminating call handling. The VMSC places CRN and GMSC address in MTC CDR and includes these elements in CAP IDP to SCP, if VT-CSI is present in VLR. The transfer of CRN and GMSC address between GMSC, HLR, VMSC and SCP (2×) allows correlation of the following CDRs:

- roaming call CDR in GMSC;
- MT call CDR in VMSC;
- CAMEL service CDR in SCP for T-CSI service;
- CAMEL service CDR in SCP for VT-CSI service.

CAP IDP sent from VMSC to gsmSCF, resulting from VT-CSI, also contains 'MSC address'. This MSC address is the address of the serving MSC. The CRN, however, is allocated by the GMSC. Hence, the CRN in CAP IDP for this call case (VT call) is associated with 'GMSC address' and not with 'MSC address'.

  If for a particular call the originating VMSC and the GMSC happen to be the same node and for both the MO call and the MT call a CAMEL service is invoked, then that node will generate two CRNs; one CRN for the MO call and one CRN for the MT call. Since the originating VMSC and the GMSC for this call case have the same (G)MSC address, that node will have one CRN generator, to guarantee uniqueness of the CRN + (G)MSC address pair.

7.5.2 MF Calls

For forwarded calls, the CRN and GMSC address are used. The rationale is that a forwarded call results from a terminating call attempt. Hence, the GMSC has allocated a CRN for that call. If the

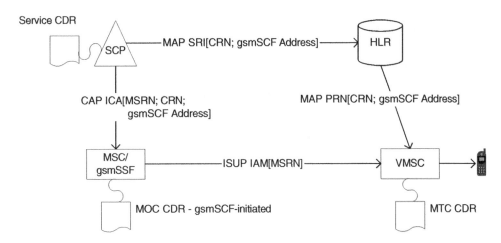

Figure 7.13 Usage of CRN and gsmSCF address for SCP-initiated calls

HLR has sent MAP PRN to VLR, and VLR has returned MSRN, then the CRN and GMSC are also available in VMSC. Hence, the forwarding MSC (GMSC for early forwarding; VMSC for late forwarding) has CRN and GMSC address available to identify the call in CAP IDP. In the case that call forwarding in VMSC is given back to GMSC (i.e. ORLCF is applied), CRN and GMSC address for this call are returned from VMSC to GMSC (in MAP RCH) and are used for the call forwarding in the GMSC.

7.5.3 SCP-initiated Calls

In the case of SCP-initiated calls (see Section 6.2.3), the SCP allocates the CRN; see Figure 7.13.

The gsmSCF address, when coupled to CRN, fulfils the same function as MSC address or GMSC address. It guarantees unique identification of the call. If the SCP sends MAP SRI to HLR to obtain an MSRN for the call, then the CRN should be included in MAP SRI; the HLR includes the CRN in MAP PRN to VMSC, as for normal MT calls. The HLR also includes the gsmSCF address in MAP PRN. In MAP SRI from SCP to HLR, the gsmSCF address takes the place of the GMSC address; HLR and VMSC treat the gsmSCF address in the same manner as GMSC address. The CRN + gsmSCF address may be used to correlate the CAMEL Service CDR, the 'MOC–CPH adapted' CDR and the MTC CDR. If the SCP does not obtain an MSRN for the call, but includes the MSISDN of the called party in CAP ICA, then the CRN + gsmSCF address is used only between SCP and MSC/gsmSSF, for correlation of CAMEL service CDR and MOC–CPH adapted CDR. The MSC/gsmSSF may behave as GMSC, resulting in regular MT call handling, including the allocation of a CRN, etc.

7.6 Global Call Reference

The CRN is coupled with GMSC address or gsmSCF address and is carried in various MAP messages and in CAP IDP. This mechanism allows for the linking of CDRs that are generated in various entities. The CRN does not, however, travel over ISUP. As a result, the call identification for a MO call differs from the call identification for the MT part of that same call. See Figure 7.14 for an example call case, consisting of an originating MSC, a GMSC-B that forwards the call, a GMSC-C for the forwarded-to party and a VMSC-C.

ITU-T has introduced a global call reference (GCR) in the bearer independent call control (BICC); ITU-T Q.1902.4 [145]. The node where call establishment is initiated generates the GCR.

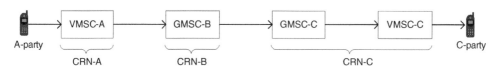

Figure 7.14 Multiple CRNs per call

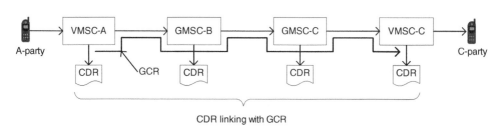

Figure 7.15 Linking of CDRs with global call reference

The GCR is transported between successive nodes in the call, through the BICC signalling that is used for call establishment. The respective nodes may place the GCR in the CDR for the call. In this manner, the CDRs for the call may be linked by post-processing; see Figure 7.15.

CAMEL does not specify the relation between CRN and GCR. A call as depicted in Figure 7.15 may be controlled by one or more CAMEL services. Each CAMEL service receives a CRN from the MSC, enabling post-processing to link CDRs between VMSC/GMSC and SCP. In that manner, the CDRs from core network (MSC, GMSC) and the CDRs from service layer (SCPs) may all be linked together.

7.7 Call Party Handling CDRs

For calls that are subject to a CPH service, the 3GPP charging specifications contain a set of CPH-adapted CDR definitions. The following CPH-adapted CDRs are specified:

- MOC record – CAMEL CPH adapted;
- MOC record – gsmSCF initiated;
- MOC record – new call segment;
- MOC call forwarding record – CAMEL CPH adapted;
- MTC record – CAMEL CPH adapted.

Although CAMEL phase 4 CPH feature is specified in 3GPP Rel-5, the CPH-adapted CDRs are specified in 3GPP Rel-6 only. Refer to 3GP TS 32.250 [120] for further details on the CPH-adapted CDRs.

8

3GPP Rel-6 and Beyond

8.1 General

CAMEL phase 4 is the last CAMEL phase; there will not be a CAMEL phase 5 in 3GPP releases later than Rel-5. However, functional enhancements are still introduced in CAMEL phase 4 in these later 3GPP releases. Hence, when referring to CAMEL phase 4, one should mention the corresponding 3GPP release: CAMEL phase 4 as specified in 3GPP Rel-5; CAMEL phase 4 as specified in 3GPP Rel-6; CAMEL phase 4 as specified in 3GPP Rel-7, etc.[1]

As described in Section 6.1, CAMEL phase 4 includes the 'CAMEL 4 sub set mechanism'. This mechanism has dual purpose: (1) GSM core network vendor and operator may introduce CAMEL phase 4 stepwise and (2) 3GPP may introduce new CAMEL features in the CAMEL specifications without the necessity to define a new CAMEL phase. One aspect of CAMEL that is fundamental to the distinction between the CAMEL phases is the application context (AC) version for the CAP protocol between gsmSSF and gsmSCF. Should the introduction of new CAMEL features require the upgrade of the CAP AC version, e.g. from CAP v4 to CAP v5, then this would have the following impact:

(1) A subscriber may have one of five different levels of CAMEL capability in the O-CSI. This increases the complexity of the CAMEL capability negotiation between VLR and HLR during location update. Similar argumentation applies for the levels of CAMEL capability in the T-CSI.
(2) The MSC/gsmSSF would need to support one additional version of CAP (CAP v5).
(3) When the SCP wants to utilize any of the new CAMEL features, it needs to support one additional version of CAP (CAP v5).

For these reasons, any enhancement to CAMEL phase 4 in 3GPP Rel-6 and beyond uses CAP 4. New parameters on CAP are placed behind the extension marker, and the (G)MSC/gsmSSF reports to the SCP which CAMEL phase 4 feature it supports. The SCP then knows which CAMEL phase 4 capability it may use for this call.

Figure 8.1 shows an example of the inclusion of new parameters behind the extension marker in CAP IDP argument.

The parameters 'enhanced dialled services allowed' (see Section 8.2.1) and 'uu-Data' (see Section 8.2.4) are introduced in 3GPP Rel-6. When an SCP does not support these new para-meters

---

[1] CAMEL phase 3 is included in 3GPP R99 and in 3GPP Rel-4. However CAMEL phase 3 in 3GPP Rel-4 is identical to CAMEL phase 3 in 3GPP R99. Hence, there is no need to mention the 3GPP release when mentioning CAMEL phase 3. The same applies for CAMEL phase 2, which is included in GSM R97 and GSM R98.

CAMEL: Intelligent Networks for the GSM, GPRS and UMTS Network   Rogier Noldus
© 2006 John Wiley & Sons, Ltd

```
InitialDPArgExtension {PARAMETERS-BOUND : bound} ::= SEQUENCE {
   gmscAddress                   [0]  ISDN-AddressString            OPTIONAL,
   forwardingDestinationNumber   [1]  CalledPartyNumber {bound}     OPTIONAL,
   ms-Classmark2                 [2]  MS-Classmark2                 OPTIONAL,
   iMEI                          [3]  IMEI                          OPTIONAL,
   supportedCamelPhases          [4]  SupportedCamelPhases          OPTIONAL,
   offeredCamel4Functionalities  [5]  OfferedCamel4Functionalities  OPTIONAL,
   bearerCapability2             [6]  BearerCapability {bound}      OPTIONAL,
   ext-basicServiceCode2         [7]  Ext-BasicServiceCode          OPTIONAL,
   highLayerCompatibility2       [8]  HighLayerCompatibility        OPTIONAL,
   lowLayerCompatibility         [9]  LowLayerCompatibility {bound} OPTIONAL,
   lowLayerCompatibility2        [10] LowLayerCompatibility {bound} OPTIONAL,
   ...,
   enhancedDialledServicesAllowed
                                 [11] NULL                          OPTIONAL,
   uu-Data                       [12] UU-Data                       OPTIONAL}
```

Figure 8.1 Extension capability for CAP Initial DP Argument. Reproduced from 3GPP TS 29.078, v6.4.0, Section 6.1.1, by permission of ETSI

in CAP IDP, then that SCP will simply discard these parameters when received from an (G)MSC/gsmSSF.

8.1.1 Capability Negotiation

Two mechanisms are used between (G)MSC and gsmSCF and between SGSN and gsmSCF to indicate the supported CAMEL phase 4 capability of that node [SGSN or (G)MSC]. These are the 'offered Camel 4 CSIs' parameter and the 'offered Camel 4 functionalities' parameter.

8.1.1.1 Offered Camel 4 CSIs

The offered Camel 4 CSIs are sent from (G)MSC or SGSN to HLR and indicate which CAMEL phase 4 CSIs are supported in that node (SGSN or (G)MSC). Figure 8.2 shows the structure of this parameter. The *bit string* definition of offered CAMEL 4 CSIs allows for the introduction of a new CSI in a later CAMEL phase 4, up to a total of 16 CSIs (including 'psi-enhancements'). However, neither 3GPP Rel-6 nor 3GPP Rel-7 includes new CSIs, compared with 3GPP Rel-5.

```
OfferedCamel4CSIs ::= BIT STRING {
     o-csi                         (0),
     d-csi                         (1),
     vt-csi                        (2),
     t-csi                         (3),
     mt-sms-csi                    (4),
     mg-csi                        (5),
     psi-enhancements              (6)
} (SIZE (7..16))
```

Figure 8.2 Structure of Offered CAMEL 4 CSIs. Reproduced from 3GPP TS 29.002, v5.8.0, Section 17.7.1, by permission of ETSI

8.1.1.2 Offered Camel 4 Functionalities

The offered Camel 4 functionalities are sent from (G)MSC/gsmSSF to gsmSCF, in CAP IDP or in CAP initiate call attempt (ICA) result. Figure 8.3 shows the structure of this parameter. The definitions for the BITs at position 15 up to 18 are new in 3GPP Rel-6. According to the SIZE definition of this BIT STRING, a total of 64 CAMEL phase 4 functionalities could be defined.

8.2 Enhancements to 3GPP Rel-6

The following enhancements are introduced in CAMEL phase 4 in 3GPP Rel-6:

- enhanced dialled service;
- handover notification criteria;
- enhancement to SCUDIF control;
- reporting of user-to-user information;
- enhancement to user interaction.

These enhancements relate to CS call control. 3GPP Rel-6 contains no enhancements to SMS control, GPRS control, mobility management etc.

8.2.1 Enhanced Dialled Service

The dialled services, as described in Section 5.2, have the restriction that they shall have a 'short dialogue'. The SCP is not allowed to arm DPs or ask for charging reports. Hence, after the processing of DP analysed information is completed, the dialled service terminates. The enhanced dialled service (EDS) resolves this issue by permitting a 'long dialogue' for the dialled service, when the following conditions are fulfilled: the serving MSC supports EDS; no other CAMEL service is

```
OfferedCamel4Functionalities ::= BIT STRING {
      initiateCallAttempt                (0),
      splitLeg                           (1),
      moveLeg                            (2),
      disconnectLeg                      (3),
      entityReleased                     (4),
      dfc-WithArgument                   (5),
      playTone                           (6),
      dtmf-MidCall                       (7),
      chargingIndicator                  (8),
      alertingDP                         (9),
      locationAtAlerting                 (10),
      changeOfPositionDP                 (11),
      or-Interactions                    (12),
      warningToneEnhancements            (13),
      cf-Enhancements                    (14),
      subscribedEnhancedDialledServices (15),
      servingNetworkEnhancedDialledServices (16),
      criteriaForChangeOfPositionDP      (17),
      serviceChangeDP                    (18)
} (SIZE (15..64))
```

Figure 8.3 Structure of Offered Camel 4 functionalities. Reproduced from 3GPP TS 29.002, v6.10.0, Section 17.7.1, by permission of ETSI

active for that call in that logical MSC; and CAP v4 is used for the dialled service. It is implied by the above conditions that the singe-point-of-control rule prevails; EDS is not allowed when the subscribed service from O-CSI is still active for the call; see Figures 8.4 and 8.5.

In the example flow in Figure 8.4, the O-CSI service is still active when the MSC/gsmSSF triggers the D-CSI service. Therefore, EDS is not allowed for the subscribed dialled service. Likewise, N-CSI may not become an EDS service for this call. In Figure 8.5, the subscribed service (O-CSI) does not retain the CAMEL relationship. When the subscribed dialled service is triggered, the MSC/gsmSSF includes in CAP IDP an indication that EDS is allowed ('enhanced dialled services allowed'). The D-CSI service may now arm events and retain the CAMEL dia-logue. When the serving network dialled service is triggered, the subscribed dialled service is still active, so the MSC/gsmSSF does not include the enhanced dialled services allowed parameter in CAP IDP.

The gsmSSF determines whether the conditions for EDS are fulfilled and informs the SCP when EDS is allowed. The subscribed service, the dialled service and the enhanced dialled service use the same application context (AC). It is the responsibility of the SCP to apply EDS only when the enhanced dialled services allowed parameter is present in IDP; this parameter is defined for CAP v4 only. Hence, EDS cannot be used in combination with CAMEL phase 3.

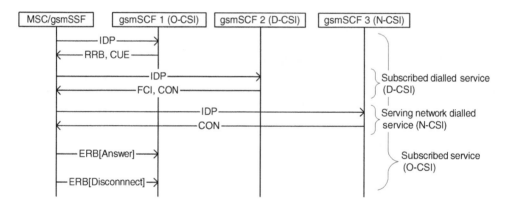

Figure 8.4   Multiple service triggering for MO call – EDS not allowed

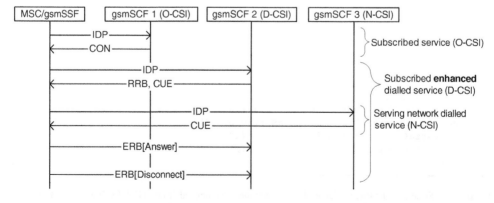

Figure 8.5   Multiple service triggering for MO call – EDS allowed

The 'offered CAMEL 4 functionalities' parameter that is included in the IDP operation may indicate either or both 'subscribed enhanced dialled services' and 'serving network enhanced dialled services'. These indications, when present in the offered CAMEL 4 functionalities, are informative for the service logic. It relates to supported capability of the MSC. However, whereas an MSC may support EDS, that does not imply that EDS is allowed for each and every call. Therefore, a dialled service (D-CSI or N-CSI) should strictly use the 'enhanced dialled services allowed' parameter in IDP to determine whether EDS is allowed for this call.

8.2.2 Handover Notification Criteria

The handover notification is introduced in CAMEL phase 4 in 3GPP Rel-5; see Section 6.2.7. The handover notification criteria were introduced in CAMEL phase 4 in 3GPP Rel-6. The notification criteria may be used to reduce the number of handover notifications that may be sent to the SCP during a call. When the SCP arms the change of position DP, it may specify notification criteria. Effectively, the SCP indicates to the MSC/gsmSSF which handover events will be reported. The SCP may specify one or more of the following events:

- The subscriber performs handover across the boundary of a specific cell (cell global identifier).
- The subscriber performs handover across the boundary of a specific service area (service area identifier).
- The subscriber performs handover across the boundary of a specific location area (location area identifier).
- The subscriber hands over between 2G radio access and 3G radio access.
- The subscriber hands over to another PLMN.
- The subscriber hands over to another MSC.

When handover is reported to the SCP, the current location of the subscriber is reported. A CAMEL service may use the handover notification criteria for dynamically adapting the notification resolution. Such CAMEL service may be a home zone-like service, where subscribers receive a special tariff when calling from (or being called in) the home zone. A subscriber's home zone (HZ) may overlap with a defined set of location areas (LA); see Figure 8.6. The HZ for this subscriber consists of cells in LA2 and LA5.

As long as the subscriber is not located in LA2 or in LA5, the CAMEL service may arm the LA criterion, specifying LA2 and LA5. As a result, the MSC/gsmSSF sends a notification only when the subscriber enters or leaves LA2 or LA5. When the subscriber is in LA2 or LA5, the CAMEL service re-arms the handover DP; this time without notification criteria. Every cell handover will now be reported to the SCP. The CAMEL service can then determine when the subscriber is in her home zone. When the subscriber is no longer in LA2 or LA5, the CAMEL service defines again the trigger criteria, i.e. notification will be sent only when subscriber crosses the boundary of LA2 or LA5.

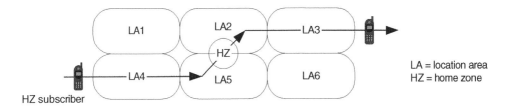

Figure 8.6 Dynamic setting of handover notification criteria

8.2.3 Enhancement to SCUDIF Control

Section 6.2.12 describes the CAMEL interaction with service change and UDI/RDI fallback (SCUDIF). When a call is established with dual bearer capability, the MSC/gsmSSF reports the 'preferred service' and the 'less preferred service'. The requested services for a SCUDIF call may be speech and video. When the call is answered, the call may be answered, e.g. as a video call. The active service at answer is indicated to the gsmSCF in the answer notification. The CAMEL service now treats the active phase call as a video call. However, should a service change take place during the active phase of the call, then this will remain unnoticed for the CAMEL service.

To overcome this dilemma, CAMEL phase 4 in 3GPP Rel-6 allows for 'service change notifications'. DP O_Service_Change and DP T_Service_Change are introduced in the O-BCSM and T-BCSM, respectively. The enhancement to the O-BCSM is given in Figure 8.7. The support of the Service_Change DP by the MSC/gsmSSF is indicated in CAP IDP, in the offered CAMEL 4 functionalities parameter.

The service change, reflecting a change between tele service 11 (speech) and bearer service 30 (64 kb/s data, used for video telephony), may be reported in the active phase of the call. The CAMEL service may arm this DP as EDP-N only. When the service change event is reported, the gsmSSF also generates a charging report. This enables the CAMEL service to adapt the call charge onwards, e.g. when the call changes from speech call to video call.

The service change may be initiated by the calling party or by the called party. The call party for which the service change is reported is determined, however, by the BCSM in which the service change is detected (Figure 8.8). For the MO call CAMEL service, a service change is always reported on leg 1. The charging report resulting from the service change is also reported on leg 1. Hence, in order to adapt the call charge to the service in use (speech vs video telephony), the CAMEL service will apply charging to leg 1, not to leg 2. For the T-CSI service and VT-CSI service, a service change, and the accompanying charging report, is always reported on leg 2.

The call party for the service change is reported (i.e. leg 1 or leg 2) does not have to be the call party that initiated the service change. In-call service change requires the use of bearer independent call control (BICC) instead of ISUP. Details of the DTAP and BICC signalling related to service change are specified in 3GPP TS 23.172 [89].

8.2.4 Reporting User-to-user Information

User-to-user signalling (UUS) allows transparent transfer of user-specific information (UUI) between calling and called party. This user-specific information is used by calling and called party only and does not affect call routing or handling. UUI may be included in regular ISUP messages for call

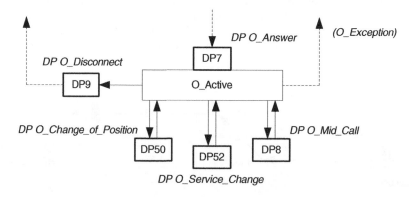

Figure 8.7   O-BCSM with DP O_Service_Change

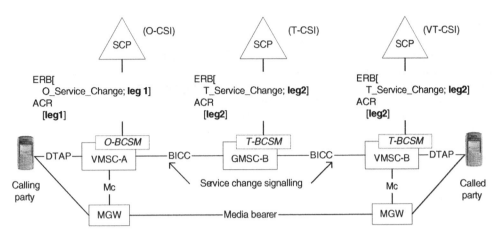

Figure 8.8 Service change notifications

establishment, including IAM, ACM, CPG, ANM, CON and REL. UUI may also be included in UUS-specific ISUP messages, such as facility or user-to-user information. Signalling and procedure details of UUS are specified in 3GPP TS 23.087. A maximum of 32 or 128 octets user data may be transported, depending on the signalling network.

Up to 3GPP Rel-5, CAP does not convey any UUI. Hence, UUS is transparent for any CAMEL service. CAMEL phase 4 in 3GPP Rel-6, however, allows for including UUI in IDP operation (Figure 8.9). UUS may be used between two (mobile) subscribers or between a mobile subscriber and network equipment. This enhancement to CAMEL phase 4 is earmarked for 'GSM-R', which is GSM-adapted for use by railways (see http://gsm-r.uic.asso.fr/).

An MSC/gsmSSF that supports this feature of CAMEL phase 4 includes the UUI that is present on the access network (DTAP or ISUP/BICC), into CAP IDP. For an MO call, UUI present in DTAP Setup is copied to CAP IDP; for an MT call in GMSC or VMSC and for an MF call, UUI present in ISUP IAM is copied to CAP IDP. Support of this feature in the MSC/gsmSSF is not indicated in the offered CAMEL 4 functionalities. The reason is that this functionality has no impact on CAMEL service logic behaviour. If the SCP does not support this parameter (the UUI) in CAP IDP, then it will ignore it. CAMEL does not allow the UUI in ISUP IAM to be modified. UUI that may be present in other ISUP messages, such as ACM or REL, is not reported to the SCP.

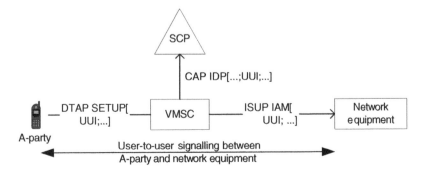

Figure 8.9 User-to-user signalling between mobile subscriber and network equipment

8.2.5 Enhancement to User Interaction

User interaction (playing announcements to the user, collecting digits from the user) entails that a traffic connection is established between the served subscriber and an announcement device. The announcement device may be internal to the MSC or external. When the announcement device is external to the MSC, a CAMEL control relationship may be established between the announcement device and the SCP (assisting dialogue; see Section 4.3.4). The CAMEL service may instruct the announcement device which announcement to play etc. The traffic connection between serving MSC and announcement device does not contain information related to the CAMEL call, such as calling party number or called party number. Of course, the ISUP/BICC IAM between serving MSC and announcement device contains called party number, but that called party number relates to the actual traffic connection to the announcement device. If calling party and/or called party, related to the CAMEL call, are needed to select the announcement to be played, then the CAMEL service will use the control relationship to indicate the required announcement to the announcement device.

To facilitate the use of autonomous announcement devices, the traffic flow between serving MSC and external announcement devices is enhanced. When the CAMEL service establishes the connection between the serving MSC and the external announcement device, it uses CAP establish temporary connection (ETC). The service may include the parameters original called party Id and calling party number in CAP ETC. The serving MSC maps these parameters on the corresponding elements in the ISUP IAM that is used to establish the connection with the announcement device.

When user interaction is done from the GMSC, then the original called party Id in CAP ETC may contain the MSISDN of the called subscriber, i.e. the subscriber on whose behalf the CAMEL service is invoked. In other words, the original called party number in CAP ETC is not related to the original called party number from the ISUP IAM arriving at the GMSC (see Figure 8.10 for a use case of this feature). The content delivery system (CDS) in Figure 8.10 requires the MSISDN of the served subscriber to select the announcement associated with this called party. The calling party number may further be required to personalize the announcement, taking the calling party into consideration.

The enhancement to CAP ETC is not indicated in the offered CAMEL4 functionalities in CAP IDP. Hence, service logic designers must take care when to use this enhancement. An MSC/gsmSSF that does not support this enhancement to CAP ETC would ignore these newly specified parameters. Prior to the availability of this CAMEL capability, service logics may use proprietary methods, like embedding the B-MSISDN into the assisting SSP IP Routing Address.

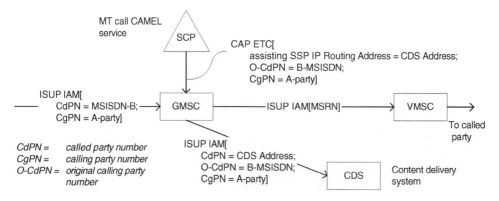

Figure 8.10 Setting the O-CdPN and CgPN in temporary connection

8.3 Enhancements to 3GPP Rel-7

The following enhancements are introduced in CAMEL Phase in 3GPP Rel-7.

8.3.1 Trunk-originated Triggering

All CAMEL services up to CAMEL phase 4 in 3GPP Rel-6 are triggered from an MSC that is serving a subscriber; this rule also applies for the dialled services. Hence, CAMEL services may be triggered from the following logical entities:

- *visited MSC (VMSC)* – a VMSC may be involved in MO, MF and MT calls;
- *Gateway MSC (GMSC)* – a GMSC may be involved in MT and MF calls.

Hence, CAMEL does not formally specify a mechanism to trigger a service from an MSC that is neither VMSC nor GMSC. Example cases where this form of triggering is required include:

- *Premium rate and free phone service* – calls to service numbers like premium rate and free phone may be routed to a designated SSP in the PLMN. Service triggering takes place from the SSP. Calls to service numbers may also originate from outside the PLMN; in that case, calls to these numbers are also routed to this designated SSP in the PLMN.
- *Remote access for VPN subscriber* – VPN subscribers need to be able to access their service using another subscriber's phone or from a connection outside the PLMN, e.g. a PSTN terminal.
- *Direct PABX connection* – a GSM operator may connect a PABX directly to an MSC in the PLMN. Calls to and from individual extensions from such PABX require service triggering from the MSC.

In the above examples, the MSC from which triggering should take place has no subscription data available, such as O-CSI or T-CSI. Neither does the MSC start a normal MO call process or MT call process. For these call cases, trunk-originated (TO) triggering may be used. Figure 8.11 shows an example of service triggering for PABX connection. In this example, the PABX is connected to the MSC through a digital subscriber signalling 1 (DSS1) connection. DSS1 is the access signalling protocol that is normally used to connect subscriber equipment to the ISDN network. The PABX may be configured to route calls to specific destinations through a trunk route to the MSC. The PABX appends a prefix to the called number for these calls; this prefix is used to identify the owner of the PABX, i.e. the company. The MSC is configured to trigger a VPN service for calls originating from this incoming trunk, and other incoming trunks. The MSC applies TO triggering to trigger a CAMEL service for this call case. All data that is required for TO triggering, such as service key, gsmSCF address, protocol (application part), etc. is provisioned in the MSC. Compare this with the provisioning of N-CSI in an operator's MSC.

CAMEL specifies that CAP v4 will be used for TO triggering. When TO service triggering is started, the MSC instantiates the CAMEL phase 4 O-BCSM. A later subsection will explain an enhancement to the O-BCSM for the TO call case. In principle, all capabilities contained in CAP v4 may be used for the TO service. The following list shows the main differences between the capability for an MO call and the capability for a TO call.

- There is no interaction with a VLR for the TO call; there is no subscriber data available. This means that no supplementary services can be performed for the call.
- Subscriber-specific data such as IMSI and location information will not be reported to the SCP. The location number, however, may be reported to the SCP, if received on the incoming trunk.

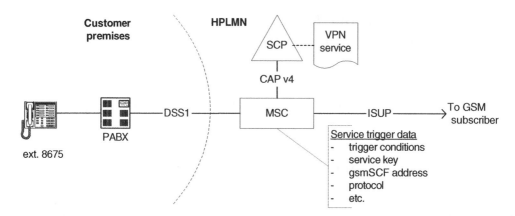

Figure 8.11   CAMEL service triggering for PABX connection

In the case of PABX connection through DSS1, no location number can be conveyed from PABX to MSC.
• Information related to MS or UE, such as IMEI and MS-Classmark, is not available.

Unlike the serving network based dialled service (N-CSI), TO triggering is triggered at DP collected information.

8.3.1.1 Collection of Address Signals

One distinctive new feature in TO triggering is the collection of address signals during call establishment. ISDN defines the following two addressing methods:

• *en-bloc sending* – the entire destination address is sent in a single call setup message to the ISDN network;
• *overlap sending* – the destination address is sent to the ISDN network in a number of successive call setup messages.

Call establishment in the GSM network normally uses en-bloc sending. When the GSM MS establishes a call, the DTAP setup message contains the entire destination number, carried in the called party BCD number parameter. The serving MSC generates a single ISUP IAM for the outgoing call.

In the case that a PABX is connected to the MSC, the PABX may be configured such that it commences call establishment when a minimum number of digits are entered. The DSS1 setup message may therefore contain incomplete address information. The MSC may nevertheless instantiate the O-BCSM and start the TO CAMEL service. The TO CAMEL service may instruct the gsmSSF to collect more address signals, until the entire destination address is received. The CAMEL service uses the following mechanism:

(1) the CAMEL service arms DP collected information as EDP-R; and
(2) the CAMEL service sends CAP operation *collect information* (CI).

The use of CAP CI has the effect that the O-BCSM makes a state transition to the O_Null state. When the requested number of further address digits is received, the O-BCSM transits back to DP collected information and reports the received address digits in a CAP event report BCSM (ERB) operation.

CAP CI is introduced in CAP v4 without application context version increase. Strictly speaking, one may argue that this enhancement to CAP (adding a new operation) is not backwards-compatible. However, when an MSC/gsmSSF triggers the TO CAMEL service, the MSC/gsmSSF indicates in CAP IDP whether it supports CAP CI. The CAMEL service takes care not to use CAP CI when the MSC/gsmSSF has not indicated the support of that operation.

8.3.1.2 Calling Party Number, Connected Number

TO triggering is a case wherein it would be advantageous if CAMEL had the capability to provide a calling party number (CgPN) to the gsmSSF. When a PABX is connected to the MSC, it may not always provide a CgPN that is suitable for the public GSM network. The PABX may provide the calling subscriber's extension number as CgPN. The PABX owner (a company) may have an agreement with the GSM operator that the network (the MSC) takes care of setting the CgPN for outgoing calls. When TO triggering is applied for a VPN service, the GSM operator may want to shift the responsibility of outgoing number translation and the responsibility of setting the CgPN, to the VPN service. However, CAP cannot set the CgPN. This may be an area where CAP v4 would need to be enhanced, to allow for setting the CgPN.

Similar agreements as described above, for setting the CgPN for direct PABX connection, may exist for setting the connected number. The PABX may return a connected number in a format that is not suitable for transportation over the public GSM network. The PABX may provide the connected subscriber's extension number as connected number. The operator may want to shift the task of setting the connected number to the VPN service. Figure 8.12 presents a call flow with useful future enhancement, for direct PABX connection.

The connected number and additional connected number would be provided to the MSC/gsmSSF in response to the answer notification. The purpose of these elements would be the following:

- *Connected number* – this element should be copied to the corresponding element in ISUP ANM; it indicates the connected line for this call. The element should contain the public number of the connected PABX extension.

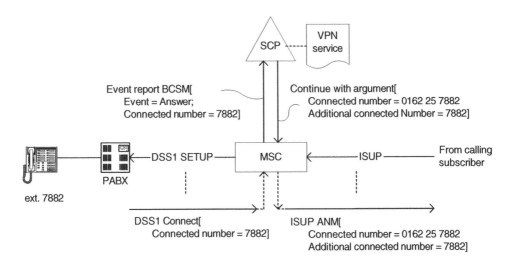

Figure 8.12 Setting the connected number and additional connected number

- *Additional connected number* – this element should be copied to generic number 5 (*'additional connected number'*) in ISUP ANM; see ITU-T Q.763 [137]. The additional connected number may be used to affect the number that will be displayed to the calling party.

The CAMEL service logic in Figure 8.12, e.g. a VPN service, may decide to restrict the use of the additional connected number to calls between subscribers of the same VPN group. The VPN service will check the calling party number here.

Appendix

A.1 Overview of CAP Operations

The following tables contain overviews of the CAP operations. The overviews are divided into call control, SMS control and GPRS control.

Table A.1 CAP operations for call control

Operation	Acronym	Phase	Class	Timer
Activity test	AT	1	3	Short
Apply charging	ACH	2	2	Short
Apply charging report	ACR	2	2	Short
Assist request instructions	ARI	2	2	Short
Call gap	CG	3	4	Short
Call information report	CIRp	2	4	Short
Call information request	CIRq	2	2	Short
Cancel	CAN	2	2	Short
Collect information	CI	4 (Rel-7)	2	Short
Connect	CON	1	2	Short
Connect to resource	CTR	2	2	Short
Continue	CUE	1	4	Short
Continue with argument	CWA	3	2	Short
Disconnect forward connection	DFC	2	2	Short
Disconnect forward connection with argument	DFCWA	4	2	Short
Disconnect leg	DL	4	1	Short
Entity released	EL	4	4	Short
Establish temporary connection	ETC	2	2	Medium
Event report BCSM	ERB	1	4	Short
Furnish charging information	FCI	2	2	Short
Initial DP	IDP	1	2	Short
Initiate call attempt	ICA	4	1	Short
Move leg	ML	4	1	Short
Play announcement	PA	2	2	Long
Prompt and collect user information	PC	2	1	Long
Play tone	PT	4	2	Short
Release call	RC	1	4	Short
Request report BCSM	RRB	1	2	Short
Reset timer	RT	2	2	Short
Send charging information	SCI	2	2	Short
Split leg	SL	4	1	Short
Specialized resource report	SRR	2	4	Short

Table A.2 CAP operations for SMS control

Operation	Acronym	Phase	Class	Timer
Connect SMS	CON SMS	3	2	Short
Continue SMS	CUE SMS	3	4	Short
Event report SMS	ERB SMS	3	4	Short
Furnish charging information SMS	FCI SMS	3	2	Short
Initial DP SMS	IDP SMS	3	2	Short
Release SMS	REL SMS	3	4	Short
Request report SMS event	RRB SMS	3	2	Short
Reset timer SMS	RT SMS	3	2	Short

Table A.3 CAP operations for GPRS control

Operation	Acronym[a]	Phase[b]	Class[c]	Timer
Activity test GPRS	AT GPRS	3	3	Short
Apply charging GPRS	ACH GPRS	3	2	Short
Apply charging report GPRS	ACR GPRS	3	1	Short
Cancel GPRS	CAN GPRS	3	2	Short
Connect GPRS	CON GPRS	3	2	Short
Continue GPRS	CUE GPRS	3	2	Short
Entity released GPRS	ER GPRS	3	1	Short
Event report GPRS	ERG	3	1	Short
Furnish charging information GPRS	FCI GPRS	3	2	Short
Initial DP GPRS	IDP GPRS	3	2	Short
Release GPRS	REL GPRS	3	2	Short
Request report GPRS event	RRGE	3	2	Short
Reset timer GPRS	RT GPRS	3	2	Short
Send charging information GPRS	SCI GPRS	3	2	Short

[a] The acronym is commonly used in signal sequence diagrams for CAP operations.
[b] CAMEL phase in which the operation is introduced
[c] The 'class' of the operation indicates whether Result and Error definitions exist for the operation. For a description of the operation class, refer to ITU-T Q.773 [141]. Table A.4 lists the class definitions.

Table A.4 Operation classes

Class	Description
1	Both Result and Errors are defined for the operation
2	Only Errors is defined for the operation
3	Only Result is defined for the operation
4	Neither Result nor Errors are defined for the operation

A.2 Overview of MAP Operations

Table A.6 lists the MAP operations that are used for CAMEL. This list forms a subset of the MAP operations that are specified in 3GPP TS 29.002 [103].

For USSD relation MAP operations, refer to Chapter 4.

Table A.5 Operation timer values

Timer[a]	Range
Short	1–10 s (call control)
	1–20 s (SMS control; GPRS control)
Medium	1–60 s
Long	1 s–30 min

[a] The operation timer indicates the maximum time that the entity that sends the operation will wait for a Result indication or Error indication. For each operation, a timer *range* value is defined.

Table A.6 MAP operations used for CAMEL

Operation	Acronym	Sender	Receiver	Description
Any-time interrogation	ATI	gsmSCF	HLR	Request for dynamic subscriber information
			GMLC	Request for location information
			MNP SRF	Request for number portability information
Any-time modification	ATM	gsmSCF	HLR	Instruction to HLR to modify subscription data
Any-time subscription interrogation	ATSI	gsmSCF	HLR	Request for subscription data from HLR
Delete subscriber data	DSD	HLR	VLR, SGSN	Instruction to delete subscription data from VLR or SGSN
Insert subscriber data	ISD	HLR	VLR, SGSN	Instruction to place subscription data in VLR or SGSN
Note MM event		VLR, SGSN	gsmSCF	Notification to gsmSCF about occurrence of mobility management event
Notify subscriber data change	NSDC	HLR	gsmSCF	Notification to gsmSCF about change in subscriber data
Provide roaming number	PRN	HLR	VLR	Request for roaming number from VLR
Provide subscriber information	PSI	HLR	VLR, SGSN	Request for dynamic subscriber information from VLR or SGSN
Restore data	RS	VLR	HLR	Request for subscription information at node refresh
Resume call handling	RCH	VMSC	GMSC	Request to GMSC to apply optimal routing at late call forwarding (ORLCF)
Send routing information	SRI	GMSC	HLR	Request from GMSC for routing instructions from HLR for terminating call
SS invocation notification	SSIN	MSC, HLR	gsmSCF	Notification to gsmSCF about invocation of a supplementary service
Update GPRS location		SGSN	HLR	Used by SGSN to request subscription information at registration
Update location	UL	VLR	HLR	Used by VLR to request subscription information at registration

A.3 Overview of ISUP Messages

Table A.7 lists the ISUP messages relevant for CAMEL call control.

Table A.7 ISUP messages used for CAMEL

Message	Acronym	Description
Address complete message	ACM	Indication that called party is alerting
Answer	ANM	Indication that called party has answered the call
Call progress	CPG	Generic call progress indication; may be used to indicate alerting state or to indicate that in-band information is available
Connect	CON	Indication that called party has answered the call. CON may be used by answering device, for example
Initial address message	IAM	Message to start call establishment
Release	REL	Indication that call is released

The complete list of ISUP messages is defined in ITU-T Q.763 [137].

A.4 Overview of CAMEL Subscription Information

Table A.8 Overview of all the CAMEL subscription data elements

Element	Full name	CAMEL phase	Description
O-CSI	Originating call CSI	1	Used in MSC for MO and MF calls and in GMSC for MF calls
T-CSI	Terminating call CSI	1	Used in GMSC for MT calls
SS-CSI	Supplementary service invocation notification CSI	2	Used in MSC to send notification to gsmSCF when supplementary service is invoked. May also be used in HLR for SS notifications to gsmCSF (CAMEL phase 3)
TIF-CSI	Translation information flag CSI	2	Indication in HLR that a subscriber may register short forwarded-to numbers; may also be used in MSC as an indication that a subscriber may deflect to short numbers
U-CSI	USSD CSI	2	Subscription data in HLR; used in HLR to relay USSD service request to gsmSCF
UG-CSI	USSD generic CSI	2	Generic data in HLR; used in HLR to relay USSD service request to gsmSCF
D-CSI	Subscribed dialled service CSI	3	Used in MSC or GMSC for subscribed dialled service invocation
N-CSI	Network dialled service CSI	3	Used in MSC or GMSC for serving network based dialled service invocation
VT-CSI	Visited terminated call CSI	3	Used in VMSC for MT call control
MO-SMS-CSI	Mobile originated SMS CSI	3	Used in MSC or SGSN for MO SMS service
GPRS-CSI	GPRS control CSI	3	Used in SGSN for controlling GPRS attached/detached state and for PDP context control
M-CSI	Mobility management CSI	3	Used in MSC for CS mobility management notifications to gsmSCF

Table A.8 (*continued*)

Element	Full name	CAMEL phase	Description
MG-CSI	GPRS mobility management CSI	4	Used in SGSN for PS mobility management notifications to gsmSCF
MT-SMS-CSI	Mobile terminated SMS CSI	4	Used in MSC or SGSN for MT SMS service
O-IM-CSI	Originating IP multimedia CSI	4	Used for subscribed services for originating IMS calls
D-IM-CSI	Dialled services IP multimedia CSI	4	Used for subscribed dialled services for originating IMS calls
VT-IM-CSI	Terminating IP multimedia CSI	4	Used for subscribed services for terminating IMS calls
TO-CSI	Trunk originated calls CSI	4 (Rel-7)	Used in MSC for trunk originated CAMEL service invocation

References

3GPP Specifications and Technical Reports

[1]	GSM TS 02.01	Principles of telecommunication services supported by a GSM public land mobile network
[2]	GSM TS 02.02	Bearer services supported by a GSM public land mobile network
[3]	GSM TS 02.03	Teleservices supported by a GSM public land mobile network
[4]	GSM TS 02.24	Description of charge advice information
[5]	GSM TS 02.30	Man–machine interface of the mobile station
[6]	GSM TS 02.31	Fraud information gathering system; service description
[7]	GSM TS 02.32	Immediate service termination; service description
[8]	GSM TS 02.33	Lawful interception; stage 1
[9]	GSM TS 02.41	Operator-determined barring
[10]	GSM TS 02.67	Enhanced multi-level precedence and pre-emption service; stage 1
[11]	GSM TS 02.72	Call deflection service description; stage 1
[12]	GSM TS 02.78	Customized applications for mobile network enhanced logic; service definition
[13]	GSM TS 02.81	Line identification supplementary services; stage 1
[14]	GSM TS 02.82	Call forwarding supplementary services; stage 1
[15]	GSM TS 02.83	Call waiting and call hold supplementary services; stage 1
[16]	GSM TS 02.84	Multi-party supplementary services; stage 1
[17]	GSM TS 02.85	Closed user group supplementary services; stage 1
[18]	GSM TS 02.86	Advice of charge supplementary services; stage 1
[19]	GSM TS 02.87	User-to-user signalling service description; stage 1
[20]	GSM TS 02.88	Call barring supplementary services; stage 1
[21]	GSM TS 02.90	Unstructured supplementary service data; stage 1
[22]	GSM TS 02.91	Explicit call transfer
[23]	GSM TS 02.93	Completion of calls to busy subscriber; stage 1
[24]	GSM TS 02.96	Name identification supplementary services; stage 1
[25]	GSM TS 02.97	Multiple subscriber profile phase 1; service description
[26]	GSM TS 03.02	Network architecture
[27]	GSM TS 03.03	Numbering, addressing and identification
[28]	GSM TS 03.08	Organization of subscriber data
[29]	GSM TS 03.14	Support of dual tone multi-frequency signalling via the GSM system
[30]	GSM TS 03.15	Technical realization of operator-determined barring
[31]	GSM TS 03.18	Basic call handling
[32]	GSM TS 03.31	Fraud information gathering system; service description; stage 2

[33]   GSM TS 03.35   Immediate service termination; stage 2
[34]   GSM TS 03.38   Alphabets and language-specific information
[35]   GSM TS 03.68   Voice group call service; stage 2
[36]   GSM TS 03.71   Location services; functional description; stage 2
[37]   GSM TS 03.72   Call deflection; stage 2
[38]   GSM TS 03.78   Customized applications for mobile network enhanced logic
                       (CAMEL); stage 2
[39]   GSM TS 03.79   Support of optimal routing phase 1; stage 2
[40]   GSM TS 03.82   Call forwarding supplementary services; stage 2
[41]   GSM TS 03.83   Call waiting and call hold supplementary services;
                       stage 2
[42]   GSM TS 03.84   Multi-party supplementary services; stage 2
[43]   GSM TS 03.85   Closed user group supplementary services; stage 2
[44]   GSM TS 03.87   User-to-user signalling (UUS); stage 2
[45]   GSM TS 03.88   Call barring supplementary services; stage 2
[46]   GSM TS 03.90   Unstructured supplementary service data
[47]   GSM TS 03.91   Explicit call transfer supplementary service; stage 2
[48]   GSM TS 03.97   Multiple subscriber profile; stage 2
[49]   GSM TS 04.08   Mobile radio interface layer 3 specification
[50]   GSM TS 04.72   Call deflection supplementary service; stage 3
[51]   GSM TS 04.80   Mobile radio interface layer 3 – supplementary services specification
                       formats and coding
[52]   GSM TS 04.87   User-to-user signalling supplementary service; stage 3
[53]   GSM TS 04.90   Unstructured supplementary service data
[54]   GSM TS 09.02   Mobile application part (MAP)
[55]   GSM TS 09.07   General requirements on interworking between the PLMN and the
                       ISDN or PSTN
[56]   GSM TS 09.78   CAMEL application part (CAP)
[57]   GSM TS 12.05   Subscriber related call and event data
[58]   3GPP TS 22.002 Circuit bearer services supported by a public land mobile network
[59]   3GPP TS 22.003 Circuit teleservices supported by a public land mobile network
[60]   3GPP TS 22.004 General on supplementary services
[61]   3GPP TS 22.016 International mobile equipment identities
[62]   3GPP TS 22.030 Man–machine interface of the user equipment
[63]   3GPP TS 22.041 Operator-determined call barring
[64]   3GPP TS 22.060 General packet radio service; service description; stage 1
[65]   3GPP TS 22.066 Support of mobile number portability; stage 1
[66]   3GPP TS 22.078 Customized applications for mobile network enhanced logic
                       (CAMEL); stage 1
[67]   3GPP TS 22.097 Multiple subscriber profile phase 2; stage 1
[68]   3GPP TS 22.101 Service aspects; service principles
[69]   3GPP TS 22.135 Multicall; stage 1
[70]   3GPP TS 22.228 Service requirements for the IP multimedia core network subsystem;
                       stage 1
[71]   3GPP TS 22.250 IP multimedia subsystem (IMS) group management; stage 1
[72]   3GPP TS 22.340 IP multimedia subsystem (IMS) messaging; stage 1
[73]   3GPP TS 23.003 Numbering, addressing and identification

[74]   3GPP TS 23.012   Location management procedures
[75]   3GPP TS 23.018   Basic call handling; technical realization
[76]   3GPP TS 23.032   Universal geographical area description (GAD)
[77]   3GPP TS 23.038   Alphabets and language-specific information
[78]   3GPP TR 23.039   Interface protocols for the connection of short message service
                          centers (SMSCs) to short message entities
[79]   3GPP TS 23.040   Technical realization of short message service
[80]   3GPP TS 23.042   Compression algorithm for SMS
[81]   3GPP TS 23.060   General packet radio service; stage 2
[82]   3GPP TS 23.066   Support of GSM mobile number portability; stage 2
[83]   3GPP TS 23.078   Customized applications for mobile network enhanced logic
                          (CAMEL); stage 2
[84]   3GPP TS 23.082   Call forwarding supplementary services; stage 2
[85]   3GPP TS 23.097   Multiple subscriber profile phase 1; stage 2
[86]   3GPP TS 23.107   Quality of service concept and architecture
[87]   3GPP TS 23.125   Overall high level functionality and architecture impacts of flow
                          based charging; stage 2
[88]   3GPP TS 23.171   Location services; functional description; stage 2 (UMTS)
[89]   3GPP TS 23.172   Technical realization of circuit switched multimedia service;
                          UDI/RDI fallback and service modification; stage 2
[90]   3GPP TS 23.205   Bearer-independent circuit-switched core network; stage 2
[91]   3GPP TS 23.228   IP multimedia subsystem (IMS); stage 2
[92]   3GPP TS 23.271   Functional stage 2 description of location services
[93]   3GPP TS 23.278   Customized applications for mobile network enhanced logic
                          (CAMEL) phase 4; stage 2; IM CN interworking
[94]   3GPP TS 24.008   Mobile radio interface layer 3 specification; core network protocols;
                          stage 3
[95]   3GPP TS 24.011   Point-to-point (PP) short message service support on mobile radio
                          interface
[96]   3GPP TS 24.228   Signalling flows for the IP multimedia call control based on session
                          initiation protocol and session description protocol; stage 3
[97]   3GPP TS 24.229   Internet protocol (IP) multimedia call control protocol based on
                          session initiation protocol and session description protocol; stage 3
[98]   3GPP TS 25.305   User equipment positioning in universal terrestrial radio access
                          network; stage 2
[99]   3GPP TS 25.410   UTRAN Iu interface: general aspects and principles
[100]   3GPP TS 25.420   UTRAN Iur interface: general aspects and principles
[101]   3GPP TS 25.430   UTRAN Iub interface: general aspects and principles
[102]   3GPP TS 26.111   Codec for circuit switched multimedia telephony service;
                          modifications to H.324
[103]   3GPP TS 29.002   Mobile application part (MAP)
[104]   3GPP TS 29.016   Serving GPRS support node SGSN – visitors location register; Gs
                          interface network service specification
[105]   3GPP TS 29.060   General packet radio service; GPRS tunneling protocol across the Gn
                          and Gp interface
[106]   3GPP TS 29.078   CAMEL application part (CAP)
[107]   3GPP TS 29.205   Application of Q.1900 series to bearer-independent circuit switched
                          (CS) core network architecture; stage 3

[108]   3GPP TS 29.228   IP multimedia (IM) subsystem Cx and Dx interfaces; signalling flows
                           and message contents
[109]   3GPP TS 29.229
[110]   3GPP TS 29.232   Media gateway controller – media gateway interface; stage 3
[111]   3GPP TS 29.278   Customized applications for mobile network enhanced logic
                           (CAMEL); CAMEL application part (CAP) specification for IP
                           multimedia subsystems (IMS)
[112]   3GPP TS 29.414   Core network Nb data transport and transport signalling
[113]   3GPP TS 29.415   Core network Nb interface user plane protocols
[114]   3GPP TS 32.005   Telecommunications management; charging management; 3G call
                           and event data for the circuit switched domain
[115]   3GPP TS 32.015   Telecommunications management; charging management; 3G call
                           and event data for the packet switched domain
[116]   3GPP TS 32.200   Telecommunication management; charging management; charging
                           principles
[117]   3GPP TS 32.205   Telecommunication management; charging management; charging
                           data description for the circuit switched domain
[118]   3GPP TS 32.215   Telecommunication management; charging management; charging
                           data description for the packet switched domain
[119]   3GPP TS 32.240   Telecommunication management; charging management; charging
                           architecture and principles
[120]   3GPP TS 32.250   Telecommunication management; charging management; circuit
                           switched domain charging
[121]   3GPP TS 32.251   Telecommunication management; charging management; packet
                           switched domain charging
[122]   3GPP TS 32.298   Telecommunication management; charging management; charging
                           data record parameter description
[123]   3GPP TS 32.299   Telecommunication management; charging management; diameter
                           charging applications
[124]   3GPP TS 33.106   Lawful interception requirements
[125]   3GPP TS 43.051   GSM/EDGE Radio access network (GERAN) overall description;
                           stage 2
[126]   3GPP TS 43.059   Functional stage 2 description of location services in GERAN
[127]   3GPP TS 43.073   Support of localized service area (SoLSA); stage 2

ITU-T Recommendations

[128]   E.164   The international public telecommunication numbering plan
[129]   E.212   The international identification plan for mobile terminals and mobile
                           users
[130]   H.248   Gateway control protocol
[131]   H.324   Terminal for low bit-rate multimedia communication
[132]   Q.118   Abnormal conditions – special release arrangements
[133]   Q.701   Functional description of the message transfer part (MTP) of
                           signalling system no. 7
[134]   Q.711   Functional description of the signalling connection control part
[135]   Q.713   Signalling connection control part formats and codes
[136]   Q.714   Signalling connection control part procedures

[137] Q.763   Signalling system no. 7 – ISDN user part formats and codes
[138] Q.764   Signalling system no. 7 – ISDN user part signalling procedures
[139] Q.769.1   Signalling system no. 7 – ISDN user part enhancements for the
                      support of number portability
[140] Q.771   Functional description of transaction capabilities
[141] Q.773   Transaction capabilities formats and encoding
[142] Q.850   Usage of cause and location in the digital subscriber signalling
                      system no. 1 and the signalling system no. 7 ISDN user part
[143] Q.1214   Distributed functional plane for intelligent network CS-1
[144] Q.1901   Bearer independent call control protocol
[145] Q.1902.4   Bearer independent call control protocol (capability set 2): basic call
                      procedures
[146] Q.series   Number portability – scope and capability set 1 architecture
       supplement 3
[147] Q.series   Number portability – call control for capability set 1 service provider
       supplement 4   portability (all call query and onward routing)
[148] Q.series   Number portability – capability set 2 requirements for service
       supplement 5   provider portability (query on release and dropback)
[149] X.200   Information technology – open systems interconnection – basic
                      reference model: the basic model
[150] X.680   Information technology – abstract syntax notation one (ASN.1):
                      specification of basic notation
[151] X.681   Information technology – abstract syntax notation one (ASN.1):
                      information object specification
[152] X.682   Information technology – abstract syntax notation one (ASN.1):
                      constraint specification
[153] X.683   Information technology – abstract syntax notation one (ASN.1):
                      parameterization of ASN.1 specifications
[154] X.690   Information technology – ASN.1 encoding rules: specification of
                      basic encoding rules (BER), canonical encoding rules (CER) and
                      distinguished encoding rules (DER)
[155] X.880   Information technology – remote operations: concepts, model and
                      notation
[156] X.881   Information technology – remote operations: OSI
                      realizations – remote operations service element (ROSE) service
                      definition
[157] X.882   Information technology – remote operations: OSI
                      realizations – remote operations service element (ROSE) protocol
                      specification

ETSI Standards

[158] ETS 300 374-1 Intelligent network capability set 1; part 1: protocol specification
[159] EN 301 140-1   Intelligent network application protocol (INAP); capability set 2
                      (CS2); part 1: protocol specification

T1 Standards

[160] ANSI T1.113   Signaling system no. 7 (SS7) integrated services digital network
                      (ISDN) user part

GSM Association Publications

[161]   PRD TD.32   Rejects and returns process
[162]   PRD TD.57   Transferred account procedure (TAP) data record format specification
                      version number 3
[163]   PRD TD.58   TAP 3 implementation handbook
[164]   PRD TD.60   General scenarios for the transferred account procedure for data
                      record format specification version 3

IETF Request for Comments (RFC)

[165]   RFC 768   User datagram protocol
[166]   RFC 1889   RTP: a transport protocol for real-time applications
[167]   RFC 2806   URLs for telephone calls
[168]   RFC 2865   Remote authentication dial in user service (RADIUS)

Other

[169]   Gonzalo Camarillo and Miguel-Angel Garcia-Martin, *The 3G IP multimedia
         subsystem (IMS): merging the internet and the cellular worlds*. John Wiley & Sons:
         Chichester, 2004.
[170]   J. Larmouth, *ASN.1 Complete*. Morgan Kaufmann: New York, 1999.
[171]   O. Dubuisson, *ASN.1 - Communication entre systèmes hétérogènes*. Springer Verlag
         et France Télécom: Berlin, 1999.
[172]   D. Steedman, *ASN1, the Tutorial and Reference*. Technology Appraisals:
         Twickenham, 1993.
[173]   I. Faynberg, L. Gabuzda, M. Kaplan and N. Shah, *The Intelligent Network Standards*.
         McGraw–Hill: New York, 1997.
[174]   U. Black, *The Intelligent Network*. Prentice Hall: Englewood Cliffs, NJ, 1998.

Abbreviations

3G	Third Generation	B2BUA	Back-to-back user agent
3GPP	Third Generation Partnership Project	BAIC	Barring of all incoming calls
		BAOC	Barring of all outgoing calls
		BC	Bearer capability
AC	Application context	BCSM	Basic call state model
ACH	Apply charging (CAP operation)	BGW	Billing gateway
		BER	Basic encoding rules
ACM	Address complete message (ISUP message)	BICC	Bearer-independent call control
ACR	Apply charging report (CAP operation)	BICR	Barring of all incoming calls when roaming
AIN	Advanced IN	BOIC	Barring of outgoing international calls
ALR	Active location retrieval		
ANM	Answer message (ISUP message)	BOIC-exHC	Barring of outgoing international calls except to the home country
AoC	Advice of charge		
AOCC	Advice of charge – charge	BOR	Basic optimal routing
AOCI	Advice of charge – information	BS	Basic service or bearer service
AP	Application part	BSC	Base station controller
APN	Access point name	BSS	Base station system
ARIB	Association of radio industries and businesses	BSSAP	Base station system application part
ARPA	Advanced research projects agency	BTS	Base transceiver station
AS	Application server	CAMEL	Customized applications mobile network enhanced logic
ASN.1	Abstract syntax notation no. 1		
AT	Activity test (CAP operation)		
ATI	Any-time interrogation (MAP message)	CAN	Cancel (CAP operation)
		CAP	CAMEL application part
ATIS	Alliance for Telecommunications Industry Solutions	CB	Call barring
		CC	Country code
		CCBS	Completion of Calls to Busy Subscriber
ATM	Any-time modification (MAP message)	CCITT	Consultative Committee for International Telegraphy and Telephony (now called ITU-T)
ATSI	Any-time subscription interrogation (MAP message)		
AUC	Authentication centre	CCS	Common channel signalling

CAMEL: Intelligent Networks for the GSM, GPRS and UMTS Network   Rogier Noldus
© 2006 John Wiley & Sons, Ltd

CCSA	China Communications Standards Association	CUE	Continue (CAP operation)
		CUG	Closed user group
CD	Call deflection	CW	Call waiting
CDMA	Code division multiple access	CWA	Continue with argument (CAP operation)
CdPN	Called party number		
CDR	Call detail record	DCH	Default call handling
CF	Call forwarding	D-CSI	Dialled service CSI
CFB	Call forwarding – busy	DFC	Disconnect forward connection (CAP operation)
CFNRC	Call forwarding – not reachable		
CFNRY	Call forwarding – no reply	DL	Disconnect leg (CAP operation)
CFU	Call forwarding – unconditional		
		DNS	Domain name server
CG	Call gap (CAP operation)	DP	Detection point
CGF	Charging gateway function	DSD	Delete subscriber data (MAP message)
CGI	Cell global identifier		
CgPN	Calling party number	DSS1	Digital subscriber signalling no. 1
CH	Call hold		
CI	Cell identifier	DTAP	Direct transfer application part
	Collect information (CAP operation)	DTMF	Dual tone multi frequency
		DTN	Deflected-to-number
CIRp	Call information report (CAP operation)		
		ECT	Explicit call transfer
CIRq	Call information request (CAP operation)	EDGE	Enhanced data rates for GSM evolution
CLI	Calling line identity	EDP	Event detection point
CLIP	Calling line identification presentation	EDP-N	EDP – notify mode
		EDP-R	EDP – interrupt mode
CLIR	Calling line identification restriction	EDS	Enhanced dialled service
		EIR	Equipment identification register
CN	Core network	eMLPP	Enhanced multi-level precedence and pre-emption
CNAP	Calling name presentation		
COL	Connected line identity	EN	European norm
COLP	Connected line presentation	ERB	Event report BCSM (CAP Operation)
COLR	Connected line restriction		
CON	Connect (CAP operation) or Connect (ISUP message)	ETC	Establish temporary connection (CAP operation)
CPG	Call progress (ISUP message)	ETS	ETSI technical specification
CPH	Call party handling	ETSI	European Telecommunications Standards Institute
CRN	Call reference number		
CS	Call segment, Capability set (as in 'CS1') or Circuit switched	FAC	Final assembly code or Facility (ISUP message)
CS1, CS2	Capability set 1, Capability set 2	FCI	Furnish charging information (CAP operation)
CSA	Call segment association	FSM	Finite state machine
CSCF	Call session control function	FTN	Forwarded-to number
CSI	CAMEL subscription information	GCR	Global call reference
CTR	Connect to resource (CAP operation)	GERAN	GSM edge radio access network

GGSN	Gateway GPRS support node	ISDN	Integrated services digital network
GMLC	Gateway MLC		
GMSC	Gateway MSC	ISIM	IMS subscriber identity module
GN	Generic number	ISO	International Standards Organization
GPRS	General packet radio system		
gprsSSF	SSF for GPRS control	ISUP	ISDN user part
GSM	Global system for mobile communication (formerly: groupe speciale mobile)	ITU	International Telecommunication Union
		LAC	Location area code
gsmSCF	SCF as specified for GSM	LAI	Location area identifier
gsmSRF	SRF as specified for GSM	LCS	Location services
gsmSSF	SSF as specified for GSM	LE	Local exchange
GT	Global title	LI	Lawful intercept
GTP	Generic tunneling protocol	LLC	Low-layer compatibility
		LN	Location number
HLC	High-layer compatibility	LSA	Localized service area
HLR	Home location register	LU	Location update (MAP message)
HPLMN	Home PLMN		
HSS	Home subscriber server		
		MAP	Mobile application part
IAM	Initial address message (ISUP message)	MC	Multi-call
		MCC	Mobile country code
ICA	Initiate call attempt (CAP operation)	M-CSI	Mobility management CSI
		ME	Mobile equipment
I-CSCF	Interrogating CSCF	MF	Mobile forwarded
IDP	Initial detection point (CAP operation)	MGC	Media gateway controller
		MG-CSI	Mobility management for GPRS CSI
IE	Information element		
IETF	Internet Engineering Task Force	MGW	Media gateway
		ML	Move leg (CAP operation)
IF	Information flow	MLC	Mobile location services centre
IFC	Initial filter criteria	MMS	Multimedia messaging service
IMEI	International mobile equipment identifier	MNC	Mobile network code
		MNP	Mobile number portability
IMEISV	IMEI + software version	MO	Mobile originating
IMS	IP multimedia subsystem	MO-SMS-CSI	Mobile originated SMS CSI
IMSI	International mobile subscriber identity	MPTY	Multi-party call
		MS	Mobile station
IM-SSF	SSF specified for IMS interworking	MSC	Mobile switching services centre
IMT2000	International Mobile Telephony 2000	MSIN	Mobile subscriber identification number
IN	Intelligent networks	MSISDN	Mobile station integrated services digital network number
INAP	IN application part		
IP	Internet protocol or Intelligent peripheral		
		MSP	Multiple subscriber profile
IPLMN	Interrogating PLMN	MSRN	Mobile station roaming number
ISC	IMS service control	MT	Mobile terminating
ISD	Insert subscriber data (MAP message)	MTP	Message transfer part
		MT-SMS-CSI	Mobile terminated SMS CSI

N-CSI	Network dialled service CSI	RCH	Resume call handling (MAP message)
NDC	National destination code		
NDUB	Network determined user busy	RD	Restore data (MAP message)
NI	Network identifier	REL	Release (ISUP message)
NSDC	Notify subscriber data change (MAP message)	RES	Resume (ISUP message) or Result
		RFC	Request for comments
O-BCSM	Originating call BCSM	RNS	Radio network system
O-CSI	Originating call CSI	RRB	Request report BCSM (CAP operation)
ODB	Operator determined barring		
OI	Operator identifier	RTP	Real-time transport protocol
OR	Optimal routing		
ORLCF	Optimal routing of late call forwarding	SAC	Service area code
		SAI	Service area identifier
OSI	Open system interconnection	SCCP	Signalling connection control part
PA	Play announcement (CAP operation)	SCF	Service control function
		SCI	Send charging information (CAP operation)
PC	Prompt and collect user information (CAP Operation)	SCP	Service control point
PCS	Personal communication system	S-CSCF	Serving CSCF
		SCUDIF	Service change and UDI/RDI fallback
P-CSCF	Proxy CSCF		
PDN	Packet data network	SDP	Service data point or Session description protocol
PDPc	Packet data protocol context		
PDU	Protocol data unit	SDS	Subscribed dialled services
PIC	Point in call	SGSN	Serving GPRS support node
PLMN	Public land mobile network	SIM	Subscriber identification module
PN	Personal number		
PRD	Permanent reference document	SIP	Session initiation protocol
PRI	Primary rate interface	SK	Service key
PRN	Provide roaming number (MAP message)	SL	Split leg (CAP operation)
		SLF	Subscriber locator function
PS	Packet switched	SLS	Signalling link selection
PSI	Provide subscriber information (MAP message)	SM	Short message
		SMLC	Serving MLC
PSTN	Public switched telephone network	SMPP	Short message peer-to-peer
		SMS	Short message service
		SMSC	Short message service centre
QoS	Quality of service	SMS-CSI	See MO-SMS-CSI
		smsSSF	SSF for SMS control
RA	Routing area	SN	Subscriber number
RAC	Routing area code	SNR	Serial number or Single number reach
RADIUS	Remote authentication dial-in user service		
		SPC	Signalling point code
RAI	Routing area identifier	SPNP	Support of private numbering plan
RAN	Radio access network		
RAU	Routing area update	SRF	Signalling relay function or Specialized resource function
RC	Release call (CAP operation)		

SRI	Send routing information (MAP message)	TS	Tele service or Technical specification
SRR	Specialized resource report (CAP operation)	TTA	Telecommunications Technology Association
SS	Supplementary service	TTC	Telecommunications Technology Committee
SS-CSI	Supplementary service CSI		
SSF	Service switching function		
SSIN	Supplementary service invocation notification	UCP	Universal computer protocol
		UDI	Unrestricted digital information
SSN	Subsystem number	UDP	User datagram protocol
SSP	Service switching point	UDUB	User determined user busy
SS7	Signalling system no. 7	UE	User equipment
STP	Signaling transfer point	UMTS	Universal mobile telephony system
SUS	Suspend (ISUP message)		
SV	Software version	URI	Universal resource identifier
		URL	Univeral resource locator
TAC	Type approval code	USSD	Unstructured supplementary service data
TAP	Transfer account procedure		
T-BCSM	Terminating call BCSM	USIM	Universal subscriber identity module
TC	Transaction capabilities		
TCAP	Transaction capabilities application part (=TC)	UTRAN	Universal terrestrial radio access network
T-CSI	Terminating call CSI	UUI	User-user information
TDM	Time division multiplex	UUS	User-to-user signalling
TDMA	Time division multiple access		
TDP	Trigger detection point	VLR	Visitor location register
TDP-N	TDP – notify mode	VMSC	Visited MSC
TDP-R	TDP – interrupt mode	VPLMN	Visited PLMN
TE	Transit exchange	VPN	Virtual private network
TIF-CSI	Translation information flag CSI	VT-CSI	Visited MSC terminating call CSI
TO-CSI	Trunk originated CSI		
T-PDU	Transfer PDU	W-CDMA	Wideband CDMA
TR	Technical report	WIN	Wireless IN
TRX	Transceiver		

Index

Page numbers followed by *t* indicate tables.

Printed and bound in the UK by
CPI Antony Rowe, Eastbourne

Printed and bound by CPI Group (UK) Ltd, Croydon, CR0 4YY

16/04/2025

14658471-0002